How to Have Stress-Free Family Holidays

How to choose, plan and enjoy travelling at home and abroad with your children

Consultant editor: Sheila Sang

BLOOMSBURY

Acknowledgements

In addition to the writers who contributed to Section 1, the publishers would like to thank the following for their help in the compilation and revision of this book.

Kate Bell, Michelle Brown, Trish Burgess, Candida Clark, Alastair Cording, Charlotte Cox, Vennetta Cox, Rachel Crasnow, Sophie Craven, Jayne Cringle, Richard Darwood M.D., Sarah Dewe, Alice Gambier, Isobel Jacobs, Daniel Jewesbury, Helen Goalen, Boyd Hilton, Mike Hirst, Emma Hudson, Suzy Lucas, Kate Newman, Anne-Lucie Norton, Vanita Parti, Elizabeth Pitman, Kathy Rooney, Shaie Selzer, Blanche Sibbald, Tracey Smith, Matthew Whyman, Theresa Wright.

All rights reserved; no part of this publication may be reproduced, stored in a retrieval system, or transmitted by any means, electronic, mechanical, photocopying or otherwise, without the prior permission of the Publisher.

Earlier editions of this book have been published as *The Family Travel Handbook*

This edition published 1993
by Bloomsbury Publishing Limited
2 Soho Square, London W1V 5DE

Copyright © 1993 by Bloomsbury Publishing

A CIP record for this book is available from the British Library

ISBN 0 7475 1376 7

IMPORTANT NOTICE
Every care has been taken to ensure the accuracy of the information contained in this publication but no liability can be accepted by Bloomsbury Publishing Limited for errors or omissions of any kind. To ensure that such errors are eliminated from future editions readers are kindly invited to notify the publishers.

Typeset by Florencetype Ltd, Kewstoke, Avon
Printed in Britain by Clays Ltd, St. Ives plc

 Contents

Preface	vii
Introduction *Sheila Sang*	ix

Section 1
    Choosing your destination   3

Section 2
    Choosing and planning your holiday   79

Section 3
    Holiday transport   189

Section 4
    Getting there
Long journey? No problem . . . *Sheila Sang*	221
Holiday reading for kids *Wayne Jackman*	227
Renting somewhere to stay *Kathy Rooney*	233
A half-term holiday break at Butlins *Sheila Sang*	236
Holiday insurance *Ernest Jones*	240

Section 5
    Family travel know-how   247

Section 6
    Medical know-how   263

Section 7
    Information and addresses   275

Section 8
International SOS	303
International food vocabulary	312

Index   325

Preface

How to Have Stress-Free Family Holidays is an essential manual for all those planning to travel with children. It is packed with down-to-earth advice and helpful tips for family travellers that may help to make the holiday journey a pleasure rather than a chore.

The book is split into eight sections, organized to guide the reader through the process of choosing the type and destination of their holiday, choosing a company to travel with, preparing for departure and coping with emergencies while away from home.

The countries featured in *Section One* are described evocatively by the personal experiences of travelling families. Each article is full of essential information for those holidaying with children. They may not tell you what a standard guide book will think important, but the authors have tried to describe exactly what it is like to travel in each country with children. We have also included a selection of recent guide books for further information.

Section Two helps you to choose the type of holiday you would like and describes the facilities provided by many hundreds of travel companies. This exclusive survey includes sample prices to help you gauge the character of the holidays offered by the company concerned.

In *Section Three*, 'Holiday transport', we describe what you can expect during your journey. This section includes an exclusive survey of facilities at airports throughout Britain. You may find your local airport far more comfortable to travel from than that suggested by the holiday brochure! For this edition we have also examined the level of amenities offered by motorway service stations throughout Britain and those of ferry ports operating major routes to and from the UK.

We have commissioned a series of articles for *Section Four* that will help you to prepare for your holiday: the editor, Sheila Sang, a highly experienced journalist and travelling mother, advises you on how to cope with your children while you are travelling and gives many handy hints picked up on her travels; Wayne Jackman, a well-known personality on children's television and capable travelling father, has compiled an extensive holiday reading list of inexpensive paperbacks graded by age, that will eliminate boredom from your journey; Ernest Jones advises families on the type of holiday insurance available; and Kathy Rooney, with many highly-successful family holidays under her belt, tells you how to avoid disaster when booking self-catering accommodation.

Section Five is full of practical holiday know-how, which we have arranged alphabetically. This will prove invaluable during the holiday, when you want to know what to do to avoid travel sickness, boredom, and other nightmare situations.

Preface

The straightforward advice of Dr Richard Darwood is well known. For *Section Six* he has compiled an A–Z of the most common holiday problems encountered by families, including such headaches as diarrhoea, insect bites and sun-stroke.

The final two sections are equally useful. *Section Seven*, 'Useful addresses and phone numbers', will help you to obtain expert advice and information on the destination you have chosen, the type of holiday you are planning and many other holiday topics.

The final section, *Section Eight*, gives a five-language vocabulary of 90 words that will help you to deal with family emergencies. It is often difficult to find medical vocabulary in a standard phrase book – you will find everything you need here, from allergy to vomiting. New for this edition, an international vocabulary of food will guide you through even the most complex of menus. At last, when the children demand sausages and chips, you will know how to order them!

Finally, we would like to appeal to our readers to help us to keep this book as up-to-date as possible. Please write and tell us of your holiday experiences with children (both good and bad). If you disagree with any advice or information in this book, please tell us. We will endeavour to include your views in the next edition and in this way, you may help us to raise the standard of service we expect when embarking on a family holiday.

Bon voyage and, above all, enjoy travelling with your family!

How to Have Stress-Free Family Holidays
Bloomsbury Publishing
2 Soho Square
London
W1V 5DE
Tel: 071 494 2111

Introduction
by Sheila Sang

Wet weather blues?

It's one thing being armed with good advice about what you and your children can expect at your holiday destination, how to get there with least family feuding, and what to take . . . but what do you do if it rains?

With children in tow the depression wet weather brings to your holiday HQ can equal any that's forecast by Michael Fish. A round of whist might work wonders with a group of consenting adults, but tip in a toddler or two and your previously surefire solutions will be thrown to the all-too-present winds.

The first thing you'll find out is that if you've planned a 'cheap' holiday in Britain, in wet weather the outlay on indoor entertainment such as butterfly farms and stately homes can quickly make two weeks in the Med look like the bargain basement. A spell of rainy weather could mean your close encounters with the cashpoint have a regularity beaten only by the patter of the rain. Ask anyone who tried it last August.

Part of your holiday planning should be to contact the local tourist office (listed under useful addresses at the back of the book) to get details of what there is to see and do near your destination. This should get you properly filled in on the area, and you can make definite plans to visit one or two of the most appealing undercover spots when rain stops play. If you've started off with such trips in mind, the whole family can look forward to these excursions instead of seeing them as a second-rate fill-in when there's nothing better to do. But if funds *aren't* unlimited, it's well worth seeking out some of the less well-advertised attractions. Bear in mind that the local people have children too, and the chances are that there will be activities laid on for them in the summer months.

Sports centres are a good place to start. Many run special courses for children during the summer months, offering anything from windsurfing to badminton. (With watersports the hire of a wetsuit is often included in the fee, so the chill factor of the local lake or bracing seashore breezes needn't be a total turn-off.) You may be able to book in for one-off sessions, or if your children are enthusiastic, by the week.

Staff at sports and leisure centres may also be able to point you in the direction of other organised schemes for youngsters and pre-school activities such as Junior Gyms with soft play mattresses or ball pools for the kids to dive into. Libraries often organise school-holiday activities, such as puppet shows and children's entertainers. There's the added bonus of being able to give your children a change from the boring old books they took from home.

Museums and art galleries are always well worth a visit. Apart from their normal cultural offerings some also run holiday programmes specifically arranged for children, with guided tours and activities geared to stimulate young imaginations. You might even learn something yourself! Many museums attract younger visitors with year-round 'hands-on' exhibits that you'll probably end up dragging the kids away from at closing time. A few phone calls will tell you what's available, and when.

The local swimming pool is also a safe bet to while away a few rainy hours. Ask around to see what's available rather than necessarily heading for the most obvious one. Leisure pools with wave machines, slides and flumes are cropping up all over the country faster than greenfly on a cabbage patch. They're great splashy fun for all the family, though the more sensational ones could seriously damage your bank balance.

Of course its not fair to imply that only British holidays can dampen our spirits (or other exposed parts). Given the notorious uncertainty of our own summer weather, we've justly earned a reputation for heading south for sunnier climes. All the more disappointing when we're met there by endless downpours or even worse – cold weather.

Most housing in warmer climates, and particularly holiday housing, is designed to maintain a level of cool on hot summer days. In the cold and rain the intended refreshing effect becomes positively chilling. Wet clothes won't dry. It can take hours to warm up after a 100 metre sprint from the local shop/restaurant/bus stop. And it quickly becomes clear why gas ovens and gas fires were designed differently (one warms a room, the other doesn't). If you've left the dozens of warm tops and trousers for the kids out of your summer packing, you're unexpectedly stuck. And no, three pairs of shorts worn one on top of the other don't give you the warmth of a decent pair of Marks & Spencer track suit bottoms.

If rain strikes late in a holiday, you're likely to be relaxed enough to take it with a certain equanimity. By then, you'll also have checked out what the locality has to offer. If it pours first thing, and the rain stays, you'll need to call on all your reserves of parental patience to prevent the onset of cries of disappointment and appalling appeals for entertainment. If you thought your children were bored on the journey, try keeping them cooped in a hotel room or rented apartment to watch the rain for hour upon hour.

Few countries have attractions themed for children in the way that we do at home. The good part of this is that you save the cash you would otherwise hand over to keep the kids dry and amused, the bad part is that you're left largely to your own resources.

Contact the tourist office and check out guide books before your holiday to get basic information about the most interesting sights to see and things to do. When you arrive, visit the information office in the most popular area as it should be well-stocked with up-to-date information and accurate opening hours of places you're interested in (there's little that is less enchanting than trudging through a downpour to find that a museum or archaelogical site is shut on that particular day of the week, or shut for four hours for siesta.)

Many hotels in resort areas will have some entertainment laid on for children – if that is the case, your problems should be taken care of for a few hours at least.

If you're not staying in a hotel you'll probably find it useful to search out the nearest cafe to find other families with children who are sheltering to brave out the storm. Resort cafes in particular welcome families – even if you do make two coffees and two lemonades last from breakfast to lunchtime. For toddlers a visit to such a cafe may be all the entertainment you need. Chasing like-minded small people around the tables appears to be a thoroughly absorbing pastime for many two-year-olds, and such behaviour isn't frowned upon in most family-orientated venues abroad.

Whatever the age of your children, there's nothing like information from others in the same boat. So while your children are making friends with the other kids you may be able to find out from Mum and Dad what else there is for families to do and (just as important) how long the rain may be expected to last! But beware of cafes with video games. From toddlerhood onwards these machines have a magnetic attraction for both children and cash. Your purse and patience will rapidly diminish if you remain in the vicinity.

If your children are of museum or sightseeing age, bad weather gives you the opportunity to profit from this when they know there's little alternative. But if bucket and spade is all that will bring a smile to your tots' faces, you needn't necessarily accept defeat at the first drop of rain. Unless you're facing a hurricane or hailstorm most little kids will happily toddle around in the drizzle on the beach. Beach umbrellas can do more than shade you from the sun!

On a more positive note, bad weather can give you the opportunity to explore without wails of protest from beach-mad babies. If you have a car, it's an ideal opportunity to tour when it isn't too hot for the kids in the back. Local wineries/factories may be happy to take you on a tour and show your children how their goods are produced. (You may even get some cheap samples.) Children will also enjoy using local buses and are likely to receive a warm welcome (if unintelligible to you) from villagers. Small village shops may offer local produce that wouldn't normally make it into tourist zones. Don't forget to check the time (and likely reliability) of the return bus!

But in the end, possibly the key to making sure the children have a good time is not to let them know that *you* see the rain as a problem. If you pretend that you planned it that way maybe they'll just get on with finding things they enjoy doing and forget to complain!

Sheila Sang

 **Section 1**

Choosing your destination

Introduction

Is it to be Norfolk again this year or should you try Thailand? The travel industry can offer you boundless choices of hotels and villas in Spain and Greece, but they also go to somewhat more exotic parts of the world *and* offer child discounts – so why not step out of the familiar? We firmly believe that it is possible to travel far and wide with children, but understand that parents will want practical information which is aimed at their special needs. As in any reference book the selection of what goes in and what does not is a difficult one. At first glance the choice of countries featured in this section may appear idiosyncratic, but there was method in our madness!

We have made the assumption for the purposes of this book that people are setting out from the UK, and therefore some countries feature because they have a special relationship with Britain. Not many people go on holiday to Saudi Arabia, for instance, but there are thousands of people working there from this country whose families want to visit them.

You will find the familiar holiday destinations here – France, Spain, Greece – but your wanderlust may be stimulated by the entries on the more unusual spots of the world – Nepal, Jordan, Bulgaria and Zimbabwe. Also this year we have 'double' entries on several popular countries such as France, Turkey and Spain. A special feature for Spain are separate descriptions of the Canaries and Majorca, the latter being the story of one family's holiday in a charming port town.

Fact boxes

The fact boxes provide essential information about each country's major cities, climate, languages, currency and time difference. The following terms are among those which are used to describe climate:

Continental Wide seasonal temperature range. Reasonably predictable.
Temperate Relatively small annual and daily temperature range. Highly variable and often unsettled.
Mediterranean Hot dry summer and mild cool winter. Summer reliable but winter more disturbed.
Tropical Hot throughout the year; often high humidity.
Monsoonal As tropical with alternate dry and (very) wet seasons.
Desert Extreme continental, with wide daily temperature range; dry.

The information given in the boxes is intended only as a rough guide. The temperatures given are the daily average for the month.

Remember, weather varies between coast and inland, and in general temperatures fall 0.6°C for every 100m increase in altitude. If you are travelling to a climate very different from what you are used to, it is advisable to do more research into what you can expect and buy the appropriate clothing to cope with it.

Antigua

See **West Indies**

Australia

Capital Canberra
Major cities Sydney, Melbourne, Brisbane, Adelaide, Perth, Darwin
Time GMT + 8–10
Currency Australian dollar (A$) = 100 cents
Language English
Climate Wide regional variation from tropical monsoon to cool temperate and deserts

Annual weather range
North (Darwin): tropical
　Temperature
　　Max 34°C (Oct–Nov)　Min 30°C (Jul)
　Rainfall
　　Max 386mm (Jan)　Min 0mm (Jul)
Southeast (Sydney): warm temperate
　Temperature
　　Max 26°C (Jan–Feb)　Min 16°C (Jul)
　Rainfall
　　Max 135mm (Apr)　Min 71mm (Oct)
Southwest (Perth): mediterranean
　Temperature
　　Max 29°C (Jan–Feb)
　　Min 17°C (Jul–Aug)
　Rainfall
　　Max 180mm (Jun)　Min 8mm (Jan)
Centre (Alice Springs): desert
　Temperature
　　Max 36°C (Dec–Jan)　Min −19°C (Jul)
　Rainfall
　　Max 43mm (Jan)　Min 8mm (Jul–Sep)

Australia is a paradise for family holidays. Once recovered from the exhausting plane journey, it's plain sailing. However, for those who have not been before, it is best to consider your mode of travel carefully. Australia is a large country and distances between places are often very considerable. For local travel, a hire car (ask for air-conditioning and, if you want one, a baby seat) is leisurely, but for interstate travel, it is advisable to use the domestic airline. You can get substantial discounts (see your travel agent for details).

Accommodation is varied and similar to the US and Canada. Motels with large family rooms are common and cots are normally available on request. The better motels have swimming pools and some have playgrounds. An alternative is the campsite or caravan park. These usually have playgrounds and sometimes swimming pools and you will probably see Ova-nite caravans which are cabins and caravans to rent at larger sites.

You will find parks in most towns with free toilets, playgrounds and picnic tables. A swimming pool, often with toddler's pool attached, is common in most towns.

Remember that the seasons are reversed in the southern hemisphere and January is hot and dusty. If your accommodation is not air-conditioned you can request a fan. I have often asked for one, sometimes with favourable results.

Supermarkets are found in every large town with fresh food supplies, long opening hours and special baby sections. In general, the cost of living is less than in the UK.

There is no national health service that covers foreign travellers and you would be well advised to include medical insurance with your travel arrangements. For minor problems, I have found the local pharmacist in small towns most helpful.

Mother's rooms and baby changing rooms are more likely to be found in department stores attached to the ladies' cloakroom – as in the UK.

If you are out and about with small children you should be aware of the hazards peculiar to Australia. You should be wary of spiders. They can be lethal in Australia so 'hands off' all creepy crawlies is the best policy to follow. Playing around long grass or wood piles is also not wise because of the snakes – there are several deadly varieties. The sun in Australia is very intense and you must monitor your children very carefully. High-factor sun-protection creams and hats are essential for the whole family.

Eating *en famille* poses no problems. Australians eat their evening meal (often called tea) from 6.00 to 6.30 p.m. Certain chain restaurants actually advertise themselves as family restaurants. Fresh fruit, vegetables, salads, grilled meat and fish are reasonably priced and commonplace. We have found cafés in small country towns excellent value for family meals. They provide good plain food and serve informal meals at all hours. Often as not the owner will be happy to prepare simpler fare for your baby. Another source of good, cheap meals is the local club where you have to be signed in. The club generally relies on revenue from visitors so the Secretary, usually found in the office, will be happy to oblige.

Recommended reading:

Australia: A Travel Survival Guide (Lonely Planet, 1989)
Insight Guides – Australia (APA Publications, 1990)
The Insider's Guide to Australia, Robert Wilson (MPC, 1989)

For further information:

Australian High Commission, Australia House, Strand, London WC2B 4LU. Tel: (071) 379 4334 (visa information)

For tourist information:

Australian Tourist Commission, Gemini House, 10–18 Putney Hill, London SW15 6AA. Tel: (081) 780 1424

Austria

Capital Vienna
Major cities Graz, Linz, Salzburg, Innsbruck
Time GMT +1
Currency Schilling (Sch) = 100 Groschen
Language German
Climate Moderate climate

Annual weather range – Vienna:
Temperature
  Max 30°C (Jul)   Min −4°C (Jan)
Rainfall
  Max 84mm (Jul)   Min 39mm (Jan)

Austria is, I would say, a good place for a 'safe' family holiday. For all its mountain ranges, forests and rivers, Austria is tame, almost cosy. Everything (or nearly everything) works. Public transport is efficient; communications are excellent. The hotels and inns are well run, immaculate and comfortable, but then Austrian hoteliers have few equals in the world. The restaurants and cafes are impeccably clean and efficient and produce delicious food – naturally, because Austrian restaurateurs have no peers (unless they be the Swiss). The water is drinkable. Medical facilities are first rate and chemists sell anything you might need. If your child is stung in the gullet while eating a slice of scrumptious Viennese strawberry flan *mit Schlagsahne* (whipped cream) his chances of survival are much higher here than in most places. In short, there is an absence of hazard which may be very reassuring. Moreover, the people are prompt, industrious, well-organized and obliging. Children are very welcome.

If you are confined to the towns and cities (e.g. Graz, Vienna, Salzburg, Innsbruck – fascinating places all), then of course you will find the kind of entertainment available for children that you would find in any other large urban area in Europe, but if your children can be profitably lured into museums, you will find more.

Beach life is confined to the shores of the lakes where the resorts are very well equipped. Frankly, I would not take small children to Austria unless they enjoy rustic life, walking, hiking, scrambling in streams, catching butterflies and so on. Then, you could well base yourself in one of the excellent *Gasthöfe* (inns) in a village where your benign host would fill your children with nourishing food. Day trips by mountain railway and cable cars will appeal to little ones.

Other possible diversions are the festivals, such as the Corpus Christi processions on boats in the Traunsee, Hallstättersee and Attersee. Another notable Corpus Christi celebration is the Samson Procession at Tamsweg in Salzburg province. A monstrous red-haired Samson (25ft tall) dances before the Town Hall. In other villages there are stage fights between Davids and Goliaths. In Carinthia, Whit Monday is the occasion for knight-like contests between boys on horses. In the Tyrol certain villages stage carnivals with traditional dances and mumming plays.

The Tyrol is well known to the English; less well known perhaps is the Vorarlberg region at the far western end of the country. The Austrians have been at some pains to develop this part for family holidays and there is a good deal of entertainment for children in delightful surroundings. It's nothing like on the scale that the Danes have devised (see description about Denmark) but the diversions are similar: playgrounds, amusement parks, 'adventure' grounds and so forth. Austria has a well-developed tourism industry and offers accommodation in private rooms, farms, campsites, inns, pensions and all types of hotel. Most of these places have special arrangements and prices for children. There is a lot of organized entertainment for little ones, including soapbox races, medieval pageants with fire-eaters, magicians and bear tamers, puppet theatres and torchlight processions.

Recommended reading:

Fodor's Austria (Fodor's Travel Publications, Inc., 1991)

Blue Guide – Austria (A & C Black, 1987)

Frommer's – Austria and Hungary, Darwin Porter (Prentice Hall Press, 1991)

For further information:

Austrian National Tourist Office, 30 St George Street, London W1R 0AL. Tel: (071) 629 0461

Barbados

See **West Indies**

Belgium

Capital Brussels
Major cities Antwerp, Bruges, Ghent, Mons, Liège
Time GMT + 1
Currency Belgian franc (Bfr) = 100 centimes
Languages French, Dutch, German
Climate Cool temperate
Annual weather range – Brussels:
Temperature
  Max 23°C (Jul)   Min −1°C (Jan)
Rainfall
  Max 95mm (Jul)   Min 31mm (Mar)

One thing all Belgians have in common is a fondness for their food. Coming a close second is their emphasis on family unity. In this way Belgium is very similar to France, so British families can expect a holiday where children are well catered for.

The Belgian welcome for young children is most evident in the coastal resorts and Antwerp rather than the main city of Brussels – notorious for being a lively capital with an extensive but rather expensive shopping centre. Even the picture-book towns of Ghent and Bruges cannot match the facilities available for children and their parents that can be found along the north-western coast of Belgium from Ostende to Knokke, which also incorporates the most fashionable beach spot of Knokke-Le Zoute.

Along Knokke-Le Zoute's winding promenade that seems to be endless, a wide variety of beach, playground and indoor sports are available for the youngest of age groups. The promenade, known as the *dijk* is a haven for the *cuistax* lover. Cuistax can best be described as pedal-driven open-topped carts; the largest ones hold twelve people and also provide baby chairs – although these do not have pedals. Alternatively, there are enclosed play areas situated between rows of outdoor coffee bars that provide special half-portions and high chairs for children. For those families who forgot to bring games for their children there are plenty of beach toy shops to be found along the *digue*.

Further into the countryside, restaurants invariably have large gardens with an assortment of slides and swings and make the celebrated *gaufres* or waffles that can appease any screaming child. The Belgians, like the Dutch, are also great eaters of *frites* – fried potato chips served only with home-made mayonnaise – and since there are stalls at many street corners all over Belgium they are a convenient between-meals snack.

Baby supplies are readily available in Belgium in pharmacies, drugstores and supermarkets, and many of the larger department stores and *galleries* (long covered arcades of shops) are very likely to contain their own baby section. For those who enjoy baby couture, 'Dujardin' – known to its clientele as 'DJ' – is the best-known children's clothes shop (for style as well as prices). Yet it is much patronized by benevolent grandparents, for the quality and design cannot be equalled anywhere else.

Larger shops are likely to provide a mother's room, rest areas and places to leave prams, while you will find that the smaller shop, often run by a family team, will be hospitable and accommodate your needs. Most hotels will be able to provide cots with little fuss.

As a member of the EEC, the usual reciprocal arrangements apply regarding medical care. Many doctors in Belgium are specialists but there are also general practitioners like those in the UK.

On the whole, you will find Belgian outdoor culture and desire for respectability and comfort make the country attractive for travelling with young children.

Belgium contains not only the bustling holiday resorts of the north coast and the old world towns of Ghent and Bruges, but also miles of unspoilt wooded countryside – in particular the forests and farms of the Ardennes. It's an area of hills and dramatic river gorges, where wild boar roam in endless tracts of forest criss-crossed by miles of beautiful and well sign-posted footpaths.

If you want a country holiday, where the children can help get the milk from the farm in the morning, try renting a gîte – addresses available from Belsud, Rue Marché aux Herbes 61, 1000 Brussels. These holiday homes, similar to those in France, can vary from small primitive farm cottages to grand chateaux with several acres of grounds.

Eating out *en famille* is common, both at midday and in the evening, with Sunday lunch particularly popular. Meals are relatively cheap; not so snacks in cafés. Jars of baby food are easy to buy and can be good, though you may consider some varieties too sweet.

Recommended reading:

Frommer's – Belgium, Holland, Luxembourg, Susan Poole (Prentice Hall, 1991)
Fodor's – Belgium and Luxembourg (Fodor's Travel Publications, Inc., 1991)
Blue Guide – Belgium and Luxembourg, John Thomas (A & C Black, 1989)

For further information:

Belgian Tourist Office, Premier House, 2 Gayton Road, Harrow, Middlesex, HA1 2XU. Tel: (081) 861 3300

Brazil

Capital Brasilia
Major cities São Paulo, Rio de Janeiro, Salvador, Belo Horizonte, Recife
Time GMT −2–4
Currency Cruzeiro (Cr) = 100 centavos
Language Portuguese
Climate Tropical with wide regional variations
Annual weather range – Rio de Janeiro:
Temperature
   Max 29°C (Jan–Feb)   Min 17°C (Jul)
Rainfall
   Max 137mm (Dec)   Min 41mm (Jul)

Brazilian life is child orientated: family life revolves around them, consumer advertising is heavily directed at them, and as a traveller in Brazil you can't fail to notice that Brazilians do regard them as a very important part of life.

So no problems when it comes to people being friendly or helpful with your children – hotels and restaurants in the major cities will provide all the necessities like cots and high chairs, and hotels will always arrange for someone (usually one of the staff) to babysit if you warn them at the beginning of the day. In the major cities too, there is no problem finding everything you need for children at fairly normal prices. Shops are open until late; most restaurants or American-style snack-bars serve some international-type food; getting around by taxi or even by bus is relatively easy too, but if you hire a car, don't expect child seats to be provided.

Outside the affluent centres of the major cities, travelling with children is another matter. Take key necessities with you and stick to food which is thoroughly cooked except for the wonderful fruit like papaya, avocado, pineapple or oranges which you should open or peel yourself. Bottled drinks will always be on offer. Never drink tap water or use ice. When travelling away from main tourist areas, malaria tablets and a vaccination against yellow fever and cholera are strongly recommended.

Distances are enormous in Brazil, and although there are some good long-distance buses (and it is an excellent way to see the country), children have limited stamina for the heat and dust. The internal air network may prove a more attractive, if a more expensive, option. The amazing Iguaçu Falls in Paraná state are a must. In Rio children love the cable-car ride up the Sugar Loaf mountain. There is a playground half way up. In Flamengo park on the waterfront there are many activities for children. And don't forget Carnival!

Wherever you are, you are likely to spend time on the beach. Take note: the sun is very strong, even if it seems very hazy, and children unused to it need very good protection and only limited exposure to start with. You may also need a strong repellent against the little beach flies which bite, particularly at dawn and dusk. Be very careful of the sea – the current or undertow can often be quite strong, and children unused to this type of sea can be easily frightened if they are tumbled by the undertow. Beware of the pickpocket teams of small children that work the main beaches.

Brazil is a magical country. Its warm-hearted people, all-pervasive music, marvellous beaches and tropical vegetation make it a very special place. Most people with small children are unlikely to feel up to seeing many of the wonderful areas of Brazil which involve travel over big distances, but living very simply, and with a bit of common sense you *can* travel with children to Brazil with a minimum of fuss and have a great amount of fun.

Recommended reading:
1990 South American Handbook (Trade & Travel, 66th edition 1990)
South America on a Shoestring (Lonely Planet, 1986)

For further information:
Brazilian Embassy, Information Department, 32 Green Street, London W1Y 4AT. Tel: (071) 499 0877

Britain

Capital London
Major cities Birmingham, Manchester, Liverpool, Bristol, Edinburgh, Glasgow, Cardiff, Belfast
Time GMT
Currency Pound (£) = 100 pence
Language English
Climate Temperate
Annual weather range – London:
Temperature
  Max 22°C (Jul)   Min 2°C (Jan–Feb)
Rainfall
  Max 64mm (Nov)   Min 37mm (Apr)

This entry was written by a visitor to Britain. It may give you an insight into how foreign tourists view our behaviour and facilities! When travelling in Britain, it is advisable to contact the relevant tourist information office for advance information and advice. The addresses for these are listed in Section 7.

For such a small cluster of islands, the UK is surprisingly complex. For a start, it's four countries in one – and don't let anyone persuade you that England, Scotland, Wales and Northern Ireland are all the same. The variety of scenery, accent and welcome can be astonishingly different and the people who live in these countries may not welcome being called English.

There are many enduring myths about Britain and no doubt you've heard them all: the bad weather, the terrible food and the frosty people. It's true that the country does not boast a predictable or especially enticing climate, but it's really not as bad as it's painted. Certain areas, like Scotland, the Lake District and Wales can be rather damp, but you'd have to go an awfully long way to beat the stunning scenery. The best months to holiday in the UK are often June and September when the sun seems to try harder than usual. If warm weather is really important to you, try the Channel Islands. Although much nearer to France they are, in fact, British territory and offer a warmer climate and a friendly welcome to families with children.

Like other capital cities, London is not really indicative of the rest of the UK. Expatriates and non-Londoners complain of the dirt and dishonesty, but foreign visitors often compare it favourably with their own cities. It is reasonably safe and compact, with excellent museums and art galleries (mostly free). Public transport is fast and frequent, although expensive by some standards. Buy a Travel Card pass for London Transport and you can hop on and off buses and tubes all day (after 9.30 a.m.).

Restaurants in London and other major cities offer every imaginable cuisine, but the variety can be somewhat restricted out in the 'sticks'. Chinese take-aways and McDonald's are in virtually every provincial town, but traditional English food can be very hard to find, and apart from fish and chips, sometimes poorly cooked. Indian cuisine, however, has taken a grip and there are ample opportunities to sample cheap and exciting menus. Eating out is not a family tradition in Britain but it is gradually becoming a more popular habit. The pub often seems to be the social centre of British life, and even the tiniest village will have one. However, some town pubs are smoke-filled male enclaves that are not particularly inviting. In Britain pubs may now stay open all day, but in some parts of Wales they may still be closed all day Sunday. Nowadays, thank goodness, the wind of change is blowing and a lot more pubs have children's rooms and/or gardens with swings. Many landlords will allow children to eat in their pubs if they have a separate eating area – but if in doubt, ask. Pub food is often only available at lunchtime and the quality varies, but it is usually hot and sustaining, if nothing else. The *Good Beer Guide* (published by CAMRA), *Good Food Guide* and *Egon Ronay* will help you to choose good places to eat.

Wherever you eat – hotel, restaurant or pub – check first whether anything is provided for children.

All major cities have big and well-kept parks. They are safe and often have special play areas designated and fenced off for children. Usually there are toilets nearby. London's major parks are especially good, and in summer there are often (free) puppet shows, concerts, plays and other entertainment. Look in local papers or listings magazines for details.

Increasingly shops and public places have rest areas for children, but they are still few and far between. On the other hand, public toilets are practically a national institution and it's rare to have difficulty in finding one. If the need is desperate, large department stores always have toilet facilities.

Hotels come in all shapes and sizes, but invariably charge per person. The National Tourist Boards for England, Scotland and Wales operate official accommodation rating schemes covering virtually all types of accommodation in Britain (look for the 'crown' scheme for hotels, guesthouses, B&Bs, inns and farmhouses). Under these schemes, every property is inspected annually, providing consumers with the reassurance they seek before they check in. Cots are available in most hotels, but do not rely on finding them in guest houses without making prior arrangements. Bed and breakfast in private houses (and sometimes pubs) is an excellent British institution. Ask at the local tourist information office for details (there are nearly 900 such centres throughout Britain – look in the telephone directory under 'tourist information' for details of the nearest or turn to our listing in section 7).

Despite the sometimes off-hand service, shopping for and with children can be a pleasure. Look out in particular for Sainsbury (supermarket grocers with shops in many towns), Marks and Spencer (hundreds of large stores with food and good quality clothes), and Mothercare (a specialist chain of clothing and equipment shops for children). All sell own-brand nappies at very competitive rates and have additional ranges of baby necessities. In a pinch, late-closing corner shops sell most essentials, if rather more expensively.

Car rental agencies are plentiful (see *Yellow Pages*), and taxis are easy to find on streets of major cities, but you can always call a 'mini cab' from a private firm if you get stuck. If you don't know the number, dial 192 and ask for details. British Rail offers an extensive and expensive network of trains which sometimes leaves a lot to be desired in terms of punctuality and comfort, though a Family Railcard or one of the frequent Awayday bargain tickets can save you a lot of money. Coaches offer a cleaner and cheaper alternative. Air travel within the UK is very expensive, and given the short distances involved, rarely worthwhile. However, international travel from the UK is the best in the world. Heathrow is London's main airport and one of the busiest. Gatwick is smaller and used by many charter companies and cheaper carriers, such as Virgin. Although the distances within the airport are not great, it is quite a long way to the direct link railway and there are very few porters. We found this extremely frustrating at the end of an overnight flight from the US.

Britain's often maligned National Health Service will provide free medical assistance, although not all hospitals have casualty departments. As the system is under severe strain, it would be prudent for overseas visitors to take out private insurance. Emergency calls (Fire, Police, Ambulance and Coastguard) are free. Simply dial 999.

It would be foolish to limit your holiday in Britain to a frantic week spent in the centre of London. By far the best places to spend a relaxing and enjoyable holiday are away from the traditional tourist attractions of the capital (which, during the holiday months are usually full of foreign tourists and may be considered overpriced).

It would be well worth your while to contact the various British tourist offices for leaflets and brochures – we have found them unceasingly helpful and eager to ply you with information on more out-of-the-way locations and unusual activities. England, Wales and Scotland are packed full of picture-postcard scenery and, with a little advance preparation, you will find it surprisingly easy to get away from the standard tourist tracks and spend time on your own in beautiful countryside. The British Travel Centre, 12 Regent Street, Piccadilly, London SW1Y 4PQ (for personal callers only) has information and a huge range of free and saleable books and leaflets on all parts of England, Scotland and Wales.

There are many unspoilt places in Britain and a great way to enjoy these is by renting a cottage by the week. Many companies offer country cottages all year round and, if booked in advance, cots and bunk beds are often available (see 'Cottages, Gîtes and Farmhouses' in section 2 for some suggestions). The National Tourist Boards also operate an accommodation rating scheme, covering every type of accommodation (further information is available in tourist board publications).

To find details of the historic buildings, monuments and gardens that are open to the public, you can contact English Heritage, Cadw (Wales) and Historic Scotland (properties owned by the Department of the Environment). The National Trust and the National Trust for Scotland are charities that own sites of historic or environmental interest and although you may have to pay a little more for entry, it is well worth contacting them for details of the buildings and gardens they administer in the area you are visiting. (The phone numbers are in 'Useful Addresses & Phone Numbers', section 7).

Local tourist offices (situated in larger towns throughout Britain) will tell you of events and places of interest in the area. You may be pleasantly surprised at what you find – there is a popular jazz festival each August in Brecon for instance and the Royal Show at Stoneleigh provides a highly enjoyable family day out.

Britain's National Parks are specially designated areas of natural beauty and you won't go far wrong for scenery if you visit one. They each run an information service and you can find their addresses from the regional tourist board offices.

All in all, the countryside of Britain provides an endless variety of activities and scenery and, with a little thought, you can have an enjoyable family holiday without meeting too many other tourists!

Recommended reading:
Blue Guide: England, Ian Ousby (A & C Black, 1989)
Insight Guides: Great Britain (APA Publications, 1985)
Let's Go: Britain and Ireland (Pan, 1990)

For further information:
English Tourist Board, Thames Tower, Black's Road, Hammersmith, London W6 9EL. Tel: (081) 846 9000
London Tourist Board, Tourist Information Service, Victoria Station Forecourt, London SW1. Tel: (071) 730 3488
Scottish Travel Centre, Scottish Tourist Board, 23 Ravelston Terrace, Edinburgh EH4 3EU. Tel: (031) 332 2433 *or* 17 Cockspur Street, London SW1Y. Tel: (071) 930 8661
Wales Tourist Board, Brunel House, Cardiff CF2 1UY. Tel: (0222) 499909

Bulgaria

Capital Sofia
Major cities Plovdiv, Varna, Ruse, Burgas
Time GMT + 2
Currency Lev (Lv) = 100 stótinki
Languages Bulgarian, Turkish
Climate Cool Mediterranean
Annual weather range – Varna:
Temperature
  Max 30°C (Jul)   Min −1°C (Jan–Feb)
Rainfall
  Max 64mm (Jun)   Min 26mm (Mar)

12 Choosing your destination

The Bulgarians are basically an orderly and tidy people so they are well organized for visitors and have made a great success of their tourist industry.

This is especially true of the Black Sea coast (no more than a long afternoon by air from Britain). Forty years ago this coast was untouched except at Varna, which between the wars developed into a resort for the members of the European middle and upper classes. Some regions were snake-infested, so the Bulgarians imported mongooses and then created a series of well-designed hotels, chalets, villas and campsites at Slanchev Briag and 'Golden Sands' and elsewhere along the many miles of superb sandy beaches. These are absolutely ideal for a family seaside holiday with small children. It is a stimulatingly cosmopolitan place. There are a lot of splendid open-air restaurants and much music and dancing in the evenings.

Most budgets and tastes are catered for and the hotel proprietors and staff are very obliging and welcoming. Numerous diversions and entertainments are available for small children and there are folklore spectacles and camel rides.

Bulgarians eat well. They enjoy a fertile country and there are copious supplies of good meat, vegetables and fruit. Meals are well prepared and well cooked. Particularly appetizing are the soups, hors d'oeuvres, goulashes and various kinds of kebab. As you would expect, the main influences on the cuisine have been Russian and Turkish. There is a plentiful range of soft drinks, and the water is very good; but, as in all Balkan countries, I would be cautious about drinking tap water. Emergency medical treatment is adequate and so is the supply of pharmaceutical goods.

Most people go on a package tour, but individual travel is perfectly feasible on public transport and by car. Self-drive car hire is well organized. It's good country for camping and there are big regions of totally unspoilt and very beautiful countryside, especially in the ranges of the Rhodope mountains. Trips on the lower reaches of the Danube can be organized and I would recommend visits to such towns as Plovdiv and Assenovgrad, as well as to the capital Sofia which is quite a sophisticated place. Nor should one miss a visit to Nessebur, a small Byzantine settlement on a peninsula which juts into the blue and seductive waters of the Black Sea. There is much to divert small children in all these places – donkey rides, playgrounds and even organized entertainment like treasure hunts.

Please note
Recent political changes are such that it is difficult to know the effect they will have on holiday makers. We would advise you to contact the relevant National Tourist Office for further details before travelling.

Recommended reading:
Bulgaria – A Travel Guide, Philip Ward (Oleander, 1989)
AA – Essential Bulgaria (AA, 1990)

For further information:
Bulgarian Tourist Office, 18 Princes Street, London W1R 7RE. Tel: (071) 499 6988

Canada

Capital Ottawa
Major cities Toronto, Montreal, Vancouver, Edmonton, Calgary, Winnipeg, Quebec, Halifax
Time GMT −3½–8
Currency Canadian dollar (C$) = 100 cents
Languages English, French
Climate Polar to warm
Annual weather range
South (Ottawa): cool continental
  Temperature
    Max 27°C (Jul)   Min −16°C (Jan–Feb)
  Rainfall
    Max 89mm (Jun)   Min 56mm (Feb)
West (Vancouver): temperate
  Temperature
    Max 23°C (Jul–Aug)   Min 0°C (Jan)

Rainfall
  Max 224mm (Dec)   Min 31mm (Jul)
Prairies (Winnipeg): continental
  Temperature
  Max 26°C (Jul)   Min −25°C (Jan)
  Rainfall
  Max 79mm (Jun–Jul)
  Min 23mm (Dec–Feb)

Canada is now one of the most popular medium haul destinations from the UK. Halifax on the Atlantic coast is only 5½ hours flying time; and a non-stop flight to the Pacific coast is 9 hours.

Visiting friends and relations in Canada has now been overtaken by tourists anxious to go further afield, and enjoy the culture and wide varieties of activities that abound in all 12 provinces. Canada is highly developed with civilized amenities which, combined with friendly people, make it an ideal place for family holidays.

The main cities such as Quebec, Montreal, Toronto, Calgary, Edmonton and Vancouver cater for every need for all ages.

Travelling with children in Canada is relatively easy despite the long distances. Motorways and other main roads have frequent service areas with eateries, often the inescapable fast food chains. It is worth leaving the main roads and trying out places to eat or stay in small towns nearby. Restaurants mostly have high chairs, children's menus and service is usually quick.

Accommodation in hotels and motels is readily available and similar to that in the USA. The charge is for the room and a family of four can usually share, making the stay an economical option. Canadian hostels also accommodate families.

Main tourist areas have information booths at the roadside which give maps and leaflets and can often provide advice on accommodation.

There are some obvious places and events which will appeal to small children. For example, there is the Wye Marsh Wildlife Centre near Midland, Ontario – a waterland sanctuary for muskrat, mink, beaver, frogs, turtles and numerous species of birds. This is a remarkable place to explore and possesses much of interest to little ones.

Niagara Falls are virtually an obligatory visit but should that awe-inspiring sight not have the hoped-for impact on your child's sensibility there are more mundane attractions at-hand such as Ripley's Believe it or Not Museum and Louis Tussaud's waxwork collection which, naturally, includes such notorious figures as Queen Victoria, Henry VIII and Mrs Thatcher.

For the technically minded 5-year-old there is a splendid aviation museum at Rockliffe Airport, Ottawa, which displays flying machines ranging from early balloons to jet fighters and the Science Museum in Toronto which is a highlight for children.

Toronto annually mounts the Canadian National Exhibition which combines water shows, air shows, bathtub races, horse shows, wilderness adventure rides, merry-go-rounds and pet zoos. It's a sort of vast amalgam of circus, fairground, tattoo, theatre and adventure playground. Also sited at Toronto is Canada's Wonderland where there are many entertainments for children.

Canada is wonderfully well suited to an adventurous open-air holiday but it has to be remembered that the country is so vast that the whole UK would fit into it over forty times.

Given a choice with small children in tow I would be tempted by the province of Alberta. In that huge area, several times the size of England, there lie the splendid Banff and Jasper national parks where there is a year-long range of outdoor activities. Calgary has its famous rodeo in July each year. This is every bit as good as a fine circus and the chuck-wagon races, steer roping and bronco busting are exciting events for children and adults alike. East of Calgary is the Dinosaur Provincial Park where the world's largest collection of dinosaurs has been preserved in rock buttes and spires. The existence of real-life Indians would appeal to small boys. There are marvellous trips possible on the Icefields Parkway between Lake Louise and Jasper.

From Alberta one could go into the 'outback' of the Northwest Territories and the Yukon. There is a good museum about Indians at Fort Smith. In the Nahanni National Park are the Virginia Falls (twice the height of Niagara).

Every year at Yellowknife the Caribou Carnival and the championship Dog Derby take place and there is another weeklong carnival at Frobisher Bay in April. Dawson City and Watson Lake in August hold parades and raft and canoe races; and at Dawson City in September there is a bizarre event: the Klondike International Outhouse Race (outhouses on wheels!).

Newfoundland and Labrador have their fair share of spectacles. A daily tattoo is held at St John's in July and August. In the Avalon Wilderness Reserve area native animals can be seen. Two big annual festivals take place in Labrador each year: the Heritage Festival in July and the Bakeapple Festival in August. Some small children might also be diverted by the settlement at L'Anse aux Meadows which recreates Viking life around 1000 AD.

In Saskatchewan, home of the Canadian Mounted Police, there are several things of interest, including good museums at North Battleford, Moose Jaw and Saskatoon. Entertaining events are Mosaic, a kind of festival at Regina in May, the Buffalo Days Exhibition in early August, and the Folkfest at Saskatoon in late August.

Manitoba is rich in diversions for little ones. Winnipeg has the splendid Assiniboine Park and Zoo, with playgrounds, picnic sites and a miniature railway. The Manitoba Museum of Man and Nature is of interest to children of all ages. At Steinbach, the Mennonite Village Museum depicts the life of settlers from Russia in the nineteenth century. Each year, in Manitoba there is a whole series of festivals and events which provide any amount of entertainment.

The provinces of New Brunswick, Nova Scotia, Prince Edward Island, British Columbia, Quebec and Ontario also provide a wide range of diversions for the whole family especially excellent fishing and whale watching trips.

Recommended reading:
Fodor's – Canada (Fodor's Travel Publications, Inc., 1991)
Canada – A Travel Survival Kit Mark Lightbody (Lonely Planet Publications, 1989)
Insight Guides – Canada (APA Publications, 1990)
Frommer's Canada (Prentice Hall, 1990)

For further information:
Tourism Canada, Canadian High Commission, Canada House, Trafalgar Square, London SW1Y 5BJ. Tel: (071) 930 5305

Canaries

See **Spain**

China

Capital Beijing (Peking)
Major cities Chongqing, Canton, Guilin, Harbin, Nanjing, Shanghai, Shenyang, Wuhan, Xian
Time GMT + 8
Currency Ren min bi (yuan) (RMB Y) = 100 fen
Language Chinese
Climate Wide variation, predominantly monsoonal
Annual weather range
Northeast (Peking):
 Temperature
  Max 37°C (Jun–Jul) Min −1°C (Jan)
 Rainfall
  Max 243mm (Jul) Min 3mm (Dec)

China 15

Coast (Shanghai):
  Temperature
    Max 37°C (Jul–Aug)
    Min −1°C (Jan–Feb)
  Rainfall
    Max 180mm (Jul)   Min 36mm (Dec)

China is a huge country, well over a million square miles larger than the US, and thus has wide variations in climate, language, custom, tradition, attitude and food. You may be surprised by how few children you see considering the fact that you are travelling in a nation of over 1000 million people. This is because of the rigorous birth-control laws that encourage the number of children to be kept to one per family in the cities and no more than two children per family in the rural areas.

This restriction is a severe hardship because the Chinese are deeply interested in children, *any* children. Babies are adored, coddled and spoiled in China. Their wardrobes are very fancy and varied, more so than any other family member. They get the choicest bits at mealtimes, the best spot to sleep in and are the focus of most family outings and get-togethers. In no country have we seen children better cared for, healthier or happier.

Children from the West are especially fascinating to the Chinese. They are welcome everywhere and likely to receive a lot of attention. There is, anyway, *intense* curiosity in all 'Western' visitors.

Foreigners are likely to find themselves in the main urban centres (Beijing, Canton or Shanghai) where they will probably be accommodated in the increasing number of European and American style hotels, often of Hilton-esque proportions and similar style; for the most part extremely boring establishments. It is *much* more interesting to stay in the less ostentatious places (often well appointed and comfortable) which are actually used by the Chinese themselves.

China provides very little in the way of specialized baby equipment. Babies are usually carried by their mothers in simple cotton yokes (newborn are carried in front, toddlers at the back).

Outside the very large and sophisticated hotels entertainment for small children is somewhat limited. No child can be expected to appreciate Chinese opera or drama, but for a few pence (or *yen*) a Peoples' Theatre will provide an acrobatic and conjuring show of the highest international standards. The performers are droll, witty and marvellously skilful, which explains why they are in great demand world-wide (for example on the Paul Daniels TV show in Britain). Nor should one fail to see the superb jugglers and puppet shows. Parks and zoos are other attractive diversions.

Toys and novelties are few, given that China exports toy products to many countries, but some homemade products such as dolls are beautiful and absurdly cheap. An hour or two in the Friendship Stores (there are several of these in most of the main towns) is time well spent. Small children will enjoy themselves in them and be surprised how different they are from those at home.

Children also enjoy the many different kinds of open markets, street markets and 'bazaars' which are common. The variety of goods is amazing, as is the variety of unfamiliar goods. Restaurant and hotel food should provide no problems for your children. Chinese food and cooking are excellent and differ from region to region. Most people in the West now know something of Chinese food, if only from the local take-away. Much of it in China will be easily recognizable except for, in some areas, delicacies such as cat, dog, watersnake, sea slugs, fish lips and some species of caterpillar. Parents with children who cannot cope with Chinese cuisine will have to confine themselves to the more sophisticated hotels and restaurants where Western-type food can be bought. On the whole staff are most helpful and obliging. You can get babysitters. Care is assured to be gentle and solicitous although few can be expected to speak anything but Chinese.

Travel by public transport is a hazardous business unless you speak some Chinese and is really to be avoided. Chauffeur-driven cars can be hired but self-drive hire is virtually unknown. However, most visitors will be involved in some kind of organized tour. If the tour includes train travel so much the better. Nearly all small children love train journeys and China is one of the places for this. It's a splendid adventure, comparable to the big rides on the transcontinental trains of North and South America, the CIS and India. The trains move at a thoroughly gentlemanly pace and afford an unfolding view of other ways of life, which any four- or five-year-old will be fascinated by.

Courtesies are as important in China as in any other country. Public displays of affection involving physical contact should be avoided and modesty in dress is desirable. Visitors are often welcomed by applause and the correct response is to applaud back. Patience is vital in the East and you should not lose your temper, for that is to lose face. The best passport is a big smile, and that applies to children.

Any parent taking children to China needs to remember that the great majority of the population survive at subsistence level. A few hundred metres from the luxurious 'Hilton' spreads the labyrinth of *utongs* (alleyways) where, in drab and primitive hovels, without running water or sewage system, tens of millions live in congestion.

Remember, too, to take all the pharmaceutical goods you think you may need (including remedies for coughs, colds and grazes) and before going study the climate of the region(s) you are visiting. Many areas are stupefyingly hot (and humid) in summer, bitterly cold in winter. There are also monsoon regions. A carefully balanced wardrobe may be necessary.

Recommended reading:
Introduction to China, John Lowe (John Murray Ltd., 1988)
China – A Tourist Guide (Foreign Language Press, 1989)
Insight Guides – China (APA Publications, 1991)
Culture Shock – China, Kevin Sinclair and Iris Wong Po Yee (Kuperard, 1990)
China – A Travel Survival Kit (Lonely Planet Publications, 1988)
Coping with China, Richard King and Sandra Schatzky (Basil Blackwell, 1991)

For further information:
China National Tourist Office, 4 Glentworth Street, London NW1. Tel: (071) 935 9427

Cyprus

Capital Nicosia
Major cities Limassol, Famagusta, Larnaca
Time GMT + 2
Currency Cyprus pound (C£) = 100 cents
Languages Greek, Turkish, English
Climate Mediterranean
Annual weather range – Nicosia:
Temperature
   Max 37°C (Jul–Aug)   Min 9°C (Jan–Feb)
Rainfall   Max 76mm (Jan)

In general, tourists are welcomed in the Greek part of Cyprus (for details of travel to Turkish Cyprus, see below). Many British people have settled there and there are still large RAF bases so wherever you go people speak English. Children are very much included and expected to join in at restaurants and tavernas. All along the coast the usual chips and fast foods Greek-style, are available at tourist restaurants. Most are very happy to give small portions or special combinations to suit children.

Living in a villa we found it easy to get a take-away from local restaurants in the evenings to feed the children – plenty of salad, chips and kebabs. Baby supplies and all modern commodities are easily available and inexpensive in the towns. Inland, however, away from the touristy and sometimes seedy coast, life has changed very little. The traditional hill farming communities remain as you imagine they have always been; and should you get off the beaten track you are always welcomed, particularly the children.

Public transport is easy in the towns but it is also easy and cheap to arrange a self-drive rental car – with child seats – to get about in. Like many of the villa packages, we had a car (in fact a minibus) thrown in. This meant we were able to take all six of our (combined) children to deserted beaches and villages. For all of us this was the best part of the holiday.

The spring and autumn climates are ideal for children, but you should be prepared for very hot summers. The beaches are lovely for children, sandy and relatively pollution-free but during our stay there were a few storms which produced heavy deposits of oil all round the coast and eventually all over the children and their clothes. I suppose, though, this could happen anywhere but the incidence of oil spills is not high in the Eastern part of the Mediterranean.

On the whole I would say that Cyprus is a very easy (if bland) place to holiday with children. One can combine a seaside holiday with visiting easily understood Greek mythological/archaeological features, such as the birthplace of Aphrodite on the southwest coast. The castles are certainly a must especially if your children are the hanging-over-the-edge kind!

Food and supplies are no problem and at the right time of year, the climate is perfect. The people certainly are most friendly and usually keen to please. Medical facilities are similar to those you would find travelling in other eastern Mediterranean countries and you should take out insurance to cover emergencies. Accommodation is the usual range of resort hotels, self-catering flats and villas.

Several companies now offer holidays in the northern Turkish part of Cyprus. Scenically beautiful and much more old-fashioned than the Greek area, it is indisputably very attractive and families with children are undoubtedly welcome. However, travellers should be aware that there are no direct flights (all planes are routed via Turkey), that the self-declared Republic is not recognised by the British Government (so consular assistance is not available), and that a North Cyprus stamp in a passport may prevent further travel in Greece. The latter problem can be avoided by requesting that the North Cyprus stamp is marked on a separate sheet of paper folded into the passport. The Ercan Airport (in the occupied area), is not recognized by the international aviation authorities (IATA).

Please note
Recent political changes are such that it is difficult to know the effect they will have on holiday makers. We would advise you to contact the relevant National Tourist Office for further details before travelling.

Recommended reading:
Landscapes of Cyprus (Sunflower Books, 1986)
Days out in Cyprus, Robert Bulmer (Bulmer Publications, 1987)

For further information:
Cyprus Tourism Organisation, 213 Regent Street, London W1R 8DA. Tel: (071) 734 2593 or (071) 737 9822

Denmark

Capital Copenhagen
Major cities Aarhus, Odense, Aalborg
Time GMT + 1
Currency Danish krone (Dkr) = 100 öre
Languages Danish
Climate Temperate

Choosing your destination

Annual weather range – Copenhagen:
Temperature
  Max 22°C (Jul)   Min −3°C (Feb)
Rainfall Max 71mm (Jul)   Min 32mm (Mar)

You can hardly imagine that the native land of Hans Christian Andersen would be in any way allergic to children, or vice versa. Indeed, Danish children are happy, outgoing, and much is done for them. Much for the visitor, too.

For example, there is Sommerland West: a vast 'leisure paradise' or adventure playground which will keep little ones occupied for days. It includes a zoo and islands on lakes, cafés and picnicking grounds. For diversion there are sandpits, mini-trains, mini-go-carts, a giant air cushion, pedaloes, canoes, aqua bikes, pony riding, a climbing castle and mini-toboggans. All the activities are included in the admission fee.

There is also the Varde Sommerland in Varde Engpark whose entertainments include excursion boats, moon cars, air cushions, canoes, aviaries, pets and pony rides. Similar recreations are available at Farup Summerland, and at Dyrehavsbakken which lies in the middle of a forest ten kilometres from Copenhagen. There is a big water dipper plus tombolas, adventure playgrounds and performances by clowns. There is no admission charge.

Nor should you miss the famous Legoland Parken. Thirty million Lego bricks have gone into its creation. Each day there are Lego building competitions with prizes. The park boasts what is claimed to be the world's largest collection of dolls, plus Titania's miniature palace and a collection of antique mechanical toys. Should boredom appear then there are puppet theatres, 'activities' and the 'Wild West' of Legoredo Town.

I would also recommend Knuthenborg, which is a kind of safari park but includes a 'playland' – Smaland where attractions include a miniature steam train. There is a big zoo at Givskud, plus a large play area with a children's farm and you should not miss the Fisheries Museum and Sealarium at Esbjerg. There is a vast pool for the big seals, and a smaller one for baby seals, plus an aquarium. A visit includes a tour of a fishing vessel.

Another favourite place is Djurs Sommerland. As well as many of the diversions already mentioned, there is a cowboy land with gold mining and shooting rinks, a Red Indians' island and a waterland with spiral slides and paddling pools.

The capital Copenhagen boasts the famous Tivoli. Apart from the fairground there are performances in the children's theatre as well as a pantomime theatre. Four days a week, in the evening, there is a fireworks display.

The more precocious five-year-old will find the Viking Ship Hall at Roskilde interesting. There are five restored ships here.

Getting about in Denmark is very easy. Communications are excellent and every conceivable form of accommodation is available. Danish inns are particularly attractive. Hoteliers are friendly and welcoming and children will find Danish food sustaining and appetizing. The water is safe.

Most Danes speak English fluently, but, like other peoples, they appreciate some attempt by visitors to say a few words in Danish. Besides, it's a language which is quite fun for children to learn a few words of. After all, we have common roots.

Recommended reading:
Baedeker's Denmark (AA, 1987)

For further information:
Danish Tourist Board, Sceptre House, 169 Regent Street, London W1R. Tel: (071) 734 2637

Egypt

Capital Cairo
Major cities Alexandria, Giza, Hurghudo, Luxor, Aswan
Time GMT + 2
Currency Egyptian pound (E£) = 100 piastres
Languages Arabic, English, French
Climate Dry and humid
Annual weather range
Alexandria:
  Temperature
    Max 31°C (Aug)   Min 11°C (Jan–Feb)
  Rainfall
    Max 56mm (Dec)
    Min 0mm (May–Sep)
Cairo:
  Temperature
    Max 36°C (Jul)   Min 8°C (Jan)
  Rainfall
    Max 5mm (Dec–Mar)
    Min 0mm (Jun–Oct)

Egyptians, like Arabs everywhere, adore children – including other people's. Whole families enjoy a promenade along the Nile on a hot summer evening, taking advantage of a cooling breeze off the river.

Public holidays and the Friday Sabbath are picnic days, and every available green patch in the capital, Cairo, is occupied by parents, assorted cousins, aunts and uncles, and children playing, eating, talking or listening to a blaring radio.

On these days, too, the zoological garden overflows with humanity, to the intense annoyance of its permanent residents. The foreign visitors may attempt these entertainments (on a Monday to Thursday basis), or else, with innocent boldness saunter through the grounds of a private 'sporting club' whose shaded park areas are a bequest of a bygone era of British imperialism. Tourists should not be put off, therefore, by the genuinely friendly stares and comments directed at their own children. Blond, blue-eyed youngsters are especially admired. The indulgence of Egyptians towards children in general should ensure a sympathetic response to your requests regarding your own. However, you must not expect the same facilities for children that you would find at home.

The tourists' short-term stay will centre around the major hotels, such as the Hilton and Sheraton, and smaller hotels which accept package tours. The 'package' is the only sensible way to see Egypt if you are a first-timer. European languages are spoken in some fashion by most employees in hotel receptions, while universally recognized hand signals work wonders with porters and maids. Special portions of food for children can be requested (even if they don't appear on the menu) and the waiter will warm a bottle of milk for your child. Sad to relate, McDonald's, Wimpy and Colonel Sanders all have outlets in Cairo; convenient they may be, but they are yet another obstacle between the tourist and nourishing, genuine Egyptian food. Nappies and some baby foods are available, at a price, from many pharmacies. Bottled mineral water (imported) is also available, but the intense summer heat can best be alleviated by a superb local beer (*Stella*), which even European children have enjoyed with no known ill-effect.

The best months to visit Egypt are October to April; July and August should be avoided. Public transport in urban areas is often difficult. The new Metro system serving Cairo, Old Cairo and Helwan, is very well organized and there is a regular and inexpensive rail service. There is also an underground service. Taxis can be found parked in convoys around the big hotels and daily rates can be negotiated (with the help of the hotel doorman – who will inevitably be a cousin of the cab-driver). Taxis are painted black and white in Cairo and the fare should always be agreed in advance as meters are not to be relied on. There are also limousine services.

For the less adventurous, or those with very young children, tours in luxury air-conditioned coaches can be arranged through the hotels. The

Cairo–Alexandria road is well served this way as well, while train or plane is the usual way to visit the archeological sites of Upper Egypt at Luxor and Aswan. Shopping in the 'native' markets (*souk*) is crowded, noisy, stuffy and fascinating. Small babies are best carried in slings Indian-fashion; somewhat older youngsters should be put in a harness and led so you don't lose them.

Public toilet facilities are virtually non-existent.

There are also main streets packed with small specialist shops selling shoes, cloth and souvenirs. All items except the last usually have fixed prices, but you never lose face by trying to bargain. And finally, remember the ancient Egyptian adage: once you have drunk from the Nile waters, you are destined to return to Egypt.

Recommended reading:

Essential Egypt (AA Publishing, 1990)
The Visitor's Guide – Egypt, A.G. Gravette (Moorland Publishing Co. Ltd., 1990)
Blue Guide – Egypt (A & C Black, 1988)
Egypt Handbook, Kathy Hansen (Moon Publications Ltd., 1990)
Egypt and the Sudan – A Travel Survival Kit (Lonely Planet Publications, 1990)
Fodor's – Egypt '91 (Fodor's Travel Publications, 1991)

For further information:

Egyptian State Tourist Office, 168 Piccadilly, London W1. Tel: (071) 493 5282/3. Fax: (071) 408 0295

Eire and Northern Ireland

Capitals Dublin, Belfast
Major cities Cork, Galway, Limerick
Time GMT
Currency Irish pound (I£) = 100 pence
Languages English, Irish
Climate Temperate

Annual weather range – Dublin:
Temperature
Max 20°C (Jul)   Min 1°C (Jan)
Rainfall
Max 74mm (Aug–Dec) Min 45mm (Apr)

The first thing you will notice about Eire is that being a parent is the norm; Eire has more children, in proportion to its population than any other European country and families will find babies and children welcomed, accepted and catered for everywhere they go. This doesn't mean that the Irish are, like the Italians or Chinese, a nation of child worshippers, just that children are seen everywhere and not regarded as noisy nuisances. Hotels and restaurants expect to provide small portions, early teas, high chairs, cots and single beds in parents' rooms as a matter of course. In a number of restaurants, children are not allowed after 8 p.m., and in most hotels babysitting facilities exist. Standards of food and hygiene are similar to British ones, though health-conscious parents may be horrified at the cholesterol level of the Irish fry-up. Baby foods and disposable nappies are available in the smallest village.

The Irish pub is a national institution and, in the country, often doubles as the village shop; they're friendly places and no one banishes families to a chilly garden. However, children are not allowed in most pubs after 7 p.m.

Most large shops have good facilities for feeding and changing babies – one large supermarket chain, Superquinn, even provides free crèches in some of its larger stores. A word of warning on breastfeeding – Ireland is a deeply conservative and religious country and women do not bare their breasts in public, so if you can find somewhere private to feed, do. Standard forms of contraception are available in Eire. Condoms and spermicides may legally be sold to anyone over the age of eighteen but IUDs, diaphragms, caps and the Pill are only available on prescription. In practice, however, all contraceptive devices may be difficult to obtain and travellers are advised to take their own supplies, though Family Planning Clinics in larger towns can usually be relied on.

On holiday, don't miss Dublin, the most cosmopolitan and intimate of capital cities, small enough to push a buggy round the centre, full of green spaces, good friendly, child-centred shops and watering holes. Phoenix Park is especially lovely – acres of parkland and a smashing zoo. The whole of the west coast is breathtakingly beautiful, with dreamlike, empty beaches; the only joker in this otherwise perfect pack is the weather – that well-known soft Irish day of fine misty rain – so pack your wellies and raincoats.

Most people would probably put Northern Ireland at the bottom of their holiday list – understandable, but a great pity; you're still in the UK, but also unmistakably in Ireland, and the North contains some of the most beautiful places in these islands – the Antrim Coast and Lough Erne. You can spend a holiday there without any sight of the troubles, but it's only fair to say that slightly older children can be both excited and frightened if they see armed soldiers and police, so some prior, simple explanation would probably be a good idea.

Of the four ferries operating on the Irish Sea, three are new with facilities for children.

Recommended reading:
Fodor's Ireland (Fodor's Travel Publications Inc., 1989)
Holiday Ireland 1990, Katie Wood and George McDonald

For further information:
Irish Tourist Board, Ireland House, 150 New Bond Street, London W1Y 0AQ. Tel: (071) 493 3201
Northern Irish Tourist Board, 11 Berkeley Street, London W1. Tel: (071) 493 0601

England

see **Britain**

France

Capital Paris
Major cities Lyons, Marseilles, Lille, Bordeaux, Toulouse, Nantes, Strasbourg
Time GMT + 1
Currency Franc (Fr) = 100 centimes
Language French
Climate Temperate in north, mediterranean in south
Annual weather range
North (Paris):
 Temperature
  Max 25°C (Jul) Min 1°C (Jan–Feb)
 Rainfall
  Max 64mm (Aug) Min 35mm (Mar)
South (Marseilles):
 Temperature
  Max 29°C (Jul) Min 2°C (Jan)
 Rainfall
  Max 76mm (Oct) Min 11mm (Jul)

France is probably the most civilized country in the world and therefore, for the potential traveller with very small children, it can provide virtually everything that might be needed. All French cities, major towns and most of the minor ones are equipped with every conceivable kind of resource. However, pharmaceutical goods are more expensive in France so it is wise to take the obvious basics such as analgesics, plasters and mosquito repellent.

The French are very family-minded, more so than most peoples and the family is a strong unit. Parental discipline tends to be somewhat stricter than we are used to in Britain and a good deal more so than in the US. Although expectations of children's behaviour are somewhat higher, they are usually given a great deal of attention. Your small children are likely to be welcomed and well looked after anywhere and everywhere.

The French eat out a lot (especially during the holidays). They do so *en famille* and children are allowed to stay up later than in other countries. It is very common to see large family parties in restaurants and cafes. They like everyone to be involved, from granny to the baby.

French cuisine is the best in the world and it is the best also in the sense that in any decent cafe, restaurant or hotel they will do their utmost to provide what their customers like. There is no need to go by the menu which, anyway, will often be varied and copious. Do not be afraid to ask for what you would like to eat. If you are staying in a place for a few days all the meals can be discussed and planned in advance. Proprietors and patrons are particularly good at providing the most desirable food for children. If you cannot get the kind of food you want and eat well, then you will never eat well anywhere. There is something for everyone. A *relais routier* – favourite stopping place of long-distance lorry drivers – will provide better food (and much more of it) than many a London restaurant and for a quarter of the price. My wife and I and our three children have been to France several times for several weeks each time in the last six years and the children have rejoiced over the food. As you can eat pretty well whenever you like in France there is little problem about mealtimes.

You can get almost any kind of hotel accommodation and hotel proprietors are flexible and helpful in making particular arrangements to suit your needs. It is still possible to get a comfortable room with a double bed for as little as £20 per night.

Plumbing has never been a strong point among the French, though they are improving. In some hotels lavatories and washing facilities are inferior to English standards, and sometimes they may seem a little squalid. Pay careful attention, therefore, to the hygiene in such installations – be careful about drinking water. I would not drink water from any old tap (especially in rural districts). Ask for fresh drinking water. Or better still, bottled. Tap water is considered perfectly safe in restaurants.

Do remember that in central and southern France it is very much hotter than it is in Britain during the summer. The sun may be fierce and the temperatures high. In June, July and August you should be very careful about exposing children to the sun. In an hour or less your child may be badly burned.

Medical facilities in any emergency are among the best in the world and French doctors are renowned for their skills. However, fees for private patients are high, so you must be sure that you have either taken out a comprehensive insurance policy or filled in the necessary forms (the E111) in Britain so you can benefit from the reciprocal National Health arrangements that exist.

Basically, you are most unlikely to encounter any inconvenience in any part of France when you travel with small children. One final point – no visitor should fail to take advantage of the clothes that are available for small children. They design and make excellent, inexpensive clothes that are original, colourful and stylish.

France is deservedly popular with English families, for obvious reasons: a short ferry ride and you're there. Couples who have been there before having children will have discovered their own favourite part. However, it is worth considering that the holiday is going to be different with children, so don't try to cram in too much travel.

Many people drive to the south by car, camping or staying in hotels en route. Children usually love camping, and French hotels are certainly more accommodating to children and babies than English ones. Long car rides are more of a problem, particularly if your children keep you awake at night, leaving you too tired to concentrate on driving.

There are two ways of coping. Some people travel at night while the children sleep, and hope they can get there in one stretch. This can be quite an endurance test. Others take a more leisurely pace, and travel by day, stopping frequently and expecting to spend two or three nights en route. Arthur Eperon or Michelin guides can be helpful in finding hotels. If you have young children and a long journey you should consider using the motorail or arranging a fly-drive holiday (contact your travel agent for the best available deals).

If your baby is at the travel cot stage and sleeps badly if you are in the same room, try asking for a room with a bathroom; you may be able to squeeze the cot in there. Don't feel you *have* to go to the seaside with young children – it can be very busy in August. River bathing and paddling is common in France (Loire camp sites can be a splendid choice for stops en route). Bathing is also allowed in reservoirs like Lac de Salagou. Small towns and even villages have their own outdoor pools, and sandpits and play areas are quite common.

Food will play a large part in any holiday in France and children are welcome in cafes and restaurants. French children stay up late and eat with the adults, but seem to behave remarkably well; wild behaviour from children may be frowned on.

For those used to restrictive English pubs it is a boon to know you can always find a cafe where the children can get a drink or snack at any time of the day.

Jars of savoury baby food are excellent, with delicacies such as puréed artichokes and green beans. If you are trying to keep your baby off sweet things, however, you may find it difficult to find an unsweetened baby cereal.

Nappy changing and toilets can be a problem. Public toilets are difficult to find and often smelly, though toddlers seem to find this less of a problem than older children. Camp-site loos are particularly variable.

Ferry companies have recently started doing more to accommodate children, and you should enquire when booking whether there is a playroom on the boat. My children particularly liked the Sally Line *Sea of Balls*.

Recommended reading:
Insight Guides – France (APA Publications, 1991)
Blue Guide – France, Ian Robertson (A & C Black, 1988)
Frommer's – France, Darwin Porter (Prentice Hall Press, 1991)
Fodor's France (Fodor's Travel Publications Inc., 1991)
France – The Rough Guide, Kate Baillie and Tim Salmon (Harrap Columbus Ltd., 1991)
The Holiday Which Guide to France (Hodder and Stoughton, 1991)
AA Baedeker's France (AA Publishing Division, 1991)

For further information:
French Government Tourist Office, 178 Piccadilly, London W1V. Tel: (071) 499 6911 (24 hour recorded message)

Germany

Capital Bonn (seat of Government), Berlin
Major cities Berlin, Hamburg, Munich, Cologne, Düsseldorf, Hannover, Dresden, Leipzig, Frankfurt, Stuttgart
Time GMT + 1
Currency Deutsch Mark (DM) = 100 Pfennige
Language German, English widely spoken
Climate Temperate continental
Annual weather range – Munich:
Temperature
 Max 23°C (Jul–Aug) Min −5°C (Jan–Feb)
Rainfall
 Max 139mm (Jul) Min 47mm (Dec)

Germany is a country of great variety ranging from the sandy, windswept beaches of the North Sea to the industrial centre and the mountains of the

Black Forest and Bavaria. The Germans are friendly and like children.

Children are welcome in most hotels and cafeteria-style restaurants; the more formal restaurants may turn children away. There are plenty of cafes in Germany which sell hot and cold meals all day. In family-style restaurants children's portions and special children's meals are likely to be available and high chairs will not be a problem.

Shopping hours are restricted by law. The shops open early in the morning, some at 8 a.m. and stay open until 6.30 p.m. Late-night shopping now operates in most German towns on Thursdays, when shops are open until 8.30 p.m. The shops close on Saturday at midday, except on the first Saturday of the month when they will be open all day.

Nappies, baby food and all the other paraphernalia are readily available, either in the 'drugstores' or larger supermarkets. You can buy beautiful children's clothes in Germany. There are wonderful toys, including of course the famous 'Steiff' range of soft toys.

Germany in general is not a country with many parks in the city centres, although on the outskirts of town and in the countryside there are many walks, amenities and picnic places. Playgrounds are generally well and imaginatively equipped with a variety of amusements.

Germany is a well-organized country and if you are organized, travel with children will be pleasurable and easy. Most major hotels will provide cots and high chairs, and baby seats in rented cars can be easily arranged – if booked well in advance. Hotels generally charge per person, but additional beds are usually available at a nominal charge.

The food in Germany in both restaurants and shops is usually very good. As in most countries travel with young children can be fun – if you organize it and are philosophical about the mishaps. In Germany this is as true as anywhere else, perhaps even more so.

Gute Reise!

Please note

Recent political changes are such that it is difficult to know the effect they will have on holiday makers in what was formerly East Germany. We would advise you to contact the relevant National Tourist Office for further details before travelling.

Recommended reading:

West Germany – The Rough Guide, Gordon McClachlan and Natascha Norton (Harrap-Columbus, 1990)

The Shell Guide to Germany, John Ardagh (Simon and Schuster, 1991)

Frommer's – Germany, Darwin Porter (Prentice Hall Press, 1991)

Coping with Germany, John A.S. Abecasis-Phillips (Basil Blackwell, Inc., 1991)

Baedeker's Germany (AA Publishing Division, 1991)

Guide to East Germany, Stephen Baister and Chris Patrick (Bradt Publications Ltd., 1991)

Berlin – The Rough Guide, Jack Holland and John Gawthorp (Harrap-Columbus, 1991)

AA Tour Guide – Germany, Adi Kraus (AA Publishing Division, 1991)

Cadogan City Guides – Berlin, Andrew Gumbel (Cadogan Books Ltd., 1991)

For further information:

German National Tourist Office, 65 Curzon Street, London W1Y 7PE. Tel: (071) 495 3990

Greece

Capital Athens, Patras
Major cities Thessaloniki, Larissa
Time GMT + 2
Currency Drachma (Dr)
Language Greek
Climate mediterranean
Annual weather range
Inland (Athens):
  Temperature
    Max 33°C (Jul–Aug) Min 6°C (Jan)
  Rainfall
    Max 62mm (Jan) Min 7mm (Aug)

Aegean Islands (Naxos):
  Temperature
    Max 32°C (Jul–Aug)
    Min 10°C (Jan–Feb)
  Rainfall
    Max 91mm (Jan)   Min 1mm (Aug)

The Greeks adore children and always make a fuss of them. They normally become the centre of attention and older people will often offer advice to what they consider an errant parent if a child is not protected against the sun or the chill of the evening.

Young children participate in all the family outings, so you are very welcome with them in restaurants, cafes and shops. It is a good idea to follow the Greek way and put your children to sleep at midday when it is usually very hot. During June, July and August it is best to take your children out (including swimming) in the mornings until noon and after 5 p.m. During this 'siesta' period (between 1 p.m. and 5 p.m.) quiet should be observed. Greek children will generally be up until 10 or 11 p.m., so feel free to enjoy your meal at restaurants if your little ones can stay up that late. You can order children's portions in most restaurants, but high chairs are uncommon. Restaurants stay open late, so they do not open before 7 p.m. in most places. The normal lunch opening hours are 12.30–3 p.m.

Shops open at 8.30 a.m., stay open unil 2 p.m. close for siesta time until 5.30 and open again 8.30 p.m. Monday, Wednesday and Saturday afternoons shops are closed, except those in tourist areas near the beach. There is always a pharmacy that opens late in every town and this is indicated on the shop itself.

Nappies can be found almost anywhere, even in the smallest village shop. Baby food, however, is found only in supermarkets and some chemists and it is more expensive than in the UK. Fresh milk is difficult to find in the villages, but tinned or long-life is available. Babies' formula milk can only be found in big supermarkets and is very expensive, so take your own.

It is important to boil water of any kind (even bottled) in most areas for making baby food in any form. It might be a good idea to take along sterilizing tablets.

Holiday equipment is plentiful and costs about the same as in the UK, however sun creams and toys are much more expensive.

Avoid travelling by public buses in summer, but if you must, catch the early morning ones. Hiring a car with a child seat should not prove difficult with a reputable firm. If you have booked a holiday through a UK tour operator ask them to arrange that the child seat be installed before the car is delivered. Some tour operators will arrange for high chairs and playpens to be available when you arrive at your hotel or villa.

Greek food is richer than English and often cooked in olive oil. Remember that you are welcome to visit the kitchen and can order something plain, without sauce. Restaurants will always cook chips for the children. Fruit and vegetables are excellent value and available anywhere in the summer months.

Greece is a member of the EEC so British citizens are entitled to emergency medical treatment. However, the DOH advises that you take out private medical insurance. This advice is very sensible since you have to pay a doctor to call at your hotel or villa. You will be pleased to know there is no shortage of paediatricians in every town.

Greek shops carry a large variety of mosquito-repelling devices and the best is the small electrical appliance which you put tablets into and plug in the wall socket. (*See* 'Coping in Hot Weather', Section 5.)

Athens is not overblessed with parks, but there are a few for those who would like to see some of Greece's ancient glory. The best of these is the zoological gardens (National Gardens) near the Parliament buildings. Other useful parks are at Lykabettos, Philipapou and Areos. Travel within Athens is greatly eased by using the underground railway system (pronounced *'eelektriko'*) and the double-decker express-bus service that links the airports with central Athens and Pireaus (details of this service can be obtained from Olympic Airways and the Greek Tourist Office).

Recommended reading:

Fodor's – Greece (Fodor's Travel Publications Ltd., 1991)
Guide to Greece, Michael Haag (Michael Haag Limited, 1986)
Blue Guide – Greece, Robin Barber (A & C Black, 1990)
Berlitz Blue Print – Greece (Berlitz Guides, 1990)
The Holiday Which Guide to Greece and the Greek Islands (Hodder and Stoughton and Consumers Assoc., 1989)
AA Baedeker's – Greece (AA Publications, 1990)

For further information:

The National Tourist Office of Greece, 4 Conduit Street, London W1R 0DJ. Tel: (071) 734 5997

Holland

See **The Netherlands**

Hong Kong

Time GMT + 8
Currency Hong Kong dollar (HK$) = 100 cents
Languages English, Chinese
Climate Tropical monsoon
Annual weather range
Temperature
 Max 31°C (Jul–Aug)
 Min 13°C (Jan–Feb)
Rainfall
 Max 394mm (Jul) Min 31mm (Dec)

Although the busy and congested urban spaces may not at first seem hospitable to children, Hong Kong boasts not only the Sung Dynasty Village and Middle Kingdom, two historical theme attractions, but there is also Ocean Park. This, the biggest theme park in Asia, has white-knuckle rides, a 191-metre observation tower, acrobats, clowns and daredevils, and Water World, a marine adventure park with water slides, pools and a shark tower. Hong Kong has dozens of museums including the hands-on Science Museum, Space Museum and the fascinating Museum of History.

The Hong Kong Tourist Association also distributes, free, *The Great Hong Kong Dragon Adventure*, a fantasy guided tour of Hong Kong for children.

And if you want to escape from the city centre there are 235 outlying islands, the rural New Territories, country parks, bird sanctuaries, paddy fields, duck farms and beaches. In fact, only a third of Hong Kong is built-up area, a fact that constantly surprises visitors.

Hong Kong is a largely Chinese place and Chinese culture, which developed quite independently of the West, so it has naturally developed its own unique theories and methods of child-rearing. However, the colony's 150-year-old history as a British possession provides a full array of British child-rearing customs and accoutrements. Disposable nappies are available cheaply, and you can get prepared baby foods even in small neighbourhood shops. There are cheap prams, Chinese-style cloth slings (for front and back), glorious clothing, good, fresh pasteurized milk (from the local 'Dairy Farm' brand, as well as a competitor from China), fresh biscuits in endless variety and sweets galore. Tap water is not drinkable (except at the major hotels) but cold, bottled beverages of every description are available everywhere.

Hong Kong must certainly rank with New York and Paris as one of the great places in the world for dining out. Virtually every major international cuisine is represented here, from northern Italian to kosher-style delicatessens including, of course, all the major regional Chinese cuisines, many of which are arguably better prepared here than in their places of origin. All this to say that children visiting Hong Kong will never be far from the taste of something familiar. Local Chinese frequently take all their children out to dinner, so restaurants are well-equipped and reassuringly tolerant. The full panoply of fast-food emporia is here if you need them.

While not yet qualifying as one of the cleanest cities in the world, Hong Kong would rank pretty well within Asia. Public conveniences and changing places are relatively numerous and nicely kept up (most are attended). Public transport, especially the sparkling MTR underground system are well kept and have fine safety records. Children's fares apply throughout the system. Chemists are well-supplied and reliable (the big Western-style chain is Watson's).

Babysitters of Western nationality are available (or English-speaking Chinese) from the hotels or from agencies. Standards are quite high, with substantial fees charged. Chinese nannies called *amahs* can be hired on a longer term basis. They are certain to be stern and conscientious although their English may be wanting.

One final note: Hong Kong hotels in general now charge by the number of occupants, with nominal fees charged for extra beds or cots in the room.

Recommended reading:

Fodor's – Hong Kong '91 (Fodor's Travel Publications, 1991)
Frommer's Hong Kong, Beth Reiber (Prentice Hall, 1990)
Hong Kong – A Travel Survival Guide, Carol Clewlow and Robert Storey (Lonely Planet Travel Publications, 1989)

For further information:

Hong Kong Tourist Association, 125 Pall Mall, London SW1Y. Tel: (071) 930 4775

Hungary

Capital Budapest
Time GMT + 1
Currency Hungarian Forint = 100 fillers
Language Hungarian
Climate Continental with mediterranean and Atlantic influences
Annual weather range
Temperature
Max 22°C (Jul)   Min −1°C (Jan)
Rainfall
Max 69mm (May)   Min 39mm (Jan)

Hungary is not a grey Eastern European country. It has an almost mediterranean feel to it. The weather is several degrees hotter than Britain, the people are friendly, and the buildings beautiful. The main holiday area for Hungarians is Lake Balaton, a very large inland lake, but you can't go to Hungary and not visit Budapest.

If you always enjoyed city holidays but thought that the advent of children in the family had put an end to that, then try Budapest. It's a beautiful *fin de siècle* city but with the added bonus of a playground in every square. Where else does the National Museum have slides, swings and a sandpit in the grounds?

Budapest has the advantage of being small so that even with young children you can walk almost everywhere and there is excellent public transport (with a flat fare of about 15p at current exchange rates). The metro is clean and not very deep so there are few escalators to negotiate and buggies can go on the trains unfolded. Trams and buses are more difficult as the steps are steep and they are often crowded, but people are helpful and will offer a seat if you're carrying a child.

As well as the metro and trams you could take a boat trip along the river or try the funicular railway (good views over the Pest side of the river), or the cogwheel railway (a steep climb into the Buda hills). There is also a narrow-guage children's railway.

There are two large parks in the city; one has a zoo, a castle, the Transport

Museum and, of course, a playground. The other is Margit Island in the middle of the Danube. The island has sporting facilities, swimming pools and thermal baths. There is a city farm and also bikes of all sizes for hire including tandems.

Children seem to be welcome everywhere. In the Fine Arts Museum, the Postal Museum and the National Museum, the attendants were more interested in talking to them (especially the baby) than in reprimanding them.

Food is plentiful and there are many street stalls selling fresh vegetables. The bread is freshly baked every day (it doesn't last much longer than that, which is also true of the fresh milk). The ice creams are very good and at 10p a scoop, you can afford to try lots of combinations.

There are a huge number of restaurants in Budapest. Torok's excellent guide book lists a variety, none of which we tried. However, the two we did visit were good and neither seemed to be put out by the children or the buggy. The West has arrived in the form of McDonalds (amazing decor but long queues), Burger King (shorter queues) and Pizza Hut (very long queues).

Baby food was available in supermarkets. It was mostly Hungarian but we did see some Gerber. This is likely to change as the influence of Western marketing increases. Chemists stocked both Pampers and the Hungarian equivalent, though we had taken an ample supply with us. First Aid is free but any more complicated medical treatment has to be paid for – make sure you are covered by your insurance.

Self-catering accommodation is plentiful and usually consists of flats that are let out while their owner is away in the country or visiting friends. Two large rooms, kitchen and bathroom in a turn of the century block seems to be the norm in the centre of the city. Our family of five fitted comfortably into this set-up and we paid £25 per night. Since these apartments are generally family homes, cots should be fairly easy to get hold of. As well as self-catering accommodation, there are plenty of hotels of various sizes.

Although the streets are almost litter-free, years of pollution have left the city fairly dirty, so don't expect your children's clothes to stay clean for more than a day.

Lake Balaton is about 130 km from Budapest. There are plenty of hotels here and some private accommodation. The lake has a bed of fine sand and stays shallow for a long while. Where we stayed the edge of the lake had been shored up, making it more like a swimming pool – too deep for very little children but fine for slightly older ones. On the whole, the water was not very clear.

The main problem when visiting Hungary is the language which is totally unlike any other (except Finnish). Don't be surprised to find your halting Hungarian phrase is answered in fluent German. It is assumed that all foreigners are German. Some people do speak English and they seem keen to practise. So just get yourself a phrase book, learn 'please', 'thank you', 'how much' and 'two beers and three lemonades' and enjoy yourselves!

Recommended reading:
Hungary: A Comprehensive Guide (Corvina, 1992)
Budapest: A Critical Guide, Torok (Corvina, 1992)

India

Capital New Delhi
Major cities Bombay, Calcutta, Delhi, Madras, Hyderabad, Ahmenabad, Bangalore
Time GMT + $5\frac{1}{2}$
Currency Indian rupee (IR) = 100 paise
Languages Hindi, English, local languages
Climate Tropical; wide variation with region and altitude

Annual weather range
Northern plains (Delhi):
   Temperature
      Max 41°C (May)   Min 7°C (Jan)
   Rainfall
      Max 180mm (Jul)   Min 3mm (Nov)
West coast (Bombay):
   Temperature
      Max 33°C (May)   Min 19°C (Jan–Feb)
   Rainfall
      Max 617mm (Jul)   Min 0mm (Apr)

'Please, you must go to the front of the queue; your children look tired,' insisted about twenty Indians in an early-morning, seething queue at Bombay Airport. Those were the first words we heard on Indian soil and continued to hear throughout our month's holiday there – in busy Bombay, on the overflowing buses in Goa, on rickshaws at Agra, on shikaras (small boats) in Kashmir.

We do tend to be the sort of family who hoist small children on backs and then take pot luck, but as India carries its fair share of health horror stories we were more careful for this trip. For a start we planned our itinerary in detail and I must admit in luxury: we had several nights at The Holiday Inn in Bombay though that did not stop us eating in very basic restaurants.

'Eat the cheapest food as long as it is cooked on the spot there and then,' advised a doctor friend of ours who knew India well. Why? Because that has the quickest turnover. Of course we avoided salads, fruit without peel, unsealed water and ice, and ice-cream (the children did not miss it).

Goa. Well, most people do not even see paradise once; it is a land of unspoilt beaches, unspoilt Portuguese architecture and archetypal paddy fields and was, childwise, plain-sailing. Even the water from the taps at the Taj Holiday Village was purified; for food we went lock, stock and child to various rustic restaurants where the fish was fantastic and so was the child-welcome. As I can remember, the children got more fun out of swinging on the basket chairs in one of the restaurant gardens than actually eating, but such was life in Goa.

Our plane journey to Delhi was rather late so we checked in hastily at the Imperial only to get up very early to take a booked train journey to Agra the next morning. But nobody starved; air snacks, even on short journeys, are quite good and safe.

The Imperial in Delhi seems quite imperial – toast and tea was served on a trolley in our rooms by old retainers. We never actually learnt if they appreciated our amused relief at reading on a huge surgical strip stuck across the loos the message 'sterilized for your use'.

But it was at Mr Butt's Clermont Houseboats on the Dal Lake in Kashmir when the children really got their feet under the table or more pedantically, their feet in the drawer. Our youngest, aged one and a bit, had to sleep in a huge drawer, not that she cared. Solemnly, at very small people's bed time, the boy (aged about 60) entered the houseboat with hot water bottles and again later for the rest of us. The junior team also got into the habit of sauntering along to the cook-house to order what they thought fitting for an early supper – French toast, Kashmiri doughnuts and maple syrup.

For a child brought up on Saudi sand, making daisy chains in Mr Butt's Moghul Garden was what India was all about but we managed to drag her out with the rest and walk for hours round the lake and up the mountains. The village children insisted on teaching ours how to bowl rubber tubes; another family pulled us out of the pouring rain and would not let us go till we had warmed up under a pile of blankets and a gallon of tea. We could not understand each other's language except the common one that we all had to look after our own and each other's children.

I mentioned precautions. We always knew our destination though would prefer to be more happy-go-lucky next time. Healthwise we followed the school of thought which avoids any extra jabs except the usual triple for all British children but we took malaria pills and ate natural yoghurt daily for the month before our trip. (Our elder children's boarding school was asked by us to do likewise for them.) We carried a couple of emergency packets of Milupa and Farex for the one year-old but only used a few spoonfuls. For the same member we had 200 nappies; we cannot remember whether there were disposables in the shops, only that we had judged the right number by the skin of her bottom.

Oh yes, and the chicken pox. After four weeks of wishing for 20 eyes to catch the sights of Bombay life turned inside-out on the streets, of listening to echoes in the Taj Mahal, of bussing, carting, boating, training, flying and walking round a country, one rather forgets matters like spots. But on our return to Saudi we learnt that there had been an epidemic. The rash that our fourth child had suffered in Kashmir was not due to daisy chains after all – so much for the diagnosis of the local pharmacist-*cum*-quack.

You may be cautious about spending part of your holiday in Kashmir. For advice on the current situation it is advisable to ring either the High Commissioner of India on (071) 836 8484 or the Indian Tourist Office on (071) 437 3677.

P.S. The jabs issue: we personally were advised that anti-hepatitis jabs can give a traveller a dangerously false sense of security. Add the fact that a jab appropriate for one bug in one area of India is not necessarily useful elsewhere – and we decided to be very careful instead about water, eating and so on. My advice is to talk to a doctor you really trust and decide on the situation as it stands at the time.

Recommended reading:

Insight Guides – South Asia (APA Publications, 1988)
Coping with India, Robert Wood (Basil Blackwell Ltd., 1990)
India by Rail, Royston Ellis (Bradt Publications, 1989)
The Insider's Guide to India, Kirsten Ellis (Hunter Publishing, Inc., 1990)
Cadogan Guides – India, Franka Kusy (Cadogan Books Ltd., 1989)
South Asian Handbook 1992, ed. Robert Bradnock (Trade and Travel Publications Ltd., 1991)
India – A Travel Survival Kit (Lonely Planet Publications, 1990)
Insight Guides – India (APA Publications, 1990)

For further information:

India Government Tourist Office, 7 Cork Street, London W1X 1PB. Tel: (071) 437 3677

Israel

Capital Jerusalem
Major cities Nazareth, Haifa, Tel Aviv, Beersheba, Eilat, Netanya
Time GMT + 2
Currency Israeli Shekel (NIS) = 100 agorot
Languages Hebrew, Arabic, English
Climate Mediterranean
Annual weather range – Haifa:
Temperature
   Max 30°C (Aug)   Min 8°C (Jan)
Rainfall
   Max 185mm (Dec)   Min 0mm (Jun–Aug)

Israel is a country of striking contrasts, where you can ski on Mount Herman in winter and sunbathe on the sandy Mediterranean beaches and see many unique historical and religious sites. It is also a country where children are loved and cherished and no matter where you go there are facilities for them.

Summer lasts from April until October and is usually very hot. August is the hottest month. Winter is from November to March but it is still pleasantly sunny with occasional rain. Whichever time of year you go it is wise to make sure your children are well protected from the sun. Cotton clothing for day is best but remember to take something warmer for the evenings. Protective creams, sun lotions, nappies and all other items you might need for a baby are easily available from local pharmacies.

Israel is a small country and easy to get around. There are several reputable car hire firms and children's seats are available. Although the railway system is limited and slow it does have some unusual scenic routes which children enjoy. The buses are reasonably efficient and there is a good network of them throughout the country. *Sherut* is a taxi service which runs between the main cities. Individual seats can be purchased at reasonable prices and there are plenty of local taxis which have regulated fares. There are many places to visit in addition to Jerusalem, Tel Aviv and Haifa and there are special organized tours available which are suitable for children – and often they go free.

Shops are generally open Sunday to Thursday from 8 a.m. until 1 p.m. and then from 4 until 7 p.m. Fridays the shops are open from about 8.30 a.m. until 2 p.m. and close for Sabbath until sunset on Saturday.

Medical care is extremely good and available all the time. Most doctors speak English and there are hospitals and clinics throughout the country. First-aid and emergency care is provided by the Magen David Adom, similar to the Red Cross. The English daily newspaper, *The Jerusalem Post*, lists emergency hospitals and pharmacists. You should take out medical insurance before you go since treatment can be expensive.

Fruit and vegetables are plentiful and of good quality. Restaurants and cafés offer an incredible variety of international cuisines and Western tastes are catered for in most hotels, which are often kosher. Kosher food conforms to Jewish dietary laws. This means that dairy food (milk, cheese, butter) will not be served with meat; pork and shellfish are prohibited.

Israel is an outdoor paradise for children – there are sandy beaches, all kinds of sports facilities and opportunities for walking and birdwatching. There is also an excellent range of cultural activities in every city – concerts, theatre, dance and museums, many of which have special children's areas. Just a few are the Ramat Gan Biblical Zoo, Jerusalem's Israel Museum, the Nature Museum in Haifa and the National Maritime Museum.

Finally, there are excellent hotels of every standard throughout the country as well as many kibbutz inns. You can also rent villas and flats and there are good campsites and youth hostels with family accommodation, or there is the option of staying in a Christian hospice. You will get a friendly welcome wherever you go.

Shalom, and enjoy.

Recommended reading:

Traveller's Guide to the Middle East, ed. Pat Lancaster (IC Publications Ltd., 1988)
Self Guided Israel (Langenscheidt Publishers, 1990)
Insight Guides – Israel (APA Publications, 1990)
Israel – At Cost, Fay Smith (Little Hills Press, 1990)
Blue Guide – Jerusalem (A & C Black, 1989)
The Traveller's Key to Jerusalem, Martin Lev (Harrap-Columbus, 1990)
Fodor's – Israel (Fodor's Travel Publications, 1991)

For further information:

Israel Government Tourist Offices, 18 Great Marlborough Street, London W1V 1AF. Tel: (071) 434 3651

Italy

Capital Rome
Major cities Milan, Naples, Turin, Genoa, Palermo, Bologna, Venice, Florence
Time GMT + 1
Currency Lira (L) = 100 centesimi
Language Italian

Climate Mediterranean
Annual weather range
Peninsular Italy (Rome):
   Temperature
     Max 35°C (Jul–Aug)   Min 5°C (Jan)
   Rainfall
     Max 129mm (Nov)   Min 21mm (Aug)
Great Northern Plain (Milan):
   Temperature
     Av 24°C (Jul)   Min 2°C (Jan)
   Rainfall
     Max 125mm (Oct)   Min 44mm (Jan)

Italy is the perfect holiday place for children, but it also has much to offer adults. Choose from mountains, lakes, seaside resorts or cities of great historical and artistic interest, all with endless sun and the warmest welcome in Europe. The scenery varies from dramatic mountain panoramas in the north to the fertile plains of the Po Valley, so temperatures vary enormously. In the mountains the heat is never overwhelming whereas it can be stifling on the plains at midday. Cities, of course, tend to trap the heat and Florence is notorious for this in summer.

Italians are naturally friendly, but become even more so if you have children with you since they play a central part in Italian life and are always a focus of attention.

There are many campsites, especially in the popular resorts, and hotels are plentiful and very welcoming. They charge by the room rather than by the person and are usually willing to put up an extra bed for a small child, if necessary. Cots are usually supplied by the hotels and the price of a double room is not higher. At self-catering places they are not always available. The staff are always willing to heat up baby's bottles and help in any way they can, even offering to dry baby clothes or babysit in some cases.

Eating out with children is easy and fun in Italy. Restaurants generally have high chairs, and children are accepted without question – even quite late at night. Staff are generally willing to serve half portions or to supply extra plates so that children can share a parent's portion.

Italian food is very popular with most children. Pasta heads the menu in every restaurant and is one of the cheapest dishes. Fast food and convenience foods are available throughout Italy. Takeaway pizza and pasta are sold in larger towns and sandwiches are quite easy to get hold of. Otherwise look out for *tavola calda* – the Italian version of a self-service restaurant. The noise level in restaurants tends to be high, so there is no need to worry about your children causing disturbance. Eating places are always very relaxed and as Italy's climate allows you to eat outside, it is a common sight to see children playing together while their parents finish a leisurely meal.

Bars are not quite so well adapted to children's needs. They often have no seating, and when they do they charge much more for waiter service – although you can sit for a long time with just one drink.

Everything you might need for children is readily available, even in small towns, but prices vary considerably. For instance, nappies (*pannolini*) cost far more in a local shop than in a supermarket where the price would be much the same as in Britain. If your child needs typical British foods, it is probably a good idea to take them with you, as although foods such as breakfast cereals are becoming more popular they are still quite expensive. On the other hand, ice creams and ice lollies are very cheap and very good. Shops tend to close for three or four hours in the middle of the day, but stay open until 8 p.m.

Trains and buses are efficient and cheap in Italy, with reduced fares for children. Travelling on a crowded bus in the heat of the day with small children, however, is not to be recommended. The Italians themselves complain of the heat and organize their lives accordingly. If you decide to hire a car, it can be expensive and children's safety seats are compulsory now.

Most towns have a municipal park with a special play area for small children. Sometimes they are supervised in the morning and late afternoon, and charge a small fee for entry, but many are free and open to children at all times. They are always shady and offer a pleasant retreat from the heat of the day, particularly if your child doesn't take to the idea of a siesta. Bicycles may be hired for a nominal sum.

As a member of the EEC, Britain has a reciprocal health agreement with Italy. On completion of the relevant form – E111 – (available from the DoH), British citizens are entitled to free medical treatment in Italy; but you may decide to take out private health insurance as well.

Sicily

It is easy to be evangelical about Sicily – it has so much to offer: wonderful scenery, marvellous beaches, terrific food and wine (far more varied than, say, Tuscany, thanks to the Greek, Arab, French and Spanish influences from its chequered past), and some of the finest archaeological remains in the world. Visit Agrigento, for example, on the south coast and you will see the Valley of Temples which surpasses anything you will find in Greece.

If you take a package trip to Sicily, chances are you will end up in one of the two main resorts on the island. Cefalu, on the north coast, is a small town with a lovely cathedral, numerous hotels, a stony beach and a sprinkling of discos; Taormina, on the east coast, is a former fishing village that has been rather self-consciously prettified. It has too many souvenir shops and tourist buses for my taste and is packed to the gills in summer, but it sits on a beautiful bay and it is easy to see why D.H. Lawrence rented a villa overlooking it.

For more intrepid visitors, I strongly recommend hiring a car and setting off in search of undiscovered Sicily. If you travel outside the peak season (July and August, when mainland Italians flock in), it is very easy to find accommodation without booking in advance, although it is a good idea to get a list of hotels and pensione from the Italian Tourist Office so you can plan your route. Note that the south coast has the best sandy beaches; elsewhere they tend to be pebbly. Don't be frightened of driving in Sicily; the natives may be noisy and flamboyant road-users, but they are no worse, in my experience, than London taxi-drivers. If driving really is out of the question, buses and trains are cheap and efficient.

Family life is paramount in Sicily, so the welcome that all visitors receive is particularly warm if you have children. In fact, producing a baby can often gain you entrance to previously full hotels or restaurants, and hotel proprietors gladly offer to babysit.

If you and your children require little entertainment beyond uncrowded beaches and good food, try the southern coast. Selinunte is a delightful place to stay, the 'modern' fishing village next to the tumbled ruins of the ancient city creating a curious feeling of time warp.

Should the coasts prove too hot, head for the cooler air of the mountains inland. Mount Etna is snow-capped all year round and you will need warm clothing at the summit, no matter what temperature it is below. The medieval hill-town of Enna, bang in the middle of the island, is well worth a visit, but its isolation makes it rather expensive to stay in. The nearby village of Pergusa has a small, modern hotel which is much more affordable, although it does not boast the atmosphere or views of the Hotel Belvedere in Enna itself.

Families, particularly those with very young children, would be well advised to avoid holidaying during July and August when temperatures, inflated by the sirocco wind from North Africa, are often uncomfortably hot and make sleeping at night difficult. If you cannot avoid visiting during these months, make sure you ask for a room with a balcony. And do remember, whether you sleep indoors or out, to cover all exposed flesh with insect repellent.

While you are in Sicily, it is worth considering a visit to the islands. The Aeolians, reached by boat or hydrofoil from Milazzo (the main port to the Aeolians and Messina), are extraordinarily varied, ranging from smelly Vulcano (full of sulphur springs which can be quite overpowering) to the desolate Stromboli (volcanic black beaches and a high cost of living as everything, including drinking water, has to be imported). The Egadi Islands, reached by boat from Trapani, are particularly splendid. They have dozens of beautiful, safe beaches, clear water, caves and grottoes. There are few tourist facilities, but rooms can be rented.

For those who think the Mediterranean has nothing left to reveal, Sicily will be a pleasurable surprise. See it before the rest of the world catches on.

Recommended reading:

Birnbaum's – Italy (Houghton Mifflin, 1991)
Frommer's – Italy (Prentice Hall Press, 1991)
Fodor's – Italy (Fodor's Travel Publications Inc., 1991)
Insight Guides – Italy (APA Publications Ltd., 1991)
AA Baedeker's Italy (AA Publishing Division, 1991)
Berlitz Blue Print – Italy (Berlitz Guides, 1989)
Italian Country Inns and Villas, Karen Brown (Harrap Columbus, 1990)

For further information:

Italian Tourist Board, 1 Princes Street, London W1R. Tel: (071) 408 1254

Jamaica

Capital Kingston
Time GMT –5
Currency Jamaican dollar (J$) = 100 cents

Language English
Climate Tropical
Annual weather range – Kingston:
Temperature
 Max 32°C (Jul–Sep) Min 20°C (Jan–Feb)
Rainfall
 Max 180mm (Oct) Min 15mm (Feb)

Jamaica is nine hours away by plane – something to be endured rather than enjoyed by parents and babies. But the warmth of the tropics smooths the kinks of the journey. Jamaica is everything the song says and the air has an effect on adults and children alike.

The northern winter is the best time to visit the island but it is also the busiest time. The summer is hot and humid, and though most places claim to have air-conditioning, this should not be taken too literally. There is always the odd power cut or water shortage and the islanders seem to have their own time and ways to do things.

Jamaica may be billed as a honeymooners' paradise but Jamaicans love to spoil children. To have a baby is positively an advantage. There are plenty of houses of all sizes to rent on or near the beach on Jamaica's north coast between Negril and Port Antonio. For a family they can be more economical than a hotel, and most come with a housekeeper or maid. It's worth checking, however, if cots are provided.

Nannies or childminders are available everywhere so parents can enjoy some time off too. The housekeeper will usually do the cooking (hot and spicy Jamaican food) but will go easy on the pepper for the children if requested. Let the housekeeper take you shopping in the local market and see the wonderful – and healthy – tropical fruit, vegetables and fresh seafood.

Baby food and care products are not always available and are expensive. Nappies are extremely expensive so it is preferable to bring your own. Since it is so hot there is also an opportunity for the children to wear very few clothes, which eases the packing. No nappies during the day is the rule because the odd accident is usually no disaster. Both our children were potty trained in Jamaica.

Eating out can be expensive, high chairs are rare and the service is sometimes slow. Not much fun for impatient little ones who would rather be at their favourite playground, the beach with its clean, shallow water.

'Higglers', local mobile traders, are everywhere and some require friendly but firm handling. Bargaining (higgling) is a way of life because few prices are fixed. The best price guide is what it is worth to you. Sensible offers are seldom refused. Higglers can offer a worthwhile service when the next shop is a mile down the beach. Most have fresh-pressed orange juice and fruit. No child will miss sweets when offered pieces of sugarcane to chew or milk fresh from the coconut.

Since you will spend a lot of time on the beach the sun has to be taken seriously. Repeated application of sunblock is essential for children and grown-ups; so are sunhats or some kind of shade. A shower in the afternoon and an aloe vera rub at night should cool everybody down. If you want to watch the picture-postcard sunset you will need some effective mosquito repellent; for a good night's sleep local burning coils called 'fish' or good old mosquito nets (bring your own) are recommended.

A day's rest from the beach is usually provided by the odd tropical downpour, which is exciting in itself. A cloudy day should be used as a reason for a journey into the mountainous and cool interior of the island, only a short drive away. Car rental is expensive, as is petrol, and baby seats are unheard of. Local drivers know the beauty spots and the sometimes hazardous roads. Hidden waterfalls and old plantation Great Houses are interesting, but young children are usually bored by long hot car rides and viable alternatives are trips in glass-bottomed boats to view the varied underwater life; rafting down rivers keeps them still for a while too.

There is no free health service in Jamaica so adequate medical insurance is needed. Vaccinations are not required but general rules of hygiene should be observed.

Jamaica is a friendly place for children, even though established play areas are rare and facilities somewhat lacking, but by bringing a pushchair, folding babydiner and travel cot we've felt quite comfortable. When our children were safely sleeping next door we would relax and watch the fireflies illuminate the scenery.

Recommended reading:
Insight Guides: Caribbean (The Lesser Antilles), ed. David Schwab (APA Publications, 1989)
Zellers 1990 Caribbean, Margaret Zellers (Fielding-Morrow, 1990)
The Penguin Guide to the Caribbean 1990, ed. Alan Tucker (Penguin, 1990)

For further information:
Jamaica Tourist Board, 1–2 Prince Consort Road, London SW7 2BZ. Tel: (071) 224 0505. Fax: (071) 224 0551.

Japan

Capital Tokyo
Major cities Kyoto, Kobe, Nagoya, Osaka, Sapporo, Yokohama
Time GMT + 9
Currency Yen (Y) = 100 sen
Language Japanese
Climate Temperate monsoon
Annual weather range – Tokyo:
Temperature
 Max 30°C (Aug)   Min −2°C (Jan)
Rainfall
 Max 234mm (Sep)   Min 48mm (Jan)

Japan must be one of the easiest (if expensive) countries to travel to with a child. It is safe, clean and efficient. Baby food, supplies and disposable nappies can be found everywhere. The Japanese, old and young, adore babies, and will come up to talk to yours. Don't worry about not understanding what is being said. Most probably it is only the word *kawaii* (cute) expressed over and over again.

If your first destination is Tokyo, be prepared for the commute into the city centre from the international airport at Narita. There is now a direct rail link from the airport to Tokyo Station which takes less than 1 hour. Limousine bus services to central Tokyo and a variety of downtown hotels are also available. If you are travelling around the country it is advisable to avoid travelling by air which is expensive and airports are usually some distance from the cities. The best way, especially with children, is to use the super-efficient rail system which is fast and convenient. An economic way to travel long distance is to purchase a Japan Railpass which gives you unlimited travel on the Japan Railways network for 7, 14 or 21 days. Children under 12 travel for half price, children under 6 travel free. If you are using the bullet train *shinkansen* get ready for your station as it will stop for no more than one to two minutes. Japanese children take off their shoes if they want to stand on the seats of trains and adults give up their seats for children!

In Tokyo, avoid travelling by train between seven and nine in the morning unless you want your child to experience what it is like to be a sardine. It can be an unnerving experience. If you have a baby, come with sensible shoes and a pushchair as platforms are long and there are several flights of stairs to manage, often with no escalators. Travel by taxi is very expensive and please note that the taxi doors are controlled by the driver so make sure children are not standing too close to the passenger door which will swing open automatically.

If you are staying in an international hotel it will, of course, have all the international amenities. If you are not, there will be no space in the rooms for extra beds or cots, no babysitting service and little English will be spoken. It would be more interesting to experience staying at Japanese-style inns called *ryokan*. There will be no problem about extra beds as everyone will be sleeping on *futons* on the *tatami* floors. Breakfast and dinner will be covered in the cost of accommodation and served in your room when the *futons* are out of the way, cleverly hidden in the wall cupboards. The problem then is that there is no choice of menu. You eat what you get – which is usually good.

Eating out in local restaurants is easy as there are often plastic replicas of dishes on the menu displayed in the windows and you need only point. There may not be high chairs but there are tatami mat floors which may not be so comfortable for you but great for babies to lie on or toddlers to romp around. Of course the usual fast food outlets like McDonalds, Mr Donut, Pizza Hut and the whole range of Western family-style restaurants can be found in most of the cities. You can always get yoghurt and fruit juices at shops and supermarkets.

Local shops and big stores open only after ten o'clock in the morning but stay open till late. They are open on Sundays and public holidays but close one day in the middle of the week. There are many 24-hour supermarkets as well. The department stores are the best places to head for to take a rest, feed or change your baby, or entertain your child. Go to the floor for children's things. There is a nursing room with cots provided and a baby menu available. There is a small play area for toddlers and there are samples of the toys for the children to test. Also there is usually a large playground on the roof of the store. A city like Tokyo can be very tiring for parents and children and, surprisingly, the department store can be a refuge.

Of course there are parks, zoos and museums but they are hardly just round the corner. Then there is Tokyo Disneyland but it is away from the city centre so be prepared for the long queues to do or go on anything. It will be much worse if it is a Sunday, public holiday or school holiday. You might have to wait for an hour or more to get into anything. We took our toddler on an ordinary day and even so the entire outing was a torture for her and us.

If you are at your wit's end trying to keep your child entertained, an outing to the local public bath might prove to be the most interesting cultural experience for you and your child. It is a communal bath, the water is very hot and everything is extremely clean. Ask where the nearest public bath-house is (it's probably just round the corner), and stay in there as long as you like. Just observe what the Japanese do and follow them. Bath-houses are usually open from 4.00 to 11.00 p.m.

The main problem you will encounter is language. You will just have to ask for help. There are police boxes called *koban* outside major train stations and at big road intersections. The tourist information centres (TIC) at Narita, Tokyo and Kyoto are very helpful and the staff speak English. Use their 'Travel Phone' service which is ready to help you solve a language or travel problem. When you are outside Tokyo or Kyoto, dial 0120 222800 for information on eastern Japan or 0120 444800 on western Japan.

Recommended reading:

The Insider's Guide to Japan, Peter Pepham (Moorland Publishing Co. Ltd., 1989)
Japan Handbook, J.D. Bisignani (Moon Publications, Inc., 1988)
Frommer's – Japan and Hong Kong, Beth Reiber (Prentice Hall, 1990)
Fodor's – Japan '91 (Fodor's Travel Publications, 1991)
Japan – A Travel Survival Kit, Ian L. McQueen (Lonely Planet Publications, 1989)

For further information:

Japan National Tourist Organization, 167 Regent Street, London W1R 7FD. Tel: (071) 734 9638/9. Fax: (071) 734 4290.

Jordan

Capital Amman
Major cities Zarka, Irbid
Time GMT + 2
Currency Jordanian dinar (JD) = 1000 fils
Language Arabic
Climate Desert
Annual weather range – Amman:
Temperature
 Max 32°C (Jul–Aug)
 Min −4°C (Jan–Feb)
Rainfall
 Max 74mm (Feb)   Min 0mm (Jun–Sep)

If you are not already in possession of a young child on entry into Jordan you might be well advised to hire one! Jordanians adore children who are the instant bond between tourist and the would-be host. Travelling in an Islamic country without my husband, as I was, my children were an essential passport to respectability.

We spent ten weeks there basing ourselves at Irbid, a town in the beautifully unspoilt countryside of northern Jordan. We bussed our way to historical sites, untouched villages, Amman and its surroundings and finally way south to Aqaba and Petra.

I am not qualified to comment on five-star hotels. Except in Irbid where we rented rooms, we had recourse to the cheapest hotels which featured along with the five-stars in the invaluable 'Classified Hotel Price List', courtesy of the Ministry of Tourism in Amman. (Staff there bent over backwards to help, apparently surprised, dare I say it, that they had visitors.) Bedsheets and bathrooms in no-star hostelries were tacky. The Ministry of Tourism Resthouses are excellent value – very clean, with good, cheap food.

There are a number of Western-style fast-food restaurants in the larger cities in Jordan. However, the place is still a haven for simple, wholesome cheap food. The children ate like kings in cheap restaurants or in friends' homes, on a diet of fresh vegetables, rice, a little meat, olives, sesame bread, yoghurt and honey. During the month of Ramadan, however, when most people fast between sunrise and sunset, midday eating was a problem if you were not self-catering. Children are not expected to fast, but in the towns there was nowhere private to consume food; in villages there was space but no food, unless you count stale biscuits and wrinkled oranges gathering dust in dubious looking village stores.

In the early days we were careful to drink bottled water but we soon had to lose our inhibitions about drinking well water; there was no alternative when we stayed with chance acquaintances in the middle of nowhere. The children survived it until our final two weeks, when we had to beetle off to a local government-run hospital. I do not know whether the water had upset them, but I mention the incident simply because we were impressed by the friendliness of the hospital staff and, incidentally, by the practical help we met in ordinary chemists in towns throughout Jordan.

Transport by bus or 'service taxi' is efficient, cheap and enchanting. The buses serving outlying villages are mobile gossip columns – everyone knew everyone else and insisted on knowing who we were too. Then, once our credentials had been established, we were invariably invited home, entertained, fed and watered. Service taxis were a boon in towns. They follow a prescribed route, picking up and dropping off passengers on the way for a minimal fare. Children on laps rode for free. We made the five-hour bus journey from Amman to Aqaba by Jett bus for 3 dinars (£6) for the lot of us. (Again the two-year-old rode free.) The bus was air-conditioned, with a loo, hostess service and video. The latter was welcome as the King's Highway route is dull, scrubby, grey desert with the odd dusty town thrown in.

Petra, which we reached by means of an overflowing mini-bus (a two-hour journey from Aqaba), was a test in ingenuity and infant stamina. Any normal human being would hire a horse for the day at a mere 2 dinars and enjoy wandering round the ancient city at leisure. However, my toddler refused to mount so we footed it into Petra, which was fine for the first few hours before the sun was really hot. After rest and refreshments at the resthouse in the middle of the old city the children could not face the several kilometres walk back to new civilization. A boy on a donkey saved the day, but that was a close one.

Would our children recommend Jordan to their friends? I think so, with one special caveat: 'Don't get excited about swimming in the Dead Sea until you have grown out of falling over – the salt stings the wounds.' And what did they like best? Eating okra and having a go on Bassim's homemade swing.

Please note

Recent political events are such that it is difficult to know the effect they will have on holiday makers. We would advise you to contact the relevant National Tourist Office for further details before travelling.

Recommended reading:

Fodor's Jordan and the Holy Land – a Practical & Historical Guide, Kay Showker (Fodor's Travel Publications Inc., 1989)
Jordan and Syria: A Travel Survival Kit, Hugh Finlay (Lonely Planet, 1987)

For further information:

Jordanian Embassy, 6 Upper Phillimore Gardens, London W8. Tel: (071) 937 3685/ 9611

Kenya

Capital Nairobi
Major cities Mombasa, Kisumu
Time GMT + 3
Currency Kenya shilling (Ksh) = 100 cents
Languages Swahili, English
Climate Subtropical; variation between coast and higher altitude inland

Annual weather range

Mombasa:
  Temperature
    Max 31°C (Jan–Mar)
    Min 22°C (Jul–Sep)
  Rainfall
    Max 320mm (May)   Min 18mm (Feb)

Nairobi:
  Temperature
    Max 26°C (Feb)   Min 11°C (Jun–Sep)
  Rainfall
    Max 211mm (Apr)   Min 15mm (Jul)

Not many British or American people take children to East Africa for the simple reason that it is a very expensive trip. Given the opportunity of a holiday in East Africa – or any form of trip which may include a holiday – I doubt whether it is worth taking children under four years old.

Most people go to these regions to visit the game reserves and look at the wildlife. For the most part the wildlife will mean little or nothing to small children because they have little sense of scale. It is a basic error on the part of adults to suppose that children will respond like adults, though of course adults very much want them to do so. I have taken my children (aged four and nine) to several splendid zoos and game parks. The smaller the animals the more they were interested in them. In short, they have no capacity for being awe-inspired when they behold a herd of 100 elephants, a pride of lions or a column of wildebeeste 5000 strong. On the other hand, a tame hyrax (they resemble large rabbits but are related to the elephant) or a tame mongoose (kept to deter snakes) pottering round a fixed camp or the grounds of a game lodge will be a source of great fascination.

If you do go to East Africa with little children various precautions are essential. They must have inoculations against tropical diseases and remember that apart from Nairobi and Mombasa no town has adequate supplies of anything you might take for granted in Britain and Europe. Remember, too, that in summer (November to March) the sun is very strong; no child should be exposed to it for long.

If you travel through or camp in game reserves the rules must be rigorously adhered to: after dusk you must not venture outside the designated camping area; in daylight stay in or very near your vehicle; avoid getting too close to wild animals. Lions and elephants often show complete indifference to vehicles, but rhinoceroses are dangerous. Their sight is poor but their hearing and smell are excellent. Moreover, they are as nimble as ballet dancers and will charge at high speed on very little pretext. Two-and-a-half tons of rhino travelling at 30 m.p.h. will make 'Whiskas' of you and your vehicle. The African buffalo is perhaps the most dangerous of all. Lastly, you must never attempt to approach any wild animal on foot however innocuous or tame it may appear to be.

Colonies of baboon are a common feature of campsites. They wander in when there are few people about and look for titbits. Never leave your tent unattended for any length of time. Baboons are extremely inquisitive and given the chance they will have the tent (however strong or secure it may be) down in a flash and will create total havoc among your belongings. Baboons often look friendly enough but they are wild animals. A male baboon may weigh 350lb and could have Muhammad Ali in his prime completely incapacitated in five seconds flat. Never leave your children unattended anywhere in a game reserve.

In the game reserves you must use the appointed camping sites and obey any local rules. These sites are supplied with fresh water and, in some cases, with brick-built installations which provide shower/washing/lavatory facilities.

The game lodges are extremely well run and very comfortable. The cuisine in all these places is absolutely first-class by any standards, including those that prevail in France. Children will lack for nothing in such establishments.

Unfortunately, children are not allowed to enjoy the beauty of the Tree Tops hotel due to the potential danger of wild animals who may be disturbed by noise or sudden movement. But for the adults both the Governor's Camp in the Mara reserve and Tree Tops hotel high in the Aberdare Mountains in Kenya afford fantastical views of the wildlife in its natural surroundings.

But don't forget the attraction of the beach on the coast of Kenya, which also has a host of historical monuments and other environmental attractions well worth a visit. The Fort Jesus on Mombasa island, the Gedi Ruins near Malindi as well as the Bamburi Nature trail and Mamba Village are but to mention a few.

Finally, most important of all, the Africans. They love children and understand them. You may be sure that your children will receive VIP attention. They will lack for nothing here.

Recommended reading:

Traveller's Guide to East Africa and the Indian Ocean (IC Publications Ltd., 1990)

Kenya – The Rough Guide, Richard Trillo (Harrap-Columbus, 1991)

Kenya – A Travel Survival Kit (Lonely Planet Publications, 1991)

The Insider's Guide to Kenya, Michael and Peggy Bond (Moorland Publishing Co. Ltd., 1989)

Spectrum Guide to Kenya (Camera-pix Publishers International, 1989)

For further information:

Kenya Tourist Office, 25 Brook's Mews, London W1. Tel: (071) 355 3144

Majorca

See **Spain**

Malaysia and Singapore

Capital (Malaysia) Kuala Lumpur
Time GMT + 8
Currency Malaysian dollar or ringgit (Ma$) = 100 cents
Singapore dollar (S$) = 100 cents
Languages Bahasa Malaysia English, Chinese, Tamil
Climate Equatorial
Annual weather range – Kuala Lumpur:
Temperature
Max 33°C (Feb–Jun)
Min 22°C (Dec–Feb)
Rainfall
Max 159mm (Jan)   Min 119mm (Jun)

Children are seen, heard and taken everywhere in Malaysia and Singapore so it is very easy to travel with them there. The two national airlines – MAS and SIA – are well known for the attention and care they offer babies and children. Even domestic travel between the major towns is best done by air as it is both cheap and convenient. Of course the trains and buses are cheaper but can be very tiring and trying for children in hot and humid climates. Car rental is easily available in Malaysia and Singapore. In the towns don't hesitate to use taxis. They are not expensive.

The big hotels have excellent facilities and offer very good rates. They usually provide cots at no extra charge and babysitters can be arranged. Baby food can also be requested without much trouble and tap water is safe for drinking.

All restaurants, even very small ones, provide high chairs and no one minds children being taken out to meals anywhere and at anytime. Children are allowed to stay up late and are included in most activities. Life happens a lot outdoors and people love eating out at the hawkers' centres where there is such variety of food, fruit and fresh drinks. Let your child experience all the different tastes of Malay, Chinese and Indian dishes and local fruits like papayas and mangoes are delicious and nutritious.

Shops and restaurants stay open till late. English is easily understood everywhere and is on all labels and menus. All doctors speak English well and most doctors' surgeries are open in the evenings.

Baby food and supplies, including disposable nappies, can be bought at most places. Don't confine yourself to the supermarkets. Venture into the little grocers' shops. As it is very hot and humid throughout the year, do take precautions against sunburn, nappy rash and mosquitoes.

Besides the time spent on the beaches in these two countries, which can easily fill up all your holiday, in Singapore you might consider a trip to the zoo (considered to be the best in southeast Asia), the Jurong Bird Park and a ride on the cable car to Sentosa. The Haw Par Villa might not be such a good idea for sensitive children with vivid imaginations. In Malaysia, if you are in Penang, the butterfly farm is unique and a ride on the funicular railway up to Penang Hill could be interesting.

Recommended reading:

Malaysia, Singapore and Brunei; A Travel Survival Kit, Geoff Crowther and Tony Wheeler (Lonely Planet, 1988)

Fodor's – Singapore (Fodor's Travel Publications, 1991)

Malaysia, Singapore and Brunei – The Traveller's Guide, Stefan Loose and Renate Ramb (Springfield Books Ltd., 1990)

A Traveller's Guide to Asian Culture, Kevin Chambers (John Muir Publications, 1989)

For further information:

Malaysia Tourism Promotion Board,, Malaysian House, 57 Trafalgar Square, London WC2. Tel: (071) 930 7932

Singapore Tourist Promotion Board, Carrington House, 126–130 Regent Street, London W1. Tel: (071) 437 0033

Malta

Capital Valletta
Time GMT + 1
Currency Maltese lira (LM) = 100 cents
Languages Maltese, English
Climate Mediterranean
Annual weather range – Valletta:
Temperature
   Max 29°C (Jul-Aug) Min 10°C (Jan-Feb)
Rainfall
   Max 110mm (Dec) Min 0mm (Jul)

Malta's long tradition of association with Britain makes it an ideal holiday destination for people with young children. The island is small enough to explore without spending too much time travelling and the Maltese are very pro-British; they also adore small children and babies which makes life far more relaxed for the potentially harassed parent of hot and bothered infants.

The climate is very hot and dry during the summer months and it would probably be advisable to avoid booking for July and August if at all possible. June and September are still hot enough for the most dedicated sunbather and even early October should be warm with the added benefit of the sea still beautifully warm.

All the sandy beaches in Malta are concentrated towards the western tip; the rest of the coastline is rocky and swimming would have to be off the rocks. If your hotel has a good pool this is not a problem, although most small children enjoy playing in the sand. Many of the small hotels situated in Valletta and Sliema (unless you are heading for one of the more secluded luxury establishments) are likely to be in fairly crowded neighbourhoods and thus rather noisy. If you enjoy a more peaceful lifestyle, the fishing villages and the area around Mellieha would perhaps be more suitable.

Another area worth considering is Gozo, Malta's sister island. Gozo is quieter and greener than Malta and also has several sandy beaches. It can be reached by ferry in about 30 minutes from Malta. Hand-made lace and knitted garments are a speciality and are good value. There are also good buys to be made in the way of jeans and baby clothes, many of which are manufactured on the mainland.

A self-catering holiday is easily organized as most English goods and supplies, disposable nappies and baby food are readily available and reasonably priced even in the local shops. Eating can be inexpensive. Chicken, steak, pizzas and salad are standard but the fish, particularly swordfish in season, is excellent. Pasteurized milk is good but it is advisable to boil drinking water.

Children's portions are happily supplied. You may not find that many restaurants go in for high chairs but children are so welcome and readily accommodated that you won't feel guilty if they slip off their seats to explore.

Most Maltese children keep late hours so if your children aren't too tired after a day at the beach there is no reason for not eating out together in the evenings.

Car hire is well-priced and it is always convenient to have your own transport for visiting some of the more inaccessible areas of the island. Child seats are not common however. Buses run regularly and are cheap but crowded; don't be surprised if the local passengers offer to hold your child or even just plonk one on to their lap. My seven-month old son was happily nursed by countless ladies, but I was never offered a seat with him.

All Maltese speak English, especially those in the shops. Shops close for a long lunch from 1 p.m. until around 4 p.m. but are then open until later in the evening. There is a pharmacy in every town. Beach umbrellas and chairs can be hired at many of the beach shops fairly inexpensively. Mosquito repellents and fly sprays are necessary for comfort, especially in the evenings.

Malta is a strongly Catholic country and there are many associated feast days in different towns and villages, especially throughout the summer months. The processions are colourful and great fun and are usually accompanied by fireworks.

Recommended reading:

Blue Guide: Malta (A & C Black 1990)
The Travellers' Guide: Malta and Gozo, Christopher Kininmonth, revised by Robin Gordon-Walker (Jonathan Cape, 1987)
Discover Malta, Terry Palmer (Heritage House, 1988)

For further information:

Malta National Tourist Office, Mappin House, Suite 300, 4 Winsley Street, London W1N 7AR. Tel: (071) 323 0506

Mauritius

Capital Port Louis
Major cities Beau Bassin-Rose Hill, Curepipe
Time GMT + 4
Currency Mauritius rupee (MR) = 100 cents
Languages English, French
Climate Subtropical
Annual weather range – Port Louis:
Temperature
 Max 30°C (Jan) Min 17°C (Jun- Sep)
Rainfall
 Max 221mm (Mar) Min 36mm (Sep)

Getting to Mauritius is likely to be the worst part of your holiday: some flights involve over 24 hours' travelling time and that's no fun in a crowded aircraft, for adults or infants, so try to use the fastest route. Flights to Mauritius non-stop by Air Mauritius take 11½ hours and by BA about 17 hours.

On the ground, however, Mauritius is delicious: an island in the middle of the Indian Ocean surrounded by warm blue seas and a subtropical climate that's best, I think, in the European summer months when Mauritius is hot and sunny with cool breezes and short freshening showers.

Tourists stay mainly at 'beach hotels' rather than inland local hotels; there are also self-catering apartments and a Club Med. The beaches are well-designed for small swimmers: mostly shallow, sandy lagoons inside coral reefs, barely deep enough for grown-ups to swim and safe for children – except for the ubiquitous sea urchins which make plastic swimming shoes (sold on the beach) obligatory. The hotel where I stayed employed people to sweep the beach and shallows clean of small spiky pieces of coral but this does not seem essential: on the public beach nearby, Mauritian children of all sizes were playing in and out of the water – which is bright and clear. Hotel beaches have pedaloes and glass-bottomed boats for viewing the corals and fish, as well as sailboarding, water-skiing and snorkelling facilities. Boat excursions for swimming, diving, fishing and visiting coves can also be arranged, but do make sure they provide lifejackets for younger passengers.

The people of Mauritius are welcoming and fond of children who, locally, are expected to behave well but join the extended family's outings, so the response of hotels and restaurants is relaxed. My hotel provided an open-air 'playschool' service in the mornings with games, songs and stories for children, and a babysitting service in the evenings. On Fridays, when local people came to the hotel casino, toddlers were playing in the foyer or snoozing on the sofa with granny (children aren't admitted to gambling areas).

Meals are formal or informal according to choice with both a restaurant and a buffet/barbecue service where everyone selects their own platefuls, sort of suit-yourself-servings which could contain only what you fancy, from oysters to shredded carrot. The food however is chiefly European, while the local cuisine is part-Indian, part-Creole. Fresh fruit and vegetables are splendid, especially the small, sweet pineapples.

Mauritius is a smallish island, and excursions and tours are available by mini-bus, taxi and hire car. No distances are great, so in my view the rough state of some roads (which are currently being upgraded) isn't important although some people complain. Public transport – by bus – is cheap and frequent but they are short distance only and regarded locally as extremely uncomfortable. Self-drive cars may be hired, but it can be useful to have a local driver, who will also act as information source and negotiator when deciding where to go, shopping or searching for special services.

I didn't need to do much everyday shopping, but my impression is that most necessities are available and, if not immediately visible, can be obtained by asking. On the beach itinerant hawkers (who can be a nuisance) will undertake to supply various other items – at a price to be negotiated.

There are several European-style stores in the capital, Port Louis, and other main towns, including a Kentucky Fried Chicken if any rising-fives are feeling deprived. Don't miss the markets – for fruit, spices, knitwear and plastic toys.

The main public parks are gardens in and around Port Louis including the huge Pamplemousses botanical garden with pools and streams and palm-shaded avenues. Everyone is sure to be helpful and sympathetic, as long as you surround your request with smiles and courtesy – Mauritians are not taken with arrogant, demanding behaviour. The local language is Creole, but nearly everyone speaks and understands both English and French. Many Mauritians have visited Britain and have relatives living here – don't be surprised if they turn out to know London well – and professional people, such as the doctor or nurse on duty at your hotel, will probably have studied and trained in Britain, which could be useful in an emergency. No special immunization or health procedures are required for visitors coming from Europe since the island is free of tropical diseases, including malaria.

On the whole, children should enjoy Mauritius. At my conference delegates from Japan, Denmark and Madagascar had brought husbands and children with them, who appeared to spend the days very happily. So it should be possible to combine a business trip and family holiday on this beautiful tropical island.

Recommended reading:

Guide to Mauritius, Royston Ellis (Bradt Publications, 1990)
Visitors Guide to Mauritius, Rodrigues and Reunion, K & E Roberts (Moorland Publishing, 1992)

For further information:

Mauritius Government Tourist Office, 32/33 Elvaston Place, London SW7 5NW. Tel: (071) 584 3666. Fax: (071) 225 1135.

Mexico

Capital Mexico City
Major cities Guadalajara, Monterrey, Puebla de Zaragoza, Ciudad Juárez, Léon de los Aldamas, Tijuana
Time GMT −6–8
Currency Mexican peso (Mex$) = 100 centavos
Language Spanish
Climate Warm temperate/tropical
Annual weather range
Centre (Mexico City 2275m):
 Temperature
  Max 26°C (Mar) Min 6°C (Dec-Feb)
 Rainfall
  Max 170mm (Jul) Min 5mm (Feb)
Southeast (Mérida):
 Temperature
  Max 41°C (Apr) Min 17°C (Jan-Feb)
 Rainfall
  Max 173mm (Jul) Min 18mm (Feb)

The tourist road to Mexico is, by now, comparatively well-travelled – both because it is a physically beautiful and interesting land and, once you have got there, cheap to live and travel in. Whether you want to go to enjoy the ritzy west coast around Acapulco or to explore the Mayan ruins in the Yucatan, you can be confident of always being within reasonable reach of those things that make travelling with a baby comfortable, while not, at the same time, losing your awareness of being in a very foreign part of the world. Mexicans are warm and friendly and they like both visitors and babies.

Health is bound to be every parent's main concern when travelling and Mexico has as many hazards as most poor and hot countries. To start with Mexico City is at an altitude of 7000 feet which means thin air. Most children adapt well and quickly, but don't be surprised at a near total loss of appetite in the first few days. This, coupled with the loss of energy which comes while you adapt to the altitude and the fact that neither parents nor children sleep as deeply or as well, can make the first few days of arrival rather bad-tempered and fraught. Unfortunately there are no obvious remedies apart from taking it easy and waiting for bodies to adapt.

Whatever precautions you all take, and however stringent, you are bound to get tummy pains and diarrhoea at some time. The best stuff for everybody to take is a locally-made medicine called Kaopectate, although it's a good idea to carry some imported Lomotil with you (suitable for children). Don't forget also that the ubiquitous Coca-Cola is surprisingly good for unsettled tummies and is very effective against dehydration.

Mexico is well supplied with good chemists, and in the major towns and cities you can usually find someone in the shop who speaks some English. Chemists have a more active role in medicine than in many countries and, if you can describe symptoms accurately, they will usually either refer you to a doctor or prescribe a medicine themselves (if the illness is quite commonplace). In the towns that tourists are most likely to visit (and this excludes small country towns on the way to ruins or beaches well off the beaten track), chemists stock disposable nappies, powdered milk and baby food plus all the other obvious paraphernalia, so that there is no need to arrive in the country burdened down with endless amounts of babykit. The prices are much the same as at home. At certain times of the year parts of the Yucatan are potentially malarial so it is strongly advisable to start a course of anti-malaria pills before arriving.

Food and drink are the second item of major concern and Mexican food may not be universally popular in your travelling family. Restaurants are open at the usual times, and are very adaptable on the whole. They will try and provide you with the food your children will like. Eggs and chicken abound, though the latter may be stringy and sometimes expensive (by Mexican standards) and there are always potatoes and other vegetables. On the coast there is always lots of fish and in most parts of the country, beef. Don't ever risk drinking tap water, even in the poshest hotel; mineral water in sealed bottles is available everywhere. Similarly, don't risk eating fruit that has been peeled by someone else, however tempting; Montezuma will wreak formidable revenge! That apart, Mexican food and drink is spicy and delicious, and you are in the home of Tequila which means memorable margaritas.

Places to stay are no problem. Mexico has good hotels of almost every class and style. The deluxe are scarcely distinguishable from the Hiltons, Intercontinentals and Holiday Inns of the rest of the world, but good tourist hotels are clean and inexpensive and even the simplest are quite acceptable within their limitations. Loos are clean and the showers (usually) work and, unless you are travelling in the most popular season and are in the most popular places you should not have a problem finding a room. The better hotels will provide cots and all the receptionists will be able to put you in touch with someone who will come in and babysit – either in the day or evening.

Travel within Mexico is not a problem, and it is a wonderful country to explore. There is an extensive internal airline system and, if time is at a premium, it's best to use it. If you have time (and Mexico is fabled for losing it) there are good intercity bus services covering the entire country. At local level the buses are always full, colourful and fun . . . and having a baby with you will certainly make you part of things. Despite the Mexican reputation local buses do have schedules, so don't rely on their mañana attitude if you are visiting a faraway place late in the day! In the major centres you will find organized tourist buses, often airconditioned, visiting the local places of interest – either just one or several during a half- or full-day trip. The hotels usually carry the information, as will the local tourist office. Alternatively, you can check with the hotel desk about hiring a car or taxi for the day; they will certainly find you either. If there are no printed tariffs consult with the hotel or other tourists about a reasonable price and then haggle with the driver before starting. Safety awareness is not high. Mexican driving leaves a lot to be desired and there may not be seatbelts or baby seats – but on the other hand, it is the best way to travel if you don't want to be tied in to other people's schedules.

As for places to go – well the tourist travel guides say it all. The Yucatan is special because it combines a wonderful coast with the evidence of an extraordinary past. Don't miss it.

Recommended reading:

Fodor's Mexico (Fodor's Travel Publications, 1991)
Birnbaum's – Mexico (Houghton Mifflin, 1991)
The Rough Guide – Mexico, John Fisher (Harrap-Columbus, 1989)
Mexico – A Travel Survival Kit (Lonely Planet Publications, 1989)
Insight Guides – Mexico (APA Publications, 1991)
The Penguin Guide to Mexico (Penguin Books, 1991)

Fodor's – South America (Fodor's Travel Publications, 1991)
Insight Guides – South America (APA Publications, 1990)

For further information:
Mexican Tourist Office, 60 Trafalgar Square, London WC2. Tel: (071) 734 1058

Morocco

Capital Rabat
Major cities Casablanca, Marrakesh, Fez, Meknes, Tangier, Agadir
Time GMT
Currency Dirham (Dh) = 100 centimes
Languages Arabic, French, Spanish
Climate Mediterranean
Annual weather range
Rabat:
  Temperature
    Max 28°C (Jul–Aug)
    Min 8°C (Jan–Feb)
  Rainfall
    Max 86mm (Dec)   Min 0mm (Jul–Aug)
Marrakesh:
  Temperature
    Max 38°C (Jul–Aug)   Min 4 °C (Jan)
  Rainfall
    Max 33mm (Mar)   Min 3mm (Jul–Aug)

Except for more intrepid travellers, it is probably best to stick to package holidays or international-style hotels when you visit Morocco with children since conditions out of the tourist resorts can still be primitive.

Morocco can get very hot in the summer, but it is a dry heat, so it is much easier to tolerate than the tropics. The Atlantic coast has the most superb long sandy beaches, ideal for a seaside holiday. However, if you visit in the winter or spring, despite the hot sun, the ocean can be freezing cold and there may be a strong wind. So choose a hotel or apartment complex with a pool.

Self-catering apartments are becoming very popular at the coastal resorts and most travel companies will provide detailed information on where to shop. French is spoken by most Moroccans you will come across and English is quite common in the tourist resorts. Most towns have supermarkets of sorts where baby foods and nappies can be bought, but shopping is most characteristic at the *souk*. Wandering round the *souk* is the high spot of the holiday, but keep an eye on your children – they can easily get lost in the winding narrow lanes and among the shops and stalls crowded with people and goods. (Some sights, especially whole sheeps' heads on butcher's stalls, can be gory, but a source of fascination for children!)

Haggling is, unfortunately, the norm. Bargains can undoubtedly be had, but it's an exhausting business. There is plenty to buy for the children, local musical instruments, leather slippers, cheap shirts and teeshirts and the flowing, embroidered Moroccan shifts.

Moroccan food is not very highly spiced, although children brought up on fish fingers and baked beans may not be keen. Most hotels, however, serve international menus, together with wonderful displays of Moroccan oranges, dried dates and figs – even at breakfast. Yoghurt is also popular. It is wise to stick to bottled water and be cautious when eating in small restaurants. It's best to arm yourselves with diarrhoea medicines and other drugs before you leave home.

Morocco has some wonderful sights, ancient walled cities and palaces as well as the desert. Most travel companies run tours which are quite feasible for young and old, with regular stops at good hotels or restaurants for food and loos. Outside these stops, loos can be very primitive indeed.

For older children, the beach resorts provide lots of sporting activities, from windsurfing and fishing, to tennis and volleyball and they can join the evening out to the local Bedouin-tent-disco to watch the belly dancers. Even the younger children can enjoy the amazing spectacle of a *Fantasia*. Here, while you sip mint tea in an enormous Bedouin tent, ancient, gnarled warriors brandishing muskets charge towards you through the dust on fiery Arab steeds, yelling bloodcurdling war cries! Just as you imagine they are going to trample right through you, they wheel the horses and gallop back where they came from. In between each spectacle, you are entertained by troupes of jugglers, child acrobats and snake charmers. Just don't sit in the front row!

Recommended reading:

Fodor's '91 – Morocco (Fodor's Travel Publications, 1991)
Traveller's Guide to North Africa (IC Publications, 1988)
Blue Guide – Morocco (A & C Black, 1988)
Morocco, Algeria, Tunisia – A Travel Survival Kit, Geoff Crowther and Hugh Finlay (Lonely Planet Publications, 1989)
Insight Guides – Morocco (APA Publications, 1990)
Morocco – The Rough Guide, Mark Ellingham and Shaun McVeigh (Harrap-Columbus, 1990)

For further information:

Moroccan Tourist Office, 174 Regent Street, London W1. Tel: (071) 437 0073

Nepal

Capital Kathmandu
Major cities Patan, Morang
Time GMT + $5\frac{1}{2}$
Currency Nepalese rupee (NR) = 100 pice
Languages Nepali, Bihara
Climate Monsoonal
Annual weather range – Kathmandu:
Temperature
   Max 30°C (May)   Min 2°C (Jan)
Rainfall
   Max 373mm (Jul)   Min 3mm (Dec)

Few travellers to the Far East plan their trip without being tempted by the numerous attractions of this beautiful and, as yet, largely unspoilt mountain kingdom. Kathmandu, its capital and the only city, is now an established tourist centre offering the inevitable five-star hotels, tour operators and luxuries for those who can afford them. Do not be fooled, however, by the apparent high standards of the few top hotels. Nepal opened to foreigners only in the mid-1950s and, although rapidly expanding, its tourist trade is relatively new and restricted only to the few major towns.

Kathmandu is constantly described as 'medieval' and, although this adds to its charm and fascination, it also reflects the sanitary standards or lack of them. The water supply is the main cause for concern.

If you do decide to visit Nepal, it is vital that you get all the necessary vaccinations before you leave home and start precautions against diseases like malaria. This said, there is no reason why you shouldn't take your healthy three-year-old along – providing you are careful. The golden rule is to avoid fresh fruits, salads and all water that has not been boiled for at least ten minutes. (Tourist restaurants do boil their water, but rarely for ten minutes.) Stick to tea, boiled milk, bottled soft drinks or mineral water.

Children are welcomed anywhere in Nepal and there are very few places that would not happily accommodate a child's special needs. Local food is not as hot or spicy as Indian, but if it is a problem for your youngster you can always request rice and vegetables. Thamel is a unique remnant of the hippy era. It has pizza houses, burger bars (made of buffalo meat, not beef), ice cream parlours and all kinds of international cuisines. The best indication as to which eating establishments are safest is whether or not they are frequented by other tourists.

Imported Indian food is available including powdered milk, baby cereal (Farex) and tinned foods. You can find them at the frozen food stores and shops along the New Road, the main shopping street in Kathmandu. This is where you will find the best stocked pharmacies and even imported children's clothes. One further word of warning: pharmacists are not medically trained.

Metered taxis are available as well as rickshaws and you can hire a taxi or car for the whole day at a very reasonable price. Public transportation consists of buses which are very cheap but slow and crowded. There are no trains.

Kathmandu valley provides much to see and your children may particularly enjoy the Monkey Temple at Swayambhunath, a 20-minute walk from the centre of town. Although a steep climb, you can stop and watch the monkeys along the way. Do not feed the monkeys – or any other animals you see – they can be vicious and they do carry fleas and disease. Elephant rides are an attraction and there is a safari park at Gokarna. Although there are public conveniences available at most of the tourist attractions you will probably be safer to do as the locals and go behind bushes.

Although tour operators make trekking and climbing trips relatively safe and comfortable they discourage taking along young children. You can, however, take a leisurely trek around the Kathmandu Valley or Pokhara, which brings you nearer to the Himalayas and is a lake of outstanding beauty. Tour operators will provide all the necessary equipment including a porter to carry your child. You should take out insurance whenever you trek – and make sure it covers the cost of a helicopter. Do not take your children on longer treks even when the tour operators offer a doctor at extra cost. They are taxing and can be hazardous even for the strongest and healthiest person.

October, November, February, March and April are the best times to visit Nepal. Avoid the monsoons of June which continue through September. It is cold in Nepal from February until April so make sure you bring enough warm clothing – and note that only the top hotels have central heating.

If you decide to take the plunge and take your children to Nepal, be prepared to be restricted in your activities and pay for the best. It is foolish to travel with young children on a shoestring. This said, you can have a fascinating holiday among friendly and charming people if you plan your trip wisely.

Recommended reading:

Fodor's – India (inc. Nepal) (Fodor's Travel Publications, 1991)

Insight Guides – Nepal (APA Publications, 1991)

Nepal – A Travel Survival Kit, Tony Wheeler and Richard Everist (Lonely Planet Publications, 1990)

For further information:

The Royal Nepalese Embassy, 12A Kensington Palace Gardens, London W8. Tel: (071) 229 1594/6285

Promotion Nepal (Europe) Ltd, 3 Wellington Terrace, Bayswater Road, London W2 4LW. Tel: (071) 229 3528. Fax: (071) 243 0321.

The Netherlands

Capital Amsterdam
Major cities Rotterdam, The Hague, Utrecht, Endhoven, Arnhem
Time GMT + 1
Currency Guilder (Gld) = 100 cents
Languages Dutch, English
Climate Temperate
Annual weather range – Amsterdam:
Temperature
  Max 22°C (Jul–Aug)   Min 1°C (Jan–Feb)
Rainfall
  Max 87mm (Aug)   Min 44mm (Mar)

A fifth of the Netherlands is covered by water and about 25 per cent of the soil area of Holland comprises land reclaimed from the sea and marshes and bogs which have been drained, so the Dutch understand water. They live by it, on it and they are constantly contending with it. These facts alone make it one of the most remarkable places in the world.

It's a small country, extremely well connected by networks of roads and railways and by waterways. Distances are short so travel is very easy. Every conceivable kind of water transport is available and using it is an exciting way of seeing the country and such wonderful places as Delft and Amsterdam; and getting about by boat is something that appeals a good deal to the small child.

I could never work up much enthusiasm for the Dutch coast though it does have plenty of sandy beaches. It is well-equipped with resort facilities and does attract large numbers of holidaymakers from northern Europe. However, sunshine is often at as great a premium as it is on British shores.

There is a vast range of accommodation, from top-class international hotels down to simple inns. The small hotels are easily the best bet for a family holiday with small children. They are cosy and comfortable and the Dutch hoteliers are congenial and hospitable.

There are also some 500 bungalow parks in the country, often sited by woodland and lakes. These are well-equipped with modern comforts and make a useful base for an open-air holiday. The Dutch are also highly organized for campers; there are well over 2000 camping sites, some of which you need to book in advance. Another possibility is to take rooms in a private house or farm. This can be most enjoyable with small children.

One of the great advantages of Holland is that many Dutch people, at all social levels, are likely to speak English as well as (and, in many instances, better than) many English people. Their fluency and ease is astonishing since they have often never set foot in England.

Medical services are highly efficient. Chemists have everything you might need for small children in an emergency or in ordinary day-to-day events.

There are any number of facilities and diversions for children, including zoos, safari parks, recreation parks, fairytale parks, miniature towns, aquaria and dolphinaria.

The average child will find Dutch food appetizing. It is more or less international; everyone is catered for; variety is provided by Indonesian restaurants. Water is good and soft drinks are abundant.

The flat countryside of Holland has a special beauty of its own, and there are two obvious ways to see it – by bike or by boat.

Bikes are everywhere in Holland, and children are naturally accommodated. You can hire sturdy bikes, with clip-on child seats so that even a child of six or so could go on the back. Babies go in a basket in front, and it is quite common to see two child passengers on one adult bike.

Holland is ideal for a cycling holiday with children with miles of cycle tracks passing through beautiful countryside, with woods and lakes for variety. The tracks continue in the big cities such as Amsterdam and there are special lights for cyclists at junctions and crossings.

Canal holidays are also popular, but could be difficult with toddlers. Older children usually love boats, and there are boat trips round the old parts of Amsterdam. For a more unusual holiday, try the islands on the Waddenzee, where English people are rarely seen. Passenger ferries go from the north coast of Holland, and there are very few cars on the smaller ones, which are criss-crossed by beautiful cycle tracks. All the holidaymakers hire cycles and child seats, children's bikes, pedal cars and trikes are often available.

Center Parcs are holiday centres of a standard far superior to our conceptions of traditional British holiday camps. A similar service is offered by a company called Gran Dorado and both provide a relatively sophisticated family holiday. These organizations are to be distinguished from bungalow parks, which are holiday chalets in quiet woodland or beside lakes.

Dutch people are usually nice to children; shopkeepers often give them a sweet (worries about dental cavities don't seem to exist here) or a slice of cheese or sausage.

There are high chairs everywhere, buggies can often be hired at zoos, and *kinder* menus are common. Children are free on public transport up to the age of four.

Breastfeeding is not usual after two to three months, so feeding older babies can produce odd looks.

Amsterdam is renowned for its dog fouling; but fortunately it does not spread to children's playgrounds, which are guarded by wardens who keep dogs out. There may sometimes be a fee, but these play areas – in most parks – are wonderful value. The water playground Gaasperplas, south of Amsterdam, is free.

There is a good variety of vegetable baby food, including more unusual varieties like brown beans, but the dental cavity warning applies to the sweet varieties.

Any kind of holiday accommodation (including self-catering bungalow parks, bed & breakfast, log cabins etc.) can be booked via the Netherlands Reservation Centre in Holland (tel: 010 31 70 3202 500).

Recommended reading:

Fodor's Holland (Fodor's Travel Publications, 1991)
Blue Guide: Holland, John Tomes (A & C Black, 1989)

For further information:

Netherlands Board of Tourism, PO Box 523, London, SW1E 6NT. Tel: (071) 630 0451. Fax: (071) 828 7941.

Nevis

See **West Indies**

New Zealand

Capital Wellington
Major cities Auckland, Christchurch, Dunedin
Time GMT + 12
Currency New Zealand dollar (NZ$) = 100 cents
Language English
Climate Temperate
Annual weather range – Wellington:
Temperature
  Max 21°C (Jan–Feb)   Min 6°C (Jul–Aug)
Rainfall
  Max 137mm (Jul)   Min 81mm (Jan–Mar)

New Zealand is a beautiful country, roughly the size of the UK, but with a population of only three million people. New Zealanders are justifiably proud of it, and enjoy showing it off to visitors. The feeling of space and room to breathe, pervades attitudes to travellers and their children, which makes for a generally relaxed time (except if you are in a hurry).

Air New Zealand are very welcoming and helpful on internal flights, carrying pushchairs and car seats in the hold, and if necessary allowing pushchairs out onto the tarmac. There are adequate facilities for changing babies at the international airports, although inevitably hit and miss at the smaller airports.

Renting cars at the main centres is straightforward. Child seats are generally available, and as most cars have rear seat belts, security is no problem. New Zealand driving is anarchic, and lane discipline on the few motorways is, to put it politely, erratic although there is a speed limit on all roads which is firmly enforced in the main urban areas. The New Zealand railway system offers comfortable and efficient travel, and there are good bus services. Coaches on the main tourist routes are reasonably modern.

New Zealand is said to have a greater proportion of its land area designated as National Parks than any other country in the world. Visitors are welcomed and there are often informative displays with good graphics in the visitor centres, which interest children – as well as clean loos. There are dozens of the most wonderful sandy beaches. In the summer the main ones have lifeguards on duty with warnings where there are potential hazards. Beaches are considered crowded if you can see other families. In addition, there are nature reserves along all the main roads where travellers can enjoy a picnic and children can stretch their legs in safety. Most towns and settlements have reasonable loos and very often a recreational area equipped with swings and other apparatus. Dairies are almost everywhere. These are the equivalent of an English corner shop, open long hours and selling a range of sandwiches, fast foods and soft drinks at reasonable prices.

There are well-placed campsites in most of the holiday areas, and it is possible to get into them without booking except at peak times in January. Most have adequate washing facilities although they vary in standard. Youth hostels will also take families, but it is worth booking in the peak holiday season and at holiday weekends. The main centres and tourist spots have first-class hotels, but many people prefer to stay in motels which are extremely good value especially for families with children. Facilities vary; some offer baby listening services, swimming pools and saunas; some have their own restaurants; others simply offer self-catering. Nearly all rooms have television and radio, and are generally very clean. Tea, coffee and milk are always provided for breakfast, and there are usually fast food take-aways or dairies nearby if the motel doesn't have its own restaurant. Most restaurants have high chairs and children's menus, but service although friendly can be very erratic. New Zealand pubs are not on the whole to be recommended when travelling with children. Most of them have take-away bottle stores for thirsty parents.

Disposable nappies are easily available at supermarkets and chemists, but along with cotton wool, wipes and suchlike, are comparatively expensive. The New Zealand National Health service, although it now charges, has reciprocal arrangements with the UK. UK visitors will fall into the lowest tier of health benefits. But it is recommended that visitors have comprehensive medical and travel insurance.

Recommended reading:
Fodor's – New Zealand (Fodor's Travel Publications, 1991)
New Zealand – A Travel Survival Kit (Lonely Planet Publications, 1991)
Insight Guides – New Zealand (APA Publications, 1991)
New Zealand Handbook, Jane King (Moon Publications, 1990)

For further information:
New Zealand Tourism Office, New Zealand House, Haymarket, London SW1 4TQ. Tel: (071) 973 0360.

Northern Ireland
See **Eire & Northern Ireland**

Norway

Capital Oslo
Major cities Bergen, Trondheim, Stavanger
Time GMT + 1
Currency Norwegian krone (Nkr) = 100 ore
Language Norwegian
Climate Arctic/temperate
Annual weather range – Oslo:
Temperature
 Max 34°C (Jul) Min −7°C (Jan–Feb)
Rainfall
 Max 95mm (Aug) Min 35mm (Feb)

Norway is a country of spectacular natural beauty and, with a population half the size of London, most of it is completely unspoilt. It is not more than half an hour's tramride from any of the capital's cafés to a spot where you can pick wild berries. The coast itself is a marvel of nature with the famous ice-blue fjords and safe, sandy beaches where you can swim.

The Norwegians like children and there are many parks and open spaces for them to play in. The air is marvellously fresh and clean.

The summer is short, with June, July and August being the best months to visit the country. Autumn is also an attractive season when the mountainous terrain turns lovely shades of yellow and red. Winter starts early, at the end of October or early November. Some of the mountain roads are closed in the winter and will only open in the spring or even summer. There are few motorways but the road standard is good and well signposted. Compared with most European countries the traffic is very light, except at ferry crossing points at the weekends.

Camping and skiing are the national pastimes. There are hundreds of approved campsites around the country and they are all well-equipped, clean and welcoming. You can even rent cabins in them if you are tentless.

All public facilities in Norway can be expected to be clean and functional.

Eating out *en famille* is not a tradition in Norway and the cafeterias are more welcoming places to take children than restaurants. English people will find the food to their taste, but there will be a higher proportion of salted or preserved meats and fish. Indeed, fish of all sorts is a staple of the diet and you will find it boiled, salted, smoked, dried, cured, marinated, pickled and fried. Make sure you try *gravet laks*, a traditional Norwegian dish, which is salmon marinated in salt, sugar and spices.

Hotel chains have several discount schemes you should take advantage of if travelling with your family. In individual hotels you will find special offers covering everything from walking tours in the glacier fields and sightseeing to babysitting and windsurfing. Many double hotel rooms will have an extra sofabed in the room so you can accommodate one more person or two small children.

The Norwegian's own favourite holiday is going to a *hytte* – a wooden cottage in the mountains or by the sea. You can rent these cottages and although they vary in size, standard, location and price, most sleep four to six people and consist of a kitchen, sitting room, often with open fireplace and one or more bedrooms.

You can also expect to find every kind of item for your baby, including nappies and baby food, in the supermarkets which are well-stocked but, as with everything in Norway, expensive.

Recommended reading:

Frommer's – Scandinavia, Darwin Porter (Prentice Hall Press, 1991)

Fodor's – Scandinavian Cities (Fodor's Travel Publications, 1991)

Scandinavia – The Rough Guide, Jules Brown and Mick Sinclair (Harrap-Columbus, 1990)

Karen Brown's – Scandinavian Country Inns and Manors (Harrap-Columbus, 1989)

Drive Around Norway, Robert Spark (Trafton Publishing, 1990)

For further information:

Norwegian Tourist Board, Charles House, 5 Regent Street (Lower), London SW1Y 4LR. Tel: (071) 839 6255

The Philippines

Capital Manila
Major cities Quezon City, Davao, Cebu, Makati
Time GMT + 8
Currency Philippine peso (PP) = 100 centavos
Languages Tagalog, English
Climate Tropical
Annual weather range – Manila:
Temperature
   Max 34°C (Apr–May)
   Min 21°C (Dec–Feb)
Rainfall
   Max 432mm (Jul)   Min 13mm (Feb)

Filipinos as a rule dote on children who are almost constant targets of affectionate curiosity and cooing.

Products and services specifically designed for children are difficult to obtain outside the capital or any of the big cities in the Philippines. However, locally made items are very popular and inexpensive compared to the widely available imported range of babycare products and foods from Europe and America. Manila, of course, is the best-supplied. Shops such as Rustan's, ShoeMart or Robinson's sell disposable nappies, processed baby foods, formula, 'long-life' and powdered milks, cereals, inexpensive children's clothing and shoes (indeed, girls' fancy dresses, manufactured widely in the Philippines for export, are sold at bargain prices).

The climate is hot and humid during most of the year (heavy rains and typhoons occur, from June through October). Most hotels are well air-conditioned as some taxis. Skin rashes may occur, however, insect bites are common. A good variety of Western preparations are readily available at chemists, particularly a national chain called Mercury.

Western-style food (McDonald's, pizza and so on) is liberally strewn around Manila which could rate as Asia's 'junk food' paradise. Most hotels and several restaurants serve good Western dishes, but many Filipino dishes are only moderately spiced and are palatable and wholesome: boiled rice, roast chicken, grilled fish and prawns, fresh fruits and vegetables. Restaurants as a rule do not provide high chairs. Meals are eaten 'family-style' with everyone sharing from dishes in the centre of the table.

Water is generally drinkable in the Philippines, although as a precautionary measure it is advisable for visitors to drink boiled water or mineral water except in the main cities.

The Philippines may be one of the easiest places in the world to find competent, inexpensive babysitters who speak English. Any hotel can make the contact for you with agencies also available in large cities.

Manila is a noisy, congested city and not very conducive to getting about with children. There are a few small urban pools and the occasional playground and most hotels have pools. Large residential subdivisions in the Metro Manila area are equipped with playgrounds and their own sports facilities and complexes.

Most TV programming is in English including both middling fare imported from the US and some engaging children's programmes, including a local imitation of 'Sesame Street' called 'Batibot'.

Recommended reading:

The Philippines – A Travel Survival Kit (Lonely Planet Publications, 1991)
Philippines Handbook, Peter Harper and Evelyn Peplow (Moon Publications, 1991)
South East Asia Handbook, Carl Parkes (Moon Publications, 1990)
Philippine Guide, *Philippine Handbook*, *Philippines – Our World in Colour*, all published by Odyssey Guides, 1991
Introduction to the Philippines: Odyssey Guides, E. Peplow (The Guidebook Co. Ltd., 1991)
Insight Guides: Philippines (APA Publications, 1990)

For further information:

Philippines Department of Tourism, 17 Albemarle Street, London W1X 7HA. Tel: (071) 499 5443. Fax: (071) 499 5772.

Portugal

Capital Lisbon
Major cities Porto, Faro, Coimbra, Setubal
Time GMT + 1
Currency Escudo (ESc) = 100 centavos
Language Portuguese
Climate Mediterranean
Annual weather range – Lisbon:
Temperature
  Max 30°C (Aug)   Min 8°C (Jan–Feb)
Rainfall
  Max 111mm (Jan)   Min 3mm (Jul)

As it happens Portugal is England's oldest ally. The 600th anniversary of the Treaty of Windsor was marked in 1986 and was celebrated on a grand scale with Anglo-Portuguese events and exchanges which can only have benefited the prospective visitor from Britain.

The Portuguese are a very charming, civilized and hospitable people. They like the English and many of them have some knowledge of our language. They live in a spectacularly beautiful country which is for the most part unspoilt by the ugly manifestations of twentieth century progress. The summers are hot and the winters mild; in fact, you can take a family holiday there at almost any time. Pretty well every conceivable need is catered for and small children are very welcome everywhere.

I particularly recommend the state-owned *pousadas*, comfortable inns (sometimes in converted monasteries, castles and palaces) which are reminiscent of the convivial *gostilne* of Slovenia. There are privately run inns (*Estalagens*) and private enterprise also provides excellent holiday accommodation in manor houses and farmhouses. All these provide rural environments in splendid countryside which are ideal for small children. The Portuguese also have well organized camping and self-catering establishments.

In hotels children under eight are entitled to a 5 per cent discount if they share a room with their parents (or other adults). Many hotels provide cots for babies. Most forms of baby food are available in chemists and supermarkets but imported brands are likely to be more expensive than in the UK.

Public transport is efficient (no problems with children) self-drive cars are readily available. Railway fares are cheap and there are discounts for family tickets (*see* 'Rail Travel'). It is fun travelling by public transport and people are most obliging and courteous to visitors.

The country is justly famous for its beautiful beaches and there are said to be some 500 miles of sand on the seaboard. The resorts are extremely well equipped and there is any amount of diversion and entertainment for children. The magnetism of the shores is strong but I should certainly take the opportunity of going inland if possible. Children will find interesting rural ways of peasant life seldom seen in Britain – and look out for the silky-coated, long-legged pigs of the Algarve.

Portuguese food is plentiful, well prepared and well cooked, though for some palates it may prove a little too oily at times. As in Spain meals are leisurely and there are many delectable dishes, such as *caldo verde* (a broth made of shredded kale and potatoes), *gazpacho* soup (which is served cold), and the national dish *bacalhau* – salted, dried cod prepared and presented in many different ways. Because of the Atlantic seaboard fish and shellfish are first class, especially mullet, halibut, sole, sardines, squid, swordfish, tuna, lobster and prawns. I also recommend *caldeirada*, a fish stew somewhat reminiscent of *bouillabaisse*. Like the Turks, the Portuguese have a very sweet tooth and produce lots of delicious confectionery and sweets. One such delicacy is *barrigas de freira* (which means 'nuns' tummies') and is a mixture of egg, almonds and sugar. There is a good range of soft drinks and the water is safe but I would always be cautious about ordinary tap water.

Like the Spaniards the Portuguese have a large number of spectacular fiestas and festivals which are delightful for children and it would be a great pity not to include one of them in a trip. Some of the more famous are the Tomar Festival of Tabuleiros (a harvest festival which happens every two years); the Golega Fair of St Martin (November); the Lisboa Festival of St Anthony, John and Peter (June); the Vila Franca de Xira Red Waistcoat Festival (in July, it includes bullrunning and bullfighting); the Viseu Cavalcades of Vila de Moinhos at Montanhas (June); the Ovar Carnival of Costa de Prata (the weekend before Lent); the Porto Festival of St John (June); and the Costa Verde festival of Our Lady of Suffering in August.

Recommended reading:
Fodor's – Portugal (Fodor's Travel Publications, 1991)
Portugal – The Rough Guide (Harrap-Columbus, 1991)
The Penguin Guide to Portugal (Penguin Books, 1991)
Karen Brown's – Portuguese Country Inns and Pansadas (Harrap-Columbus, 1989)
Frommer's – Portugal, Madeira and the Azores, Darwin Porter (Prentice Hall, 1990)
Collins Traveller – The Algarve Travel Guide (William Collins, 1990)

For further information:
Portuguese National Tourist Office, 22/25a Sackville Street, London W1X 1DE. Tel: (071) 494 1441. Fax: (071) 494 1868. Telex: 265653

Russia

Capital Moscow
Major cities St Petersburg
Time GMT + 3–12
Currency Rouble (Rub) = 100 kopecks
Language Russian
Climate Continental/arctic
Annual weather range – Moscow:
Temperature
  Max 23°C (Jul)   Min −16°C (Jan)
Rainfall
  Max 88mm (Jul)   Min 38mm (Feb)

Russia is the largest state in the newly formed CIS. People in Russia are generally very fond of children, but travelling around the country with them can be very difficult. Little provision is made for children, especially infants, at hotels and restaurants because it is not customary for Russians to travel with their offspring. Little attention is paid to children on the airlines, but airports, like big stations, usually have a mother and baby room.

Accommodation at Russian hotels usually consists of twin-bedded rooms with private bathrooms and televisions. Rooms with more than two beds are unusual and cots are not generally provided.

The easiest way to visit the CIS with children is on a package tour as all transportation, including buses and trains, is then organized for you. Car travel and hiring a car may be complicated because of certain travel restrictions. Public transport, which is very cheap, is usually packed but people willingly give up their seats to children.

Feeding children can also be a problem. There are usually plenty of stalls and coffee bars selling sandwiches, buns and soft drinks, but children are only permitted to eat in restaurants during the day. High chairs are not provided. Travellers with small children are advised to take food with them since milk and other dairy products may be hard to get outside the main cities. There are foreign currency foodstores at the main tourist centres, but the selection is limited. Local supermarkets are open from 8 a.m. until 9 p.m. and usually close from 1 until 2 p.m.

More foreign currency stores are appearing in St Petersburg and Moscow and their selection of stock has increased to include things such as nappies. But disposable nappies, paper tissues, toilet rolls and waterproof pants are hard to find and bottle teats, baby wipes, plastic tie pants and plastic feeding bottles are simply not available. Baby creams are not always in stock either. No provision is made in WCs for changing babies and public conveniences should be avoided with small children anyway.

There are lots of pleasant children's playgrounds and plenty of circuses and puppet shows for older children. In winter there are opportunities for children to toboggan, ski and skate outside. However, visiting museums and art galleries with small children is frowned upon.

The winters are very cold but the buildings are well heated so light clothing is recommended indoors, however children need to be warmly dressed outside. If you are going to Russia at a very cold time of year, it is wise to take a pure lanolin cream (with a low water content) to protect your child's cheeks from the frost. In summer it can be very warm, especially in the south. Children certainly need to have sun hats in the Central Asia Republics or on the Black Sea.

Russia has a national health service, but foreigners are required to pay for medical treatment (payment must usually be made in roubles) unless some kind of reciprocal agreement exists between the respective countries. (Such an agreement exists between the UK and Russia.) Health facilities are provided at large clinics and there is a special clinic for foreigners in Moscow. Mothers with children are advised to arm themselves with antiseptic creams, anti-stomach-upset pills and cotton wool as these may prove hard to find. Bear in mind that if you need something for your child, it is sure to be out of stock!

Please note

Although the political situation is, at present, calm recent changes are such that it is difficult to know the effect they will have on holiday makers. We would advise you to contact the relevant National Tourist Office for further details before travelling.

For further information:

Intourist Travel Ltd, 219 Marsh Wall, London E14 9FJ. Tel: (071) 538 8600

Saudi Arabia

Capital Riyadh
Major cities Jiddah, Mecca, Taif, Medina
Time GMT + 3
Currency Saudi Riyal (SAR) = 100 hallalas
Language Arabic
Climate Desert
Annual weather range – Riyadh:
Temperature
  Max 42°C (Jun–Aug)   Min 8°C (Jan)
Rainfall
  Max 25mm (Apr)   Min 0mm (Jun–Dec)

As the Kingdom of Saudi Arabia does not issue tourist visas it is very unlikely that anyone would go to Saudi for a holiday with their children. However, there are many English people working there whose families either live with them for part of the year or who plan to visit. The annual pilgrimage, the Haj, also attracts hundreds of visitors.

For those already living there, the area is more varied and interesting than one would, perhaps, expect of a desert area. Just one example is the cooler mountainous region in the south-west, the Assir.

There are some fairly obvious health precautions that must be taken by everyone who visits the country. Although there are many diseases that are no longer officially present in Saudi Arabia it is wise to be protected against tuberculosis and polio, and should you be planning to leave the major cities, against cholera, typhoid and paratyphoid as well. Malaria is still present in the southern part of Saudi Arabia on both the Gulf and the Red Sea coasts. It is best to check with your doctor exactly which innoculations you need.

Although tap water is safe to drink you may prefer your children to stick to boiled or bottled mineral water. And of course all fruit and vegetables should be washed.

Shoes should be worn at all times: there is bilharzia in many pools and standing water in the mountains; stone fish and other nasties inhabit the shallows at the seaside and there is also a chance of hookworm. The ground also gets hot enough to burn the soles of the feet.

Do not touch any animals, however appealing they may look. There are few pets in the Kingdom and rabies is endemic. Take out comprehensive health insurance before you leave the UK and ensure it covers absolutely every eventuality. Hospital facilities are very good, but expensive.

The climate is severe. Your skin should be protected at all times and children should certainly wear sunhats. It is important that drinking water (bottled) is always available and children especially should be encouraged to drink it. Extra salt is not generally thought to be necessary if you eat a balanced diet.

Remember to wear modest clothes at all times. There are separate facilities for men and women. Small compartments at the rear of buses, for instance, have separate entrances and are reserved for women. Benches and restaurants are similarly segregated although there are 'family' areas in most city hotels and restaurants where men and women can sit together.

There are a few things you should know about etiquette in Saudi Arabia: don't use your left hand to point at people or to hand things to others. Don't allow the soles of your feet to show when sitting, as this can be taken as an insult. Of course, the local inhabitants will be tolerant of mistakes, but you should remember to be respectful of a somewhat more private society than our own. Most visitors quickly understand and adapt to the Saudi laws and customs.

Everything you need for children is readily available in the major towns and things like disposable nappies are in every little corner shop. The local brands are relatively cheap.

Recommended reading:
Berlitz Travel Guide: Saudi Arabia (Macmillan, 1985)
MEED Guide: Saudi Arabia (MEED, 1983)

For further information:
Royal Embassy of Saudi Arabia, Information Centre, 18 Cavendish Square, London W17 0AQ. Tel: (071) 629 8803

Scotland
See **Britain**

Sicily
See **Italy**

Singapore
See **Malaysia**

South Africa

Capital Pretoria, Cape Town, Bloemfontein
Major cities Cape Town, Johannesburg, Durban
Time GMT + 2
Currency Rand (R) = 100 cents
Languages Afrikaans, English
Climate Wide regional variation from temperate to desert to tropical
Annual weather range
Eastern interior (Pretoria and Johannesburg):
 Temperature
  Max 28°C (Dec)   Min 3°C (Jun–Jul)
 Rainfall
  Max 132mm (Nov–Dec)
  Min 5mm (Aug)
Mediterranean (Cape Town):
 Temperature
  Max 26°C (Jan–Feb)   Min 7°C (Jul)
 Rainfall
  Max 89mm (Jul)   Min 8mm (Feb)
Tropical (Durban):
 Temperature
  Max 27°C (Jan–Mar)
  Min 22°C (Jul–Aug)
 Rainfall
  Max 122mm (Feb, Nov)
  Min 51mm (May)

The climate is wonderful. In Johannesburg the heat can sometimes get unpleasant, but the Cape has a mainly Mediterranean climate, while Durban is more tropical. Remember, the seasons are reversed, and Christmas means summer. Even South Africa's winter compares favourably with summer in Britain, and along the Natal coast, winter is the height of the holiday season. You will only need light clothes, plus sweaters for winter evenings. Shield your children at first from the sun, which can be very fierce. Inevitably, such a climate means a concentration on outdoor activities. A huge variety of sports and watersports is available, so there's lots for children to do.

The country is spectacular throughout, with a huge variety of scenery. The nicest place to visit is the Cape Coastal Belt, which is incredibly beautiful. Start at Cape Town and go up the Garden Route to Port Elizabeth.

Inland attractions which would appeal to children include the Cango Caves, and the ostrich and crocodile farms of Oudtshoorn, on the edge of the Little Karoo semi-desert plain. Further up the coast, The Transkei and Kwazulu – both so-called black 'homelands' – are well worth visiting.

In Johannesburg you can visit the surface workings of a gold mine and see the mine dancers and you can go underground at Gold Reef City. Special attractions for children are Santa-rama Miniland in Rosettenville and Gillooly's Farm in Bedfordview. But take care in Johannesburg: muggings are on the increase in the city centre.

You will probably want to visit a game park. The most famous (and largest) is Kruger National Park, along the border with Mozambique, but there are many others worth considering. You can book tours, or go on self-drive visits. You can get bungalow accommodation suitable for families.

Hotels are excellent. Most of them have swimming pools and all the larger ones provide cots and high chairs on request. The game parks will provide cots if these are reserved in advance when booking. Car hire is reasonable, the roads are excellent and you drive on the left.

White South Africa is an African version of California, so life for those travelling with children is relatively easy. Most of the precautions you might need to take in other countries – Zimbabwe or Egypt for example – do not apply if you stay within the white areas. The water is perfectly safe to drink and uncooked foods don't need special treatment. Nappies of all kinds are available everywhere and the standard of health care, should you need it, is extremely high. But you will have to pay for treatment. Anti-malarials are recommended for visiting the Lowveld, Kruger National Park and Zululand. Unless you have come from the Yellow Fever area there are no other special health requirements.

If you want to see the country by train, the Blue Train is the way to go. It's South Africa's deluxe, air-conditioned express train, travelling between Pretoria, Johannesburg and Cape Town, and commonly spoken of as a memorable experience.

You won't get to see how black South Africans really live unless you make an effort. There are now organized trips to Soweto. The experience can be an uncomfortable revelation to white visitors who may be expecting fundamental change since the release of Nelson Mandela.

Please note

Recent political events are such that it is difficult to know the effect they will have on holiday makers. We would advise you to contact the relevant National Tourist Office for further details before travelling.

Recommended reading:

Central Africa: A Travel Survival Kit, Alex Newton (Lonely Planet, 1989)
The Traveller's Guide to Central and Southern Africa (International Communications, 1990)

For further information:

South African Tourism Board, 5/6 Alt Grove, Wimbledon, London SW19 4DZ. Tel: (081) 944 6646. Fax: (081) 944 6705.

Spain

Capital Madrid
Major cities Barcelona, Bilbao, Malaga, Seville, Valencia, Zaragoza
Time GMT + 1
Currency Peseta (Pa) = 100 céntimos
Language Spanish
Climate Mediterranean
Annual weather range
Madrid:
   Temperature
      Max 27°C (Jul)   Min 14°C (Dec–Feb)
   Rainfall
      Max 53mm (Oct)   Min 11mm (Jul)
Balearic Islands (Palma, Majorca):
   Temperature
      Max 29°C (Jul–Aug)
      Min 20°C (Jan–Feb)
   Rainfall
      Max 55mm (Sep)   Min 3 mm (Jul)

As a major tourist country, Spain should hold no fears for holidaymakers with children heading for the main resorts. If in doubt, ask a travel agent about special child facilities at hotels and take the normal precautions over heat, mosquitoes, sun and water.

Over 206 beaches in Spain have been awarded the Blue Flag symbol denoting that beaches, sea and facilities are of a high standard of hygiene and safety.

If you do venture into the interior, things may be a little different. Spain is a child-loving country and many couples, including the young middle-class, still have quite large families. Nevertheless, childrearing is perhaps more relaxed and you may find some differences in safety standards. Baby seats in cars are not common, however the main car hire companies will provide one on request. Not all children's playgrounds are maintained to a high standard. Also the current concern over sugary foods doesn't seem to have hit Spain yet. Expect your children to be offered sweets, biscuits and crisps as a matter of course. (However, in general, the Spanish diet is much healthier than the British.)

One major difference is the time of meals. Few Spaniards will sit down to lunch before 2 p.m. and 3 p.m. is more usual. The evening meal is from 9 p.m. onwards (10 p.m. is more common) and you will probably not be served in a restaurant, unless it is a transport or tourist restaurant, before 8 p.m. Spanish children stay up for meals and are more than tolerated in restaurants – very little bad behaviour would be unacceptable.

Spain is noted for its cuisine and each region has a local speciality. Many dishes use a combination of meat, fish, pulses and vegetables and salads are common. Children allergic to garlic or olive oil will not be happy, but typical menus usually offer pasta, grilled meat – lamb (*cordero*) and pork (*lomo*) are the most popular – chips, ice cream, fruit and the ubiquitous *flan* (cream caramel), which all but the fussiest eaters should find acceptable. Restaurant meals are nearly always freshly cooked, using the freshest ingredients, and are still very cheap. Child portions are not a special feature but few restaurants would object to providing extra plates and spoons for dividing up meals. Similarly, high chairs are not normally available.

As in most continental European countries, children are allowed in bars in Spain, where soft drinks, ice cream and coffee are served, as well as alcohol. They also usually have televisions and bar-football machines.

Shopping in Spain is a great experience – as long as you have plenty of time. Modernization is overrunning Spain very quickly, and new *supermercados* have brought with them a kind of Sainsbury mentality. But in the market, butchers and fishmongers are a law unto themselves. The best advice is to be bold – and take your own shopping basket. Children, naturally, can tire of the experience quickly, but may be revived by a stop at a *churrería*, a bar or stall serving *churros*, a delicious kind of long doughnut, often served with a hot chocolate sauce for dipping.

Shops are usually open from 9 or 10 a.m. to 1 or 1.30 p.m. and then from 3 or

3.30 p.m. to 7.30 or 8 p.m., but check locally as times can vary according to the season and area.

Disposable nappies (*panales*) and other baby equipment are available in all but the smallest villages and a chemist can usually be relied on to prescribe appropriate drugs for most holiday complaints. Most towns and large villages will have a doctor on duty for accidents and emergencies (ask at your hotel or campsite for details). There are also Red Cross clinics for emergencies.

Spanish children don't seem to be fussed over as much as their Italian counterparts, and the large families mean that older children entertain younger ones. If you are staying in a mainly Spanish area, your children will be the centre of attention and should have no problems making friends even with a language barrier. For older children, football is the universal language, but beware, very young Spanish children, as young as 11 or 12, often own small motorbikes which they use off the main roads in campsites, holiday villages and so on.

When not in swimming costumes, Spanish children are always immaculately dressed. You may find that your younger daughters with short hair will be mistaken for boys since most Spanish girls have their ears pierced at birth and are therefore easily recognized by their earrings.

The Spanish interior has some wonderful sights – Seville, Granada, Segovia, Toledo, Avila – but distances between cities are very great and it's no coincidence that cowboy films are shot in the desert landscape. So, make sure your children are good car travellers before attempting a touring holiday.

N.B. Over the recent years the road network has been much improved with a total of 6,000 km of dual carriageways and motorways having been built.

The Canaries

Like much of mainland Spain the Canaries welcome British visitors and their families. Taking your baby with you should present no particular problems provided you plan sensibly in advance for your baby's (and your own) needs. Forward planning may ultimately help to reduce the cost of your holiday. Remember, even after you have landed on the island you are going to, there may be a journey of one or two hours before you reach your final destination, so go prepared and take your own supplies with you. Even in the airports where baby facilities are advertised it is usually busy and chaotic. It is possible to buy most medication in the pharmacies.

If you have arranged car hire in advance ask for a child seat when booking because they are common and it is not rare to see local children restrained in the rear of cars.

If you are staying in a hotel you will probably be able to arrange for a cot in your room in advance for a small charge. If staying in private accommodation it is difficult to arrange the hire of one. However, most airlines will carry a travel-cot and pushchair, if tied up securely, in addition to the normal adult baggage allowance, at no extra cost. High chairs are also difficult to borrow, though some hotels may have them. A pair of reins tied on to a normal chair may be a satisfactory compromise.

Supermarkets are everywhere and most products found at home can be purchased. Fresh fruit and vegetables are particularly tempting, however tinned foods are expensive. Disposable nappies, or 'panales' are also available from pharmacies and prices are similar to the UK. Prepared baby foods, or 'potitos', cost the same as in Britain. Buy bottled water for children's drinks.

Restaurants are very tolerant of children. In tourist areas it is easy to find English-style food (if that's what you want), and most restaurants have dishes suitable for children. If you manage to find a restaurant frequented by Spanish families on a Sunday you can enjoy a break while all the children mix and are amused by the Spanish families.

Most children will love the sunshine, beaches and activities available in the Spanish islands, provided they are well protected with a hat and sun cream. Many of the beaches are black and the sand is coarse. Make sure young feet are covered as the black sand absorbs the heat readily, making it very hot. Most beaches have sunbeds and sunshades available for hire. Children are well catered for in the swimming pools and aqua-parks, although at some times of the year unheated pools may be cool for very small children.

As in most holiday destinations other attractions such as zoos, parrot parks and banana plantations welcome families.

All in all you will find the Canaries pleasant, clean and welcoming for an enjoyable family holiday at reasonable cost.

Majorca

I spent two weeks in Puerto Pollensa, Majorca in spring, 1987 with my mother and two children, Bryn aged 5 and baby Siân aged 15 months. We went during the Whitsun holidays, as Easter might have been too cold and summer can be too hot for small children and the elderly to enjoy the whole day out in the sun. We had an early morning flight, but it was a Bank Holiday weekend with threats of airport strikes and subsequent long delays, so I took piles of food, juice and nappies. The juice spilt before we even got to the check-in desk and they nearly turned us off the plane when they saw all the junk! Anyway, the journey from Manchester to the resort by plane and bus didn't take long, and we were walking along the beach by the afternoon.

It is worth mentioning that since then Government agreements with Spanish air-traffic controllers have meant that there are no longer significant delays in flights to Spain.

Puerto Pollensa was originally a small fishing village. It still has its harbour and jetty for fishing boats and leisure craft and there are frequent boat trips around the bay every day to the beach at Formentor. We used to save some bread from our meals and throw it into the sea to watch hundreds of small fish rushing in to eat. There is a long promenade of small hotels, shops, bars and restaurants, with a tree-covered walk alongside the beach at one end and huge new white sandy beaches towards Alcudia at the other. Many British people buy or rent villas and apartments here, and the town is full of interesting shops and eating places. It never gets boring. There's little drunkenness or rowdiness about, and we stayed in a small hotel with the locals and their children coming in and out all day, talking Spanish to the children and being very friendly.

The various supermarkets are crammed with good food and we picnicked well on local bread, cold meats and cheeses, fruit and salads. The juices are expensive, but would probably be too heavy to carry from home. We really liked the peach one, however. When I ran out of disposable nappies I found replacing them expensive and wished I had brought more. Siân didn't wear a nappy on the beach but drank a lot more and used about six per 24 hours. Also you never know if the 'runs' are going to attack.

The mountains are all around and when we were on the beach there was plenty of scenery there too – hills in the distance, trees, flying-boats landing and taking off from a special harbour, and people using the wind-surfing and water-ski facilities. We enjoyed the slower pedaloes and chairs or loungers were available for daily hire. Bryn was in the sea so much that he learnt to swim, but I had to put creams on both children to stop them burning. They did get an itchy rash the first few days, but it disappeared without any special treatment. There's a doctor's surgery on the main street, open in the day at various times and all are welcome. Also chemists sell most general medications.

We hired bicycles cheaply and baby seats were also available. This is a very good way of exploring the town and surrounding country. There was a another pretty cove called Cala San Vincente next to Puerto Pollensa which was well worth pedalling to. Also local buses are very cheap, and we just walked down to the bus station at the harbour and got on. There are two old towns nearby to explore, and you could go on a longer trip over the mountains into Puerto Soller. Oranges and lemons are growing everywhere. Just outside Pollensa town five miles away, we found a garden centre full of ornamental clay pots – I wanted them all! We also found plenty of shoe shops and bought a beautiful pair of leather shoes. The clothes and swimwear shops are full of stylish goods. It is almost worth taking empty suitcases and getting all your holiday teeshirts there.

There isn't a launderette so I washed a few things by hand and was allowed to borrow pegs and use the line on the hotel roof. The sun dried everything in just a few hours. I took a travel iron but didn't need it much. If required, there was a dry cleaners next to the hotel.

Just outside the port there are riding stables and there are small Shetland ponies available for children to use. On Wednesdays the market comes to the town square and the stalls are laid out with fruit, vegetables and piles of olives. There are also leather goods, clothes, pot plants and flowers everywhere.

A tip: everything the children ate stained like mad, e.g. delicious bottled chocolate drinks, local cherries and strawberries, tomato purée in the rice and pasta dishes, so a plastic bib and lots of 'wipes' are necessary.

Many people go back to Puerto Pollensa every year and I can heartily recommend it as a good place to visit with the family.

Recommended reading:
Insight Guides – Spain (APA Publications, 1990)
Fodor's – Spain (Fodor's Travel Publications, 1991)
Berlitz Blue Print – Spain (Berlitz, 1990)
The Penguin Guide to Spain (Penguin Books, 1991)
The Insider's Guide to Spain, John de St Sterne (CFW Publications, 1990)
Blue Guide – Spain (A & C Black, 1989)
Spain – The Rough Guide, Mark Ellingham and John Fisher (Harrap-Columbus, 1990)

For further information:
Spanish Tourist Office, 57 St James's Street, London SW1. Tel: (071) 499 0901. Fax: (071) 629 4257. Open 9am–4pm Mon–Fri.

Sweden

Capital Stockholm
Major cities Göteborg, Malmö, Uppsala
Time GMT + 1
Currency Swedish krona (SKr) = 100 ore
Languages Swedish, English
Climate Continental
Annual weather range – Stockholm:
Temperature
  Max 22°C (Jul)   Min −5°C (Jan–Feb)
Rainfall
  Max 76mm (Aug)   Min 25mm (Mar)

Sweden is a successful industrial nation with a small population and a very advanced welfare state structure. The standard of living is high and although visitors to the country will probably be impressed by the quality and efficiency of transport and other services, they are likely to be less pleasantly surprised by the cost. It is one of the most expensive European countries to visit but its attractions make the expenditure worthwhile. There are large areas of breathtaking scenery and one can feel completely free and at peace with one's natural surroundings.

Most Swedes own a summerhouse *stuga* so that they escape the city and can relax with the pleasures of water, sun and greenery. The shortness of the summer growing season in comparison with the long dark winter months makes the pleasure of being outside even more intense, and the first warm days of spring brings city Swedes out onto the pavements, stretching their faces to the sun like cats or lizards basking in its warmth. If you plan to visit Sweden in the winter with your family you will need extra warm clothing for outside but all Swedish houses and flats are through necessity well-heated and insulated against extreme weather. A self-catering holiday or a house swap with Swedes can be a marvellous way to see the country but some visiting families might be surprised how primitive the summer *stugas* can be, with no electricity or sanitation except a chemical toilet. Swedish families feel that the return to basic living is part of the charm of their summer holidays and so it can be, if one is prepared for it! In contrast, visitors are likely to be impressed by the high standard of cleanliness and efficiency in normal Swedish everyday accommodation.

Sweden is one of the easiest places in the world to travel with a child. Most people speak English, there are excellent facilities, plain palatable food and good medical care. The majority of Swedish mothers work outside the home and there is a strong tradition of women's opinions being listened to. This means that in the major cities at least there should be no problems with pram/buggy access, nursing and changing areas, availability of high chairs in restaurants and so on. Some of the larger city stores provide creches and play areas for children and most supermarkets have miniature trolleys allowing your young family to charge alongside you as you shop, which is entertaining for them but can be hazardous for other shoppers and your wallet!

When visiting Sweden we have always travelled by ferry with our car using Scandinavian Seaways (formerly Torline) who provide excellent facilities in the form of playareas, children's competitions, cartoon shows etc. to while away the long journeys.

When planning a family holiday in Sweden be prepared for easy but expensive living in cities and simple, basic living in the country. Distances between destinations can be very great indeed and shops few and far between. There are numerous lakes and rivers providing every form of water sport but care must naturally be taken that children are well supervised as the waters are usually very deep. Mosquitoes can also be a real pest in the summer months.

Recommended reading:

Frommer's – Scandinavia, Darwin Porter (Prentice Hall Press, 1991)
Fodor's – Scandinavian Cities (Fodor's Travel Publications, 1991)
Scandinavia – The Rough Guide, Jules Brown and Mick Sinclair (Harrap-Columbus, 1990)
Karen Brown's – Scandinavian Country Inns and Manors (Harrap-Columbus, 1989)
The Visitor's Guide – Sweden (Moorland Publishing Company, 1991)
Fodor's – Sweden (Fodor's Travel Publications, 1991)
Insight Guides – Sweden (APA Publications, 1990)
Drive Around Sweden, Robert Spark (Trafton Publishing, 1989)

For further information:

Swedish Tourist Board, 29/31 Oxford Street, London W1R 1RE. Tel: (071) 437 5816. Fax: (071) 287 0164.

Switzerland

Capital Bern
Major cities Bern, Zürich, Basel, Geneva, Lausanne
Time GMT + 1
Currency Swiss franc (SFr) = 100 centimes

Languages German, French, Italian, Romansch
Climate Varies with altitude
Annual weather range – Zürich:
Temperature
   Max 25°C (Jul)   Min −3°C (Jan)
Rainfall
   Max 136mm (Jul)
   Min 64mm (Mar, Dec)

It is not for nothing that the Swiss have a reputation for efficiency: their country is highly organized and it works. Coupled with this is the very high degree of affluence which allows the Swiss to improve, review, rebuild and convert to such an extent that it is possible to say that Swiss services – whether they be transport, telephones, banking, restaurants, postal services, facilities for the disabled or those travelling with young people – are almost without exception of an extremely high order.

Switzerland itself is the country of tourism par excellence: the Swiss have long been accustomed to people from abroad coming to enjoy and marvel at their beautiful country, initially the British, but now also from all over the world. If you speak French, German or Italian to them they will appreciate it but English is spoken to a high standard throughout most of the country and the Swiss, with four languages of their own to learn, enjoy the challenge and opportunity of practising a foreign language with visitors from overseas. But do remember that there are still, I am happy to say, many remote areas, notably in the mountains. Here life is rural, primitive and surprisingly poor: it may be difficult to find someone who understands English or to send a telegram to Bangkok – on the other hand, the mountain people are unfailingly helpful especially where children are concerned. Generally this is true wherever you go in Switzerland, the people will be welcoming, polite and helpful – and children – providing they behave in ways which are not too remote from the high standards of the Swiss themselves – are often the passport to offers of generous assistance and advice.

To come to specifics, in restaurants children are well catered for with high chairs and small portions, especially in the big chain restaurants (Migros, Mövenpick, Coop). The cost of living is high, so go for the chain stores – Migros and Coop are especially reasonable and offer a vast choice of baby goods, including Nestlé milk. The shop on the corner will be more Swiss and enjoyable but much more pricey as well. Parks and open spaces exist in abundance, and in urban areas they usually include a corner with swings and climbing frames. Public transport is astonishingly punctual (set your watch by it), clean and well organized. Car rental is easy and although I have no experience of baby seats in rented cars I would be most surprised if it created a problem.

Swiss food, unlike that in neighbouring France, tends to be straightforward, based on milk products, wholesome and excellent. If your baby enjoys cheese and chocolate then you're home and dry! Hotels are generally spotless and will often provide a cot for a child at no extra cost and many of the big supermarkets have facilities for changing and feeding young children.

Above all, don't hesitate to ask. The Swiss are not very good at putting up notices to tell you what they have to offer and if, in the unlikely event that you catch them out and they don't have what you want then they will go to immense trouble to get you something as good if not better.

Our daughter was born there and both our children were brought up there. I hope that these few notes will help you to enjoy this wonderful country as much as we did.

Recommended reading:
Blue Guide – Switzerland, Ian Robertson (A & C Black, 1989)
The Visitor's Guide to Switzerland, John Marshall (Moorland Publishing Co., 1990)

For further information:
Swiss Tourist Office, Swiss Centre, Swiss Court, London W1. Tel: (071) 734 1921

Thailand

Capital Bangkok
Major cities Chiang Mai, Nakhon Ratchasima, Khon Kaen, Udon Thani
Time GMT + 7
Currency Baht (Bt) = 100 satang
Language Thai
Climate Equatorial
Annual weather range – Bangkok:
Temperature
 Max 35°C (Apr)   Min 20°C (Jan, Dec)
Rainfall
 Max 305mm (Sep)   Min 5mm (Dec)

Thailand means the land of the free and it's important to remember when you consider going there that the country was never colonized. This means that everything about Thailand is entirely its own, even though the US has been a big influence in recent years.

The Thais love children and are unfailingly polite and welcoming to foreigners. Their own children are generally brought up to be quiet and polite. Living is very public because it is a very hot country – everyone is out on the verandah or balcony most of the time – so babies are much in evidence and people are tolerant and sympathetic towards them.

Travelling in modern Thailand is easy. There are excellent hotels providing all the facilities and conveniences you could hope for. They are quite luxurious and everything works. Bangkok's Oriental Hotel is generally thought to be one of the world's best. For the visitor with children there are one or two real bonuses thrown in. The staff in a Thai hotel will be very helpful and go to great lengths to assist you. Medical and nursing care is very good indeed. Doctors are well-trained and thoroughly responsible. You will need to take out medical insurance though, as treatment is not free to visitors. Baby supplies are available in the big hotels and are reasonably priced.

Thai food is one of the great cuisines of the world but it may be a bit of a shock to your children since it is so hot and peppery. In hotels you can get Western food and many of the coffee shops or restaurants have children's menus. Opening hours are long. But do try Thai food. I would recommend the delicious and delicate soups which are excellent for young children. Thais don't eat courses: they sit down to the whole lot at once and there will always be rice and soup available. There is an abundance of wonderful fruit. Our year-old daughter took to papaya at once.

Hotels have pools and terraces where children can play in safety. Bangkok has an excellent zoo with paths and bridges over ornamental lakes. There is a kind of Thai Disneyland outside Bangkok which is also a good outing for the children. Provided you go with recognized tours – and that means air-conditioned buses – getting around the sights should present no problem. Struggling with local taxis and the heat and the impossible traffic would not be any fun, though walking the streets with a pushchair in a town like Chiang Mai in the north would be all right. Unless you are the rugged sort keep clear of local transport with your children. Bus travel in Thailand has its charms, but is also sticky, crowded and not for the fainthearted. If you want to avoid the cities, go to the beaches of Pataya and Phuket.

After all these reassurances you should be prepared for a number of things. First the climate. It is always very, very hot and humid. Some Westerners sweat all the time, but you do get used to it. The Thais have several baths a day to keep cool. After dark mosquitoes are a nuisance, but there are creams and sprays available to repel the unwelcome creatures. There are snakes in Thailand but they don't really like people and provided you know where your children are playing, you should

Choosing your destination

have no qualms about this. Don't hesitate to take a trip to Thailand with your children – it is a beautiful country and its people are smiling, charming and incredibly tolerant.

Recommended reading:

Thailand – A Travel Survival Kit, Joe Cummings (Lonely Planet Publications, 1990)
The Insider's Guide to Thailand, Bradley Winterton (Moorland Publishing Co., 1991)
Thailand and Burma – The Traveller's Guide (Springfield Books, 1988)
Fodor's – Thailand (Fodor's Travel Publications, 1991)
Insight Guides – Thailand (APA Publications, 1991)
Culture Shock – Thailand, Robert and Nanthapa Cooper (Kuperard, 1982)

For further information:

Tourism Authority of Thailand, 9 Stafford Street, London W1X 3FE. Tel: (071) 499 7679

Tunisia

Capital Tunis
Major cities Tunis, Sfax, Sousse
Time GMT + 1 (winter), GMT (summer)
Currency Tunisian dinar (TD) = 1000 millimes
Languages Arabic, French
Climate Mediterranean
Annual weather range – Tunis:
Temperature
  Max 33°C (Aug)   Min 6°C (Jan)
Rainfall
  Max 64mm (Jan)   Min 3mm (Jul)

It takes about 150 minutes to reach Tunisia from England and thus be transported to one of the most beautiful and exotic countries in the world which has the bonus of 700 miles of beaches whose sands are the texture of caster sugar. A highly civilized country, too, and not only because it was a French Protectorate for seventy-five years.

There are several ideal holiday resorts of outstanding merit: Hammamet, Cap Bon, Nabeul, Sousse, Monastir, Mahdia and the little islands of Zarzis, Kerkennah and Jerba. These (and other places) have been well organized for foreign visitors for many years and there is a wide variety of accommodation available at reasonable prices. Some hotels even have self catering bungalows in their grounds. Most holiday sites are near small supermarkets. Many excellent restaurants cater for guests of all means and most palates.

Tunisian cuisine is an interesting blend of Arab and French. Though many children are notoriously conservative about food many, I find (and I include my own), are perfectly capable of being adventurous and experimental and Tunisia is a place to experiment in. You must try proper couscous, for example, with lamb, poultry or fish. And there is *koucha fi kolla* – a dish of young lamb baked with herbs in a jar. Avoid local national dishes that contain the peppery *harissa*. Though it is Mediterranean, the fish is often quite good, especially sea bass, bream, mullet, sardines, prawn and squid. The fruit is superb and, as Arabs like sweet things, there is a splendid range of confectionery (not so splendid for the teeth however). Standard soft drinks are in plentiful supply and the water is good – but, don't drink tap water. Stick to bottled mineral water. Incidentally, be cautious about an excess of chilled or iced drinks. These are a prime cause of stomach upsets.

Some knowledge of French stands one in good stead in hotels, shops and restaurants but many of the staff speak some English as well as Arabic and French. Menus are printed in Arabic and French and sometimes in English.

In summer when it is pretty hot children only need lightweight clothes, but if you are going in September or October, take some warmer things for the evenings. In winter take some warm clothes and waterproofs.

Communications by bus and train are fairly efficient and travel by public transport is entertaining for small children. Car rental is easy and there are good organized excursions. Immunization and vaccination are not necessary. Chemists are quite well supplied but I would play safe and take basic supplies.

There are a number of important and spectacular festivals which children would greatly enjoy. The main ones are the Sahara festival at Douz (January); the hawking festival at El Haouaria (May); the Sirens festival on the Kerkenah Islands (July); the Aoussou Festival at Sousse and the Festival of Kharja at Sid-bou-Said (August).

To the average 4- or 5-year-old a beach or a building seems to look much the same wherever they be and children of this age are not much interested in scenery *per se*. However, an alert and observant 5-year-old could hardly fail to be struck by the differences between Marks and Spencer (or Sainsbury's) and a Tunisian *souk*. In fact, such a child will find the *souks* fascinating and – dare one say it? – 'educational'.

Nor would most children be impervious to the splendour of their surroundings off the beaten track: the pine-clad hills, the orchards, orange and lemon plantations, vineyards, olive groves, date palms; the walled medinas and Berber villages, the black tents of the Bedouin encampments, the vast plains with their flocks and the long camel trains.

No trip to Tunisia would be complete without a visit to one of the great oases. These are not just large waterholes with a few dozen date palms but cover many acres and support tens of thousands of date palms, bananas, pomegranates and other vegetation. Gabes in the south is an outstanding example, and can be toured in a horse-drawn *caleche*. Moreover, it is a suitable base for excursions (which small children would greatly enjoy) through the *chotts* (vast dried salt lakes gleaming white in the summer sun) to Gafsa, Tozeur, Nefta and Douz. The oasis at Gafsa may also be toured by *caleche*. At Tozeur some 200 different springs form the oasis and there is a zoo, and at Douz there is a marvellous camel market on Thursdays. You may not wish to risk your children on a camel ride but they will certainly enjoy the camel wrestling which is a feature of the Douz festival.

In the far south are extraordinary cave dwellings, weird underground houses (even an underground hotel) and subterranean oil factories where the mills are powered by camels. Should you wish to make a trip into the desert proper (and this is a remarkable experience) you must take full precautions and have plentiful supplies of water and provisions plus, unless you are experienced, a guide.

Recommended reading:

Essential Tunisia; All You Need to Know (AA, 1990)
Discover Tunisia, Terry Palmer (Heritage House, 1988)
Where to Go in Tunisia, Reg Butler (Settle Press Hippocrene Books Inc., published in association with Thomson, 1990)
Rough Guide: Tunisia, Peter Morris and Charles Farr (RKP, 1985)

For further information:

Tunisian National Tourist Office, 77a Wigmore Street, London W1H 9LJ. Tel: (071) 224 5598 (admin) 224 5561 (enquiries)

Turkey

Capital Ankara
Major cities Istanbul, Izmir, Adana, Bursa, Antalya
Time GMT + 2
Currency Turkish lira (TL)
Language Turkish
Climate Mediterranean on coast, continental inland
Annual weather range – Izmir:
Temperature
  Max 33°C (Jul–Aug) Min 4°C (Jan–Feb)
Rainfall
  Max 122mm (Dec) Min 5mm (Jul–Aug)

Turkey is a very big country, astonishingly beautiful and varied in almost every topographical feature you can think of, and in much of it comparatively primitive conditions prevail. It is only during the last 50-odd years that the majority of it has been Westernized. In summer most of the country is hot or very hot; in winter, cold or very cold yet in the south it remains mild. The Black Sea coast is the most humid region. Spring and autumn are the ideal seasons for the visitor.

Most visitors go to Istanbul, along the Black Sea coast and down the Aegean or Asia Minor coast to the southern shores. Some go inland, to, for instance, Iznik, Bursa and Ankara; a handful visit the central and eastern regions. I would be chary of these last with very small children.

In the major cities and towns (Istanbul, Bursa, Ankara, Konya, Antalya, Izmir) and along the principal tourist routes you may expect standards of living similar to those in Britain and Western Europe and you will have little difficulty in meeting the basic requirements of babies and small children. Soon after you journey beyond such areas you may easily find essentials difficult or impossible to come by. I would play safe and take with me all the pharmaceutical goods that might be needed.

Almost any kind of accommodation is available for the holidaymaker in traditional tourist areas and proprietors are obliging and helpful in making special arrangements for small children. As ancient codes of hospitality and courtesy in Turkey have few or no equals in the world (total strangers will buy you drinks or have food sent over to you in a restaurant) you may be absolutely sure that children (as well as their parents) will be very welcome. Indeed, little ones will be fussed over, indulged and given mini-VIP treatment. British children, because they are so fair and light-skinned by comparison with the Turks, arouse much curiosity and attention.

In general Turkish hotels are clean and well run, but you may find that the lavatory and washing facilities are not what you are used to. In simpler hotels there are often several beds to a room, so, unless you don't mind sharing the room with other guests, you need to pay for all the beds in a room. Do take note that scorpions have a tendency to turn up in odd places. They particularly like a snug site such as a shoe or slipper.

Small children may not take too readily at first to the food although Turkish cuisine is generally acknowledged to be among the best in the world. Initially, they may find it a little too oily or spicy. In regions frequented by visitors there are very high hotel standards and every effort is made to ensure that guests get what they need or want. The Turks are expert at delicious cold hors d'oeuvres (*soğuk meze*) and also hot hors d'oeuvres (*sicak meze*). These are brought on trays and are a meal in themselves. Meat and fish are also very good and are usually on display in the refrigerators of restaurants. So you choose what you want. It is also standard practise for guests to go into the kitchen and select what they want. Vegetables and salads are good and the fruit is wonderful (especially melons, peaches, figs and nectarines). Always wash and peel fruit. Water, especially spring and well water, is delicious but be wary of tap water. Ask for drinking water. Mineral waters are plentiful and excellent. Soft drinks (particularly pure fruit juices) are very good. A change in diet may cause tummy troubles so take some suitable medicine.

If you value your children's teeth you will make sure they are not overexposed to Turkish confectionery. The Turks love sweet things and there is a wide range of the most delectable pastries, cakes, tarts and biscuits and many unfamiliar delicacies.

You can buy a fair variety of toys in all the main towns and any amount of standard entertainment is accessible. Children expecially enjoy the *karagöz*: witty and lively puppet shows in the Punch and Judy tradition.

Travelling by public transport is

something of an adventure, particularly in out of the way places, but most small children enjoy it very much; and it provides insights into unfamiliar ways of life that surprise children. The markets and bazaars and the old 'oriental' quarters of the cities where modes of life and work have changed little during the last 500 years are fascinating for any observant and inquisitive 4- or 5-year-olds. Traditional rural life in Turkey is equally fascinating.

Seaside resorts are well equipped for small children and I would recommend such places as Bodrum, Marmaris, Fethiye, Finike, Kemer, Alanya and Anamur.

The first time we took the children to Turkey in 1986 we booked a flight and village room accommodation with SunMed holidays. We left the resort after a week to explore on our own, as we don't particularly like organized tours and felt we weren't getting the feel of real Turkey. The next time we went we just booked a charter flight and a hire car. We went on a night flight to Dalaman airport. (We find flying through the night a good idea with young children, as they are usually asleep soon after take-off and not fidgeting around the plane for 3.5–4 hours. Adults don't feel so good the next day though!) Miraculously, the pre-booked hire car (Avis) was waiting for us at 4 a.m. and the sleeping children transferred to it. Children's car seats or restraint straps don't seem to be available in hire cars in Turkey.

A word about Dalaman airport: it is very poor on facilities, e.g. rock-hard plastic seats and limited refreshment facilities. Young soldiers with machine guns parade around the planes while passengers are disembarking.

Local driving standards aren't too bad outside Istanbul (where they are dreadful), but watch out for the odd car on the wrong side of the road. On country roads wild-looking dogs belonging to gypsies often chase after cars and can be rather frightening.

Although hire cars are convenient for travelling between resorts and for less accessible sites, short excursions are more fun on the extremely efficient and cheap minibus (*dolmus*) network. The locals are often accompanied by the odd hen or goat and lemon-scented cologne is passed around to refresh jaded passengers.

One of our main reasons for visiting Turkey was to see the wonderful archaeological sites, of which there are many. Before and during the holiday we primed the children with elementary versions of Greek myths and legends (e.g. *The Usborne Book of Legends*) so that they are familiar with places like Troy and monsters such as the Chimera whose legendary cave we passed.

Some places we would recommend: Altinkum on the Aegean coast is a relatively new and unspoiled resort with a vast sandy beach. People are very friendly and it is within easy reach of the ancient sites of Ephesus, Didyma, Miletus and Priene. Olu-Deniz near Fethiya has wonderful swimming in either sea or lagoon. The scenery is beautiful but rooms are more expensive here. Kas is an attractive former fishing village; the beach is rocky and a good walk from the centre. A fifteen-kilometre sandy unspoilt beach can be found at Patara, twenty kilometres west of Kas. The resort at Side is more commercialized, but good fun, with lots of sandy beaches and ruins.

Although they are not compulsory, check with your GP about inoculations for such things as typhoid and cholera etc. A hepatitis jab is strongly recommended. Don't let children stroke dogs or cats because of the risk of rabies. On both holidays to Turkey we have had to visit the local doctor when our youngest son developed acute ear infections. The locals were very concerned and eager to help and provided transport to the doctor. The surgeries were basic but the doctors seemed proficient, albeit rather enthusiastic with their prescriptions of antibiotics and pain-killers.

Always carry a loo roll with you. Toilet paper isn't usually provided in accommodation other than hotels and rarely in public lavatories.

The ice-cream vendors are good entertainment for children in the larger resorts. Usually dressed in local costume they play lots of teasing tricks on the children by manipulating ice-cream just out of their reach, to the accompaniment of theatrical shouts and gestures. Adults are also subject to this routine. Children in general are made a great fuss of in Turkey, especially by young men. Complete strangers will buy them ice-creams and give them kisses and cuddles. Although this sort of behaviour is frowned on in the UK, in Turkey it is a genuine expression of hospitality and friendship. Another pastime to provide unexpected fun is a visit to a carpet shop. It is very relaxed, there is no pressure to buy, exotic tea is offered and children are well tolerated. The ritual of unrolling carpet after carpet becomes a game for the children.

Recommended reading:

Insider's Guide to Turkey, Donald Carroll (CFW Publications, 1991)

Fodor's – Turkey (Fodor's Travel Publications 1991)

Turkey – A Travel Survival Kit (Lonely Planet Publications, 1990)

Turkey – The Rough Guide (Harrap-Columbus, 1991)

Blue Guide – Turkey, Bernard McDonagh (A & C Black, 1989)

Insight Guides – Turkey (APA Publications, 1989)

For further information:

Turkish Embassy Information Office, Egyptian House, 170/173 Piccadilly, London W1. Tel: (071) 734 8681

United Kingdom

See **Britain**

United States

Capital Washington DC
Major cities New York, Chicago, Los Angeles, Philadephia, Houston, Detroit
Time GMT – 5–11
Currency Dollar ($) = 100 cents
Language English
Climate Wide regional variation
Annual weather range
Northeast (Washington DC) – continental:
  Temperature
    Max 31°C (Jul) Min −3°C (Jan)
  Rainfall
    Max 112mm (Jul) Min 66mm (Nov)
Southeast (Miami) – subtropical:
  Temperature
    Max 31°C (Jul–Sep)
    Min 16°C (Jan–Feb)
  Rainfall
    Max 234mm (Oct) Min 51mm (Dec)
Southwest (Los Angeles) – mediterranean:
  Temperature
    Max 28°C (Aug) Min 8°C (Dec–Feb)
  Rainfall
    Max 79mm (Jan) Min 0mm (Jul–Aug)
Northwest (Seattle) – temperate:
  Temperature
    Max 23°C (Aug) Min 2°C (Jan)
  Rainfall
    Max 142mm (Dec) Min 15mm (Jul)

Americans generally like children and are well prepared for them. Special provisions such as children's menus and a general 'open for business' attitude are found nationwide.

Renting a car is easy (automatic transmission is the norm) and there is usually a small charge for a child's seat. If you are visiting a small town, try and give the rental company a few days' notice.

In the US distances are huge, but the roads are excellent. Highways are well signed and in many states the 55 m.p.h. speed limit is no longer strictly enforced (but be careful!). For short and medium distances driving is the easiest way to get around *en famille*, but watch out for cars overtaking on both sides of you and the 'right turn on red' convention in some states (which allows drivers to turn right when the traffic lights are red providing there is no oncoming traffic).

Main roads have frequent service stations often with a variety of restaurants and clean rest-rooms (toilets). The railway system run by Amtrak is good where it exists. The carriages ('Cars') are roomy and comfortable, although often crowded. Fares are expensive, with fewer bargain fares than British Rail offers. Children travel half-price. The cheapest way to travel the country is by bus with several bus companies offering short and long distance routes at reasonable rates. While most have toilets on board and some offer refreshments, it is probably the least comfortable option when travelling any distance with children. If you are travelling more than 200 miles, go by air if finances allow. The many domestic airlines vary in their attention to young passengers, but most will ask parents with children to board first. Do check with your travel agent and remember that there are substantial discounts for American domestic flights if you book before you leave the UK.

Hotels and motels, of which there are many, charge by the room not per person. As each room usually has two double beds, a family of three or four can easily share one room and save money. If you need a cot, ask for a 'crib'. A 'cot' (or put-u-up/camp bed) may also be available and useful if kids object to sharing a bed. Rooms have toilet, shower and television, varying from basic to luxurious depending on price. Hotels and motels sometimes have pools. It may be worth paying extra for one with a pool to let hot, tired younger travellers cool off at the end of the day while parents relax at the poolside.

Motels do not usually offer breakfast, but a diner or restaurant offering huge American breakfasts will not be far away. Kids will probably make a beeline for the fast-food variety, try to persuade them to try some of the wonderful breakfasts at family restaurants.

Eating out is a family event in the US so you can expect high chairs almost everywhere as well as children's portions and free salad in steak and salad bars (provided the adults buy steaks). It is not just the fast-food restaurants which provide meals quickly; this is a standard feature of most American restaurants so you can eat out with the family and have time to do something else. Americans eat early so it is no problem going for dinner at 6.00 or 6.30 p.m.

In general restaurant portions are extremely large, so don't over-order. This applies especially to Chinese restaurants where servings are gigantic. If you are defeated by the quantity of food, ask for a 'doggie bag' and you can take food home with you. This can be useful if children want a snack of tasty pizza in the car mid-afternoon.

Supermarkets and shops of all kinds are always open late so you are never stuck if you run out of baby food or nappies. The labelling on most food items is very good. Incidentally, nappies are called 'diapers' in the US and disposables are more expensive than in the UK.

Toilets ('rest rooms') are almost without fail clean even at the smallest roadside filling ('gas') station, those in restaurants and shopping malls can be very spacious. Many of these have fold-down nappy changing tables and readily available paper towels.

American museums and zoos are extremely good. There is a wide variety available and those focusing on natural history and the environment are best for little ones. Facilities for the disabled are usually very good and there are ramps everywhere – especially useful for people pushing prams and pushchairs.

National parks, monuments and tourist attractions are very well organized. They are likely to have lots of space with play areas and other features for children. The theme and amusement parks of the Disney organization are found in California and Florida.

Climate varies enormously from north to south and east to west. Make sure you know the temperature ranges of your destinations. Ferocious central heating and air conditioning is common so you may find indoor temperatures uncomfortable. Bring layers of clothing for your children which can be put on or peeled off as necessary. Tee-shirts are essential.

There is no National Health Service in the US so it is vital that you arrange for comprehensive medical and dental insurance before you go. If you have a serious health emergency go to a hospital 'Emergency' room (not 'Casualty'). For problems such as minor cuts or bruises, slight fever, colds or upset stomach go to a 'Primary Care Centre'. These are an alternative to overcrowded emergency rooms and offer immediate attention by trained and fully qualified medical personnel, are usually accessible in shopping areas and often at less cost than hospitals would be. They will ask you how you will pay your bill and expect to see evidence that you can – like a credit card or insurance policy. Outside the big cities the trend is towards all-in-one medical centres with on-site specialists like dentists and paediatricians. Again, you will be asked how you are paying. Credit cards are the norm. One last point – a general practitioner is usually a 'family practice' physician or an 'internist' who specializes in internal medicine.

For the British family on holiday, America offers infinite opportunities for amusement to parents, teenagers, and toddlers. With no language problem, and with airline deregulation making it cheaper to fly coast-to-coast in the US than to make a much shorter trip in Europe, America is an ideal destination for the British traveller. A sampling of America should include at least one city, one major amusement park and one national park.

Starting a visit to the US in Boston eases the British visitor into America because the older sections of the city, especially those around Beacon Hill, are reminiscent of an English city. Walk the Freedom Trail (marked with a wide red stripe in the pavement), there you'll see the house of Paul Revere, the USS Constitution, and the Old State House, from whose balcony the Declaration of Independence was read. Stroll the streets of Beacon Hill, especially at twilight, when the street lamps are lit.

A sure hit with children will be a visit to Quincy Market, where the scores of food stalls lining the central market building allow each family member to choose their own meal.

Just south of Boston's centre, you'll find Museum Wharf, housing two museums of special interest to young people – the Computer Museum, tracing information processing from the Chinese abacus to today's supercomputers, and the Children's Museum, with its many 'hands on' exhibits, including the very popular Japanese House and the Small Science Factory.

A few miles farther south, you'll reach the John F. Kennedy Museum and Library, with its mementoes of the President's life and career – and its glorious view of Boston Harbor.

A visitor to Florida's Walt Disney World, Epcot Center and MGM Studios said, 'Everything there makes you smile. The staff are so wonderful. The place is so clean. And the roller coasters are the most exciting I've ever been on.'

Don't be surprised if the family members most enthusiastic about Disney World are the teenagers and the adults. They appreciate the great humour in the exhibits – the ghosts' dining room in the haunted house, the Peter Pan ride over London. Expect to run into beloved Disney characters anywhere and everywhere. Donald Duck may turn up on water-skis, or you may find yourself breakfasting next to Mickey and Minnie Mouse.

Just a short distance away (no worries about transportation – Disney World provides it) is Epcot Center, where you can take a ride in Spaceship Earth or a

Journey into the Imagination. You can also shop and eat around the world in a village with restaurants and shops representing countries around the globe. (The homesick will find a realistic nineteenth-century London street.)

A ten-minute drive from the Disney complex brings you to Sea World, where dolphins, killer whales and sharks cavort in a 150-acre marine park. If Sea World inspires a desire for personal cavorting in the water, you aren't far from some part of Florida's 8,000 miles of coastline, sections of which sport such inviting names as The Gold Coast, The Platinum Coast, The Space Coast, and The Treasure Coast.

The wonder of Yosemite Park was captured by one of its founders, John Muir, who wrote that in the park are '... the most songful streams in the world ... the noblest forests, the loftiest granite domes, the deepest ice sculptured canyons'.

Families can choose any level of accommodation, from the luxurious Ahwahnee Hotel (where Queen Elizabeth II stayed on her California visit) to Curry Village's cabins to a campsite in one of the park's five valleys.

Visitors will find activities for all ages and lifestyles. The more sedentary can tour in one of the park's open-sided buses, each one with a National Park Service guide. The more vigorous can opt for a hike or backpacking jaunt on part of the 773 miles of trails in Yosemite. In winter, there's skiing in the Badger Pass Ski area and skating at Curry Village's outdoor rink. In warmer weather, there's fishing, swimming and boating. All this in the midst of Yosemite, with its famous peaks such as El Capitan and Half Dome, its waterfalls and its magnificent giant sequoias.

From Yosemite, go northward to the area where gold was discovered and fortunes made overnight, or travel northeast to Lake Tahoe, the world's second deepest alpine lake, ringed by mountains that are snowcapped year round. Yearning for urban life? Head for San Francisco – ride a cable car, walk across the Golden Gate Bridge, and sample one of Asia's great cultures in Chinatown.

In planning a family holiday in the US, keep in mind that many American families travel there, and facilities for children tend to be very good – convenient places for changing nappies at the rest stops on interstate highways, menus for children in restaurants and generous discounts for children at hotels and motels.

It is now possible for British citizens to travel to the US without a visa. There are, however, several conditions of entry and travellers are advised to contact the US Embassy in advance on (0898) 200290 to confirm their eligibility.

Recommended reading:

Coping with America, Peter Trudgill (Basil Blackwell, 1988)
Fodor's – USA (Fodor's Travel Publications, 1991)
Insight Guides – Crossing America (APA Publications, 1989)
Culture Shock – USA, Esther Wanning (Kuperard, 1991)
Insight Guides – Native America (APA Publications, 1991)
Fodor's – California (Fodor's Travel Publications, 1991)
The Unofficial Guide to Disneyland, Bob Sehlinger (Prentice Hall, 1991)
Insight Guides – Florida (APA Publications, 1991)
Walt Disney World (The Official Guide), Steve Birnbaum (Avon Books and Hearst Professional Magazines, 1991)
Fodor's – Disney World and the Orlando Area (Fodor's Travel Publications, 1991)
The Unofficial Guide to Walt Disney World and Epcot, Bob Sehlinger (Prentice Hall, 1991)

For further information:

US Travel and Tourism, PO Box 1EN, London W1A 1EN. Tel: (071) 495 4466.
For visa information: Telephone (0891) 200290

Wales

See **Britain**

West Indies

Antigua
Capital St John's
Time GMT – 4–5
Currency East Carribbean dollar (EC$) = 100 cents
Language English
Climate Tropical
Annual weather range – St John's:
Temperature
 Max 31°C (May–June, Aug– Sep)
 Min 21°C (Jan–Mar)
Rainfall
 Max 180mm (Nov) Min 22mm (Feb)

Barbados
Capital Bridgetown
Time GMT –4
Currency Barbados dollar (Bds$) = 100 cents
Language English
Climate Tropical
Annual weather range – Bridgetown:
Temperature
 Max 30°C (May–Sep)
 Min 20°C (Jan–Mar)
Rainfall
 Max 206mm (Nov) Min 28mm (Feb)

Bright sunshine, cool sea breezes, warm water and sandy, palm tree-fringed beaches give this part of the world a special magic all of its own. Once experienced, the desire to return to the West Indies becomes irresistible and young children need not stand in the way. December to March is high season but April and May are also lovely months to visit, with the bonus that prices drop considerably from April. The eight-hour flight, with a change of plane at Antigua for Nevis, may seem a long way with a young family, but the promise of guaranteed warmth makes it all worthwhile. There are also direct scheduled flights to Barbados with British Airways and British West Indian Airways (BWIA). For lovers of remote, far-away places island hopping is tempting, but with children, the extra (but not outrageous) expense of a private charter is desirable. In any case it is bliss to be met off the plane and whisked to a small aircraft.

West Indians love children and are welcoming and kind; babysitting and help with housework are easy to arrange. With an apartment or villa, a maid to do the cleaning, washing and ironing is usually, but not always, provided. Cots and high chairs are generally available and hotels will normally organize an early evening meal for children. Fresh food is limited; but plenty is available frozen or canned. Local fruits such as papaya, mangoes and bananas can be found and also locally-grown vegetables, but these are not always in abundance. Hotels are influenced by US cuisine and at lunchtime hamburgers and club sandwiches are always on the menu. Young children may take a few days to adjust their internal clocks and I suggest you arrive with a supply of powdered milk and cereal, so that in a hotel demands for a 6.00 a.m. breakfast can be satisfied.

The temperature ranges from 20°C to 30°C and the sun's piercing rays are made deceptively comfortable by cool breezes. It is very important to protect small children with a strong barrier cream and to make sure that they have hats and long-sleeved coverups. Children need very few clothes, so a holiday in the Caribbean can make an ideal opportunity for potty training; our boys ended their nappy days in Nevis. Disposable nappies, incidentally, are expensive and it is sensible to pack them in the luggage together with pharmaceutical items. Do take insect repellents as you can be bothered by mosquitoes in the evenings.

Car hire is always possible but baby seats are not much in evidence. This is not a real problem since distances are short and most of the roads are not made for speed. Apart from the few towns where congestion can be found, there is virtually no traffic. West Indians are so friendly that it is common practice to stop and offer lifts, especially on a small island like Nevis. This is fun for children and one of the charms of travelling in a part of the world where time has, in some ways, stood still. Although there are few buildings of great historic interest, there is a romantic awareness of the past and reminders of the days when sugar plantations flourished and naval sailing ships were based in Antigua. Horatio Nelson spent his youth in the Caribbean and met and married Frances Nesbit on the island of Nevis.

Armed with buckets and spades, young children will be happy wherever they are taken in these islands. On Nevis there are crabs to creep up on before they disappear down tunnels and pelicans to watch plunging clumsily into the sea. Nevis is a marvellous escape from modern urban life. On Antigua remote places can be found, but there is much more sophistication and a full range of water sports, plus the excitement of being taken out to a reef in a glass-bottomed boat to watch the brightly coloured fish. The famous St James beach on Barbados has everything; hotels, apartments, villas and most known water sports, but there is still plenty of room on the pink-tinged coral sand. We have been going to the Caribbean for many years and have been taking our boys ever since they took their first staggering steps. The anticipation is enough to cheer the greyest winter day.

Recommended reading:

Insight Guides – Barbados (APA Publications, 1990)
Barbados – The Visitor's Guide (Macmillan Publishers, 1988)
Barbados – A Traveller's Guide, David Milne (Roger Lascelles Publications, 1991)
Insight Guides – Trinidad and Tobago (APA Publications, 1988)
Caribbean Islands Handbook 1992 (Trade and Travel Publications, 1991)
Fodor's – Caribbean (Fodor's Travel Publications, 1991)
The Penguin Guide to the Caribbean (Penguin Books, 1991)
Cadogan Guides – The Caribbean, James Henderson (Cadogan Books, 1990)
The Outdoor Traveller's Guide – Caribbean, Kay Showker (Stewart, Tabori, Chang, 1989)
Frommer's Caribbean (Prentice Hall, 1991)
Insight Guides – The Caribbean (APA Publications, 1991)
Baedeker's Caribbean (AA Publications, 1991)
Frommer's – Bermuda and the Bahamas, Darwin Porter (Prentice Hall, 1990)
Fodor's – The Bahamas (Fodor's Travel Publications, 1991)
Insight Guides – Bermuda (APA Publications, 1991)
Crowood Travel Guides – Jamaica and the Greater Antilles (The Crowood Press, 1991)
Insight Guides – Bahamas (APA Publications, 1988)

For further information:

Antigua & Barbuda Tourist Office, 15 Thayer Street, London W1. Tel: (071) 486 7073
Barbados Board of Tourism, 263 Tottenham Court Road, London W1. Tel: (071) 636 9448/9
St Kitts & Nevis, 10 Kensington Court, London W8. Tel: (071) 376 0881

Zimbabwe

Capital Harare
Major cities Bulawayo
Time GMT + 2
Currency Zimbabwe dollar (Z$) = 100 cents
Languages English, Shona, Ndebele
Climate Warm temperate and tropical
Annual weather range – Harare:
Temperature
 Max 28°C (Oct)   Min 21°C (Jun–Jul)
Rainfall
 Max 196mm (Jan)   Min 0mm (Jul)

It's a shame that the vast majority of travellers to Zimbabwe – Southern Rhodesia in the old days – go there because of some sort of connection with the place; very few go there as independent tourists with no ties. This is very likely because Zimbabwe is still popularly thought of as a dangerous country, because of its black socialist government and the war being fought in Mozambique next door. None of these factors should put you off: Zimbabwe is an amazingly beautiful, friendly country.

Musts for visitors: obviously Victoria Falls, Lake Kariba and Hwange Safari Park. Kariba has a string of lake shore hotels, all with pools and stunningly beautiful. Hwange is without doubt one of the finest game reserves in Africa. Some of the safari tour companies in the park will accept children in their game drives – though mine simply spent the journey asleep. Locals will recommend a trip to Nyanga, which sparks off memories of the Scottish Highlands: very beautiful and slap on the border with Mozambique (but safely away from the fighting). There are some breathtaking views and a huge variety of mountain landscapes. The ruins of Great Zimbabwe are a mysterious reminder of an ancient African civilization, and of course there are the main towns, Harare and Bulawayo, both pleasant and interesting.

The climate is close to perfect, with the days mostly dry and sunny. In summer the heat rises and October is the hottest month. Small children will need some protection from the sun – sun block, and sun hat. They may be affected by heat rash when they first arrive. Keep them cool with lots of drinks, and for tiny babies, a sun canopy over the buggy. My 7-month-old adapted to the hot weather and the relaxed, luxurious lifestyle with great ease and pleasure.

Because of the historical connection with Britain, food everywhere can best be described as old-fashioned English, with some Italian, French, Greek and Indian restaurants in Harare and Bulawayo. Major hotels have standard international cuisine, welcome children and have high chairs.

The Zimbabwe Sun group of hotels, a large chain of good hotels, accommodate children free when sharing with parents. The tourist industry is in the doldrums right now, and hotels are extremely attractively priced. It's possible to travel right round Zimbabwe on a package tour for ridiculously low amounts.

Because of the distances involved, air travel is preferable, though both trains and coaches are of a reasonable standard. Car hire is expensive but the roads are completely empty and very good. Take a small battery fan to cool your child during a long hot journey.

Malaria tablets are essential for travellers even though locals will boast they never take them. They should be taken at least one week before departure, for the whole trip and for six weeks afterwards. This aftercare is particularly important to maintain protection. Take both daily and weekly tablets, but check with your doctor about the correct children's dose.

Never let children swim in rivers or dams because of the dangers of bilharzia. The water in towns, hotels and swimming pools is from purified central water supplies or boreholes and is therefore perfectly safe. Use insect repellent after sundown, and a mosquito net (easily available everywhere). Cover arms and legs. AIDS is prevalent in Africa, so if possible take an emergency supply of syringes, needles and stitches and try not to need a blood transfusion.

There are no disposable nappies or baby wipes available in Zimbabwe. Bring your own, or get them from South Africa or Botswana. All washing, including terry nappies, must be ironed before being worn. This is because the imfulu fly lays its eggs on anything damp. Once developed the larvae bore their way under the skin and can be quite painful. Ironing kills them off.

Electric plugs are three point square and the same voltage as in Britain.

Recommended reading:

Central Africa – A Travel Survival Kit, Alex Newton (Lonely Planet Publications, 1989)

Traveller's Guide to Central and Southern Africa (IC Publications, 1990)

Zimbabwe and Botswana – The Rough Guide, Barbara McCrea and Tony Pinchuck (Harrap-Columbus, 1991)

Spectrum Guide to Zimbabwe (Camerapix Publishers International, 1991)

For further information:

Zimbabwe Tours and Travel, 3 Broadway, London N14 6PJ. Tel: (081) 882 0141

Section 2

Choosing and planning your holiday

What kind of holiday do we want?

Most people take one major holiday a year. After poring over brochures and parting with hard-earned cash, the next six months are spent in anticipation of a few weeks in the sun. The summer holiday is probably the single biggest and most important yearly purchase for the average family. When you think how much your well-being can hinge on that break, it's no wonder that the decision-making can be fraught rather than fun, and through insufficient planning you can easily make expensive mistakes.

This section of *How to Have Stress-Free Family Holidays* avoids the time-consuming and sometimes confusing wade through piles of brochures. It describes what is on offer from different tour operators and travel companies. At a glance you can see the destinations available, assess what kind of holiday is on offer and judge whether their price range fits yours. Each entry focuses on the facilities available for families, and you can quickly discover if cots and high chairs are easily available, or whether there are children's clubs and entertainment.

The companies listed cover almost every country in the world, so if you fancy a change of destination, you have the facts at your fingertips. If the kind of change you are considering is in the *type* of holiday rather than the destination, this section is invaluable. We cover sixteen types of holiday, from cycling to safaris, from home-swapping to traditional hotels. There is useful information to help you choose car rental companies, or to find out about the major air, sea and rail carriers.

All the companies are listed alphabetically within each section. You can take time to consider what you want from your holiday and who is best equipped to supply it. Whether you like to have every tiny detail planned in advance, or to travel independently, this section has information relevant to you and your needs.

In general, the prices quoted are for the 1990 season. They are intended as a guide to the character of the company's programme and special offers and specific prices should always be checked before booking.

To help us get better information and more cooperation from operators and their agents please always mention *How to Have Stress-Free Family Holidays* when you make a booking or ask for a brochure.

We'd also be interested to hear of your experiences, both good and bad. Write to us at: Reference Book Department, Bloomsbury Publishing, 2 Soho Square, London W1V 5DE, marking your envelope '*How to Have Stress-Free Family Holidays*'.

Bargain breaks and short-stay holidays

If you long to forget the terrors of February or liven up the middle of a seemingly interminable school holiday, a short-stay break may be just the thing for your family.

There is a vast selection of special packages specially designed to lure families into hotels and lodges that empty of their regular business clientele as soon as office hours end on a Friday evening. And it's not just weekends these days. More and more hotels are offering mid-week breaks and special bank holiday offers.

You get the best of both worlds with short-stay offers – low prices and first-class facilities. Many of the hotels featured by the companies listed below have such amenities as luxurious health and fitness centres, indoor heated swimming pools, in-house movie channels, squash courts and even croquet lawns to keep you and your family amused. This is, of course, in addition to first-class cuisine, comfortable rooms and private bathrooms.

We decided to include chains of hotels and lodges rather than to attempt to list separate deals at separate hotels. You will quickly be able to see from the general description of what each company offers which brochures are for you.

Many companies offer extremely generous discounts or free accommodation for children. There are family rooms, baby-listening services and special children's menus. Other companies offer rooms specially designed or adapted for disabled guests and one or two even advertise non-smoking accommodation.

As in all holiday offers, prices vary according to hotel and season. We would always advise you to contact the reservations staff if you have any doubts concerning holiday rates – the restriction listed in the small print of these brochures can be extremely complicated.

One other word of advice – if you have *any* special requests, such as a cot, high chair for use in the restaurant, vegetarian food, baby food, access for a wheelchair, or even two rooms on the same floor, make sure you tell the hotel reservations staff when you are booking your holiday. It may be as well to confirm these requests in writing before arrival (and keep a copy of your letter) – the hotel companies will try their best to help you, but even the best-laid plans can be forgotten!

Prices

Unless otherwise stated 1992 prices have been quoted. These are intended only as a guide to the type of holiday offered and travellers should check with the company concerned for 1993 pricing details.

Best Western Hotels

Vine House, 143 London Road,
Kingston-upon-Thames, Surrey KT2 6NA
Tel: 081 541 0033 (Reservations)
Telex: 8814912 BWHOTL G
Fax: 081 546 1638
Brochure request: 081 541 5767

Best Western is a group of over 200 independently owned hotels throughout England, Scotland, Wales, Ireland, the Channel Islands, and the Isle of Man.

Getaway Breaks

A colour brochure giving full details of Getaway Breaks (minimum two nights stay; starting any day of the week) can be obtained from the number above. Standard Getaway rates are based on double occupancy of a room and include full traditional breakfast, a three-course dinner each night (or daily a la carte allowance), a private bath/shower, toilet, colour TV and telephone. Most rooms also offer free tea and coffee making facilities. Prices include VAT at 17½% and service.

Up to two children under 16 will be accommodated free when they share a standard double or twin room with two adults. Meals are charged as taken (ie if they don't eat breakfast, you don't pay for it). Under 12's are offered half-price children's portions, or meals from their own special menu. A few hotels carry a minimum age for dining in the restaurant – this is clearly shown in the relevant brochure entry.

It is worth asking on application whether the hotel offers family rooms or suites and whether there would be a supplement for such accommodation. Children under sixteen occupying their own rooms are charged 75% of the adult Getaway rate (inclusive of meals). However, some Best Western hotels only charge 50% of the adult rate – well worth asking about when booking.

Children Welcome

Many Best Western hotels promote their family facilities by using the label 'Children Welcome'. This indicates special features such as baby listening, special menus, play areas, etc. See brochure for details.

Single Parent Family Breaks

A particularly welcome feature of the Best Western short-stay offers is the Single Parent Family Break package. Up to two children under sixteen may stay free of charge, when sharing a standard double or twin room with one adult. Meals are charged as taken (i.e. if the kids don't eat breakfast, you don't pay for it). This offer is currently available at 69 of of the Best Western hotels, which are identified clearly in their catalogue entries.

Activity Breaks

Many hotels offer special breaks featuring such activities as golf, bridge, theatre, museum and stately home visits, walking, jazz and other music, trips to theme parks, bird watching, and 'Whodunit' weekends. It is advisable to ask about rates and suitability for children before booking.

Disabled travellers

Access for disabled guests is listed in each hotel's catalogue entry. This is based on the AA's criteria and indicates that the hotel has met the minimum standards. Best Western will always try to help guests with their particular needs if requested at time of booking.

Calotels

3rd Floor Suite, Hampshire House,
Bourne Avenue, Bournemouth BH2 6DP
Tel: 0202 297888 (Reservations)
Fax: 0202 299182

Calotels is a privately owned group of 5 hotels located in Bournemouth, Datchet (near Windsor), Wantage and Warwick. The 2 Bournemouth hotels offer extensive indoor leisure facilities, whilst the others are conveniently located for touring local attractions in their respective areas.

Prices Various special offer and short break packages are offered according to the hotel and season and you are advised to contact the reservations staff for brochures and pricing information.

Consort Hotels

Consort House, 180–182 Fulford Road,
York YO1 4DA
Tel: 0904 643151 (Reservations)
0904 620137 (Office)
Telex: 57515
Fax: 0904 611320

Choosing and planning your holiday

Consort Hotels is a consortium of over 200 hotels located throughout Britain. The organization offers a number of different types of holiday, including Consort Breaks (two-night packages), touring holidays, and, at selected hotels, Weekaways.

Consort also feature Wayfarer Inns and Consort Crown International hotels, for which brochures can be obtained from the office number listed above. The company is affiliated to hundreds of hotels throughout Europe, North America and Australia (separate brochure available).

Consort Breaks

These are two-night packages offering dinner, bed and breakfast (London hotels, bed and breakfast only). Prices are per person, based on accommodation in a twin or double room with private bath or shower. A single room supplement applies to packages where stated, and is per person per night. Many packages include an extra feature, such as special menus, complimentary wine, or guide books and free admission to local attractions. Many hotels have leisure centres which may offer such amenities as a swimming pool, sauna, solarium, gymnasium or squash court.

Consort Family Hotels

Many Consort hotels offer such features for families as free accommodation for children, free fun packs and children's meal times. Some have been designated Consort Family Hotels, and will offer special services and amenities (details available on request).

Family rates
Consort breaks and Weekaway holidays

Where children stay free of charge, the following restrictions apply:
Children (under 16) sharing a room with two adults will be free of any charge for accommodation. Payment for all meals taken (including breakfast) should be made directly to the hotel.

Children (under 16) in their own room will be charged 75% of the adult rate for accommodation and meals, payable with the holiday balance in the normal manner. Children's portion meals are provided.

Where children stay for 50% of adult charge, the following restrictions apply:
Children (under 16) sharing a room with two adults will be charged 50% of the adult rate for accommodation and meals, payable with the holiday balance in the normal manner. Children's portion meals are provided.

Children (under 16) staying in their own room will be charged at the adult rate.

Touring holidays

For up to two children (under 16) sharing a room with two adults, accommodation will be free of charge. Payment for all meals taken (including breakfast) should be made direct to the hotel. There is no reduction on touring holidays for children occupying their own room.

Entry charges to local attractions

Where entry charges are included as part of the adult accommodation package, they are not necessarily included in the free accommodation offered to children by the same deal. Entry charges for children who are being accommodated free or at a reduced rate should be paid locally to the hotel or facility operator concerned.

Forestdale Hotels

Central Sales & Reservations, Lyndhurst Park Hotel, High Street, Lyndhurst, Hampshire SO43 7NL
Tel: 0800 378640 (freefone)
0703 283003
Fax: 0703 283019

Forestdale Hotels is a privately owned group of eleven hotels across the south of England. The group operates a two-day Leisure Break programme throughout their hotels with prices and facilities varying from resort to resort. Details of children's discounts are available on request. As an example, one hotel offers a weekend break package including two nights accommodation and full breakfast during which up to two children (under 13) sharing a room with two adults will be accommodated free and need only pay for meals taken. Some hotels charge for the use of cots and all hotels ask that you book these in advance. Disabled travellers should contact the hotels direct before booking as

Bargain breaks and short-stay holidays

many Forestdale properties are country house hotels and access may be difficult. However, the group will do everything they can to help and can advise on availability of ground-floor rooms etc.

Forte Hotels

St. Martin's House, 20 Queensmere, Slough, Berkshire SL1 1YY
Tel: (Head Office) 0753 573266
Tel: (Reservations) 0345 40 40 40
Tel: (London Theatre & Rail inclusive breaks) 0345 543555

Forte Hotels operates over 700 hotels worldwide. They offer many different packages for their business customers and those wishing to travel far afield. The group produces the Forte Leisure Breaks brochure offering short breaks and holidays in over 240 UK hotels. You can also chose from weekend breaks, '7 nights for the price of 6' holidays, and special weekends such as Theatre Breaks in London. Prices start from just £41 per person per night.

There are many different offers for families in the Leisure Breaks brochure – contact Forte Hotels for the latest details. However the following offers give a good indication of what you can expect with a Forte Leisure Break package.

Children stay free in their own room:
At selected Forte Crest and Forte Posthouses up to three children under 16 may stay free in their own room when accompanied by two adults (subject to room availability). Children's meals are charged only as taken. Participating hotels are indicated in the relevant brochure entries.

Children stay free when sharing with two adults:
In hotels other than selected Forte Crests and Forte Posthouses, up to two children under 16 are accommodated free when sharing a twin room with two adults or a single room with one adult. Single parent families with more than one child are also suitably accommodated where possible. When children under 16 sleep in their own room, they receive a 25% discount off the normal Leisure Break adult price. Children's meals are charged as taken. Participating hotels are indicated in the relevant brochure entries.

Special features for families
Bumper Funcase – in all Forte Hotels, when you arrive with children, they're given a free Bumper Funcase to keep them amused while you explore your surroundings. This includes games, postcards, an activity book, etc.
Soft drinks – a soft drink will be provided in the bedroom with your tea and coffee making facilities.
Baby food – all Forte Hotels will provide Heinz baby food in tamper-evident packaging.
Children's food – Forte Hotels offer a special Hungry Bear menu. There is also a 50% reduction for child portions taken from the adult menu.
Cots and high chairs – cots and high chairs are available on request.
Baby listening and baby sitting facilities – these facilities are available at many Forte Hotels, but must be requested at least seven days prior to arrival. Check for availability and any special requests well before booking.
Weekend host families – at all Forte Posthouses a local family with a good knowledge of the area will be on hand throughout the weekend, to tell you about the hotel and to answer any questions you may have.
Children's playroom – all Forte Posthouses have a special children's playroom.
Weekend hostess – at selected Forte Crest hotels you are met on arrival by a Weekend Hostess and a welcome drink. She will be able to tell you about your hotel and help you to get the most out of your stay.

Leisure Breaks
The price of a Leisure Break includes accommodation in a twin, double or single room with private bathroom, colour TV, tea and coffee making facilities, traditional breakfast every morning, three course table d'hote dinner with coffee each evening, and VAT. For children accommodated free (see above), all meals including breakfast are charged as taken. Children may choose either from the 'Hungry Bear' menu or from the adult menu at half price.

Children in their own room receiving a 25% discount off the adult rate will find breakfast and dinner included in the price of their Leisure Break. Children under five can enjoy free breakfast and free meals at Forte Hotel carverys. A Leisure Break is for two nights or more and must include a Saturday night.

The Great British Holiday

If you book a seven-night Leisure Break holiday at any one of 200 participating hotels your seventh night is free. Hotel participation is shown in relevant brochure entries. Children may choose either from the Hungry Bear menu or from the adult menu at half price. Children in their own room receiving a 25% discount off the adult rate will find breakfast and dinner included in the price of their holiday. Children under five can enjoy free breakfast and free meals at Forte Hotel carverys. The Great British Holiday comprises seven nights for the price of six, commencing any day of the week.

Bank Holiday Bonus

A three-night break is available for the price of two nights during Bank Holidays at selected hotels (the break must begin on a Saturday night). Prices include accommodation in a single, twin or double room with private bathroom, colour TV and tea and coffee making facilities (London hotels do not always provide the latter), traditional English breakfast and a three course table d'hote dinner with coffee each evening.

London Theatre Breaks

Contact the reservations staff on 0345 543555 and reserve a ticket for you favourite London show – the rates for this package are based on a grading system that varies according to which show you choose. If there is a particular concert, opera or show you want to see, ask the reservations staff whether they can arrange it for you. The package includes your ticket, two nights accommodation in a twin, double or single room, with private bathroom (except one hotel in central London, which has no private bathrooms), colour TV, telephone, full traditional breakfast each morning, Dining Around London vouchers (redeemable at a wide selection of hotel restaurants throughout the capital) plus VAT.

Christmas and New Year Celebrations

It is not yet known what Forte will offer for Christmas 1993, but in 1992, the following offers applied. The details below will give you some idea of what will be on offer.

There are several different Christmas packages, labelled according to the style of holiday you want – Gala Celebration Christmas, Traditional Christmas, a lively Christmas Cracker Christmas and a Family Christmas. The latter includes a free room for two children under 16 when accompanied by two full-rate paying adults, a daily programme of events designed to keep the kids amused, baby-listening and babysitting services.

New Year celebration packages are split into three categories – Celebration New Year, Hogmanay New Year and Candlelit New Year. Contact Forte Hotels on 0345 404040 for details of rates and discounts.

Another feature of Forte Hotels' offers during the Christmas season is a series of special overnight rates for those visiting friends and relatives. The four graded rates apply for the period of 21 December to 3 January inclusive.

London Travel Inclusive

Forte Hotels offer a range of Leisure Break and Holiday Special deals that include rail travel from any mainline station in Britain. You should obtain a copy of the Leisure Breaks brochure for details of the travel restrictions and child discounts, but British rail carry children under five free and offer half-price travel for under 16's. These packages include a free one-day London Travel Transport Travelcard, providing unlimited transport on the London Underground and Red Bus network. For children travelling free on a rail-inclusive break, one-day London Travelcards can be obtained at the time of booking, for which a small supplement will be charged.

Touring Holidays

Forte Hotels operate a system of travel vouchers which can be used for five consecutive nights or more, allowing you the flexibility to follow your own route, using one voucher per person for each night of your holiday. Only your first night must be booked in advance. The system is not without its restrictions and you are advised to obtain an up-to-date copy of the Leisure Breaks brochure before booking such a trip. However, there are substantial benefits for travelling families, with many Forte Hotel discounts and offers of free accommodation for children. Reservation and brochure information is available by calling 0345 543543.

Other Forte Hotel Packages

Among the features not mentioned above are Forte Posthouse Stopovers (at a set rate per room per night, these offer a good deal for families, but remember that no food is included), London Stopovers (room only, but rates are still per room rather than per person), Golfing Breaks, Murder-Mystery Breaks, Music at Leisure, and European Leisure Breaks (check with reservations staff prior to booking for details of children's discounts).

See also Forte Travelodge & Welcome Lodge

Forte Travelodge

Tel: 0800 850950 (freefone)
Tel: 0345 500400 (North America)

There are nearly 100 Forte Travelodges in England, Wales and Scotland and the number is still rising. Forte Travelodge is part of the Forte Hotels group. There are also 450 travelodges in North America – contact the number listed above for further details. The lodges are designed with motorists in mind and you pay one standard charge per room per night. You pay for your first night's accommodation on arrival and can check in anytime after 3 pm – you will be welcomed even into the small hours. By using a major credit card to book, you will guarantee your room no matter how late your arrival.

This price covers accommodation in a room that will sleep a family of up to three adults, a child in a cot (please book in advance) and a child under 12 – a good deal for those travelling with children. As Forte Travelodges are usually located within close proximity of a Little Chef, Happy Eater, Harvester or Welcome Break, meals can be purchased easily. All rooms are comfortable and well equipped: the double bed has a feather duvet, there's a single sofa bed and a child's 5 foot bed if required. Cots should be requested when making your reservation. Other room features include ensuite bathroom facilities, personal central-heating controls, a colour TV with radio/alarm clock, and tea and coffee making facilities. The lodges have been designed with security in mind and the reception is constantly occupied during the hours of darkness.

Every Forte Travelodge has at least one room with special facilities for disabled people. Containing a single bed to ease wheelchair access, Forte Travelodge recommend two adults and one child as the maximum booking for these rooms (please check availability and reserve a disabled room in advance).

Four Pillars Group

The Olney Suite, Witney Lodge Hotel, Ducklington Lane, Witney, Oxfordshire OX8 7TJ
Tel: 0993 700100
Fax: 0993 700101

This small privately owned hotel group operates four hotels in the south east of England. Each hotel offers different facilities (two have a health & fitness club, one has its own pub). The brochures for two of the properties mention that special accommodation for disabled guests is available and three hotels also offer non-smoking and family rooms.

The Four Pillars Group operates a Short Break package, with rates per person, per night (minimum two nights), including accommodation, lunch or dinner, full English breakfast, VAT and service. There is usually a Dinner Dance on the Saturday night which is included in the rate, in addition to a Welcome Reception on the Friday. Check for availability on booking.

The breaks normally apply for weekends, Friday/Saturday, Saturday/Sunday and the whole of August, but the group will also allocate rooms on the Short Breaks rates for midweek stays, subject to availability.

Friendly Hotels

Consort House, 180–182 Fulford Rd, York YO1 4DA
Tel: 0800 591910 (freefone, office hours only)
0904 611662
Telex: 57515 (attn: Friendly)
Fax: 0904 611320 (attn: Friendly)

Friendly Hotels operate properties throughout England and Scotland. Many of the hotels in this group offer Short Breaks and Weekend Breaks at special rates.

The price of a Friendly Break includes three course dinner with coffee, overnight accommodation, full breakfast and tea and coffee making facilities in your room. All rooms have en suite facilities, colour TV, radio, direct dial telephone and most hotels offer in-room movies and/or satellite channels. There is free parking at most Friendly Hotels. For an extra supplement, you can upgrade your room to a Premier Plus class – this will entitle you to a free welcome drink and extra in-room facilities such as a Teletext TV, trouser press, mini bar, hairdryer and a range of personal toiletries. The Friendly Break rates only apply for Friday, Saturday and Sunday night.

There is no accommodation charge for children up to 14 when sharing their parents' room. Meals are charged as taken. Friendly Hotels also organise activity and themed weekend packages.

Granada Hotels and Lodges

Granada Motorway Services Ltd, Head Office, M1 Service Area, Toddington, Bedfordshire LU5 6HR
Tel: 0800 555300 (Central Reservations)
05258 73881 (Head Office)
Fax: 052 555 602 (Central Reservations)
05258 75358 (Head Office)

Granada Hotels and Lodges are situated on the motorways and major trunk roads of Britain, enjoying easy access to both transport systems and some of the most scenic and interesting parts of the country. All establishments have either the AA three star and RAC hotel rating or the AA Lodge Stamp of Approval. Granada operate a Weekend Budget Break package in both their hotels and lodges, in addition to offering single night accommodation.

Weekend Budget Breaks

In Granada Hotels:
Prices include accommodation in a single, twin, double or family room (sleeping two adults and two children up to the age of 16), full English breakfast and, for half board, dinner up to a standard limit (currently £13) in the hotel's restaurant. All rooms enjoy private bathroom, colour TV, radio, wake-up alarm, tea and coffee making facilities, direct dial telephone, hairdryer and trouser press. All hotels have their own fully licensed restaurant and lounge bar. Rooms specially adapted for disabled travellers are also available – check for availability before booking.

In Granada Lodges:
Prices include accommodation in a single, twin, double or family room (sleeping two adults and two children up to the age of 16), full English self-service breakfast in the Country Restaurant of the Granada Service Area. All bedrooms have private bathrooms, colour TV, radio, wake-up alarm, and tea and coffee making facilities. Rooms specially adapted for disabled travellers are also available – check for availability before booking.

Hotels of the Cinque Ports

Mermaid House, 15 Udimore Road, Rye, East Sussex TN31 7DS
Tel: 0797 223788
0797 223065
Telex: 957141
Fax: 0797 226995

This is a small group of eight hotels and historic inns located in south east England. The hotels are of medium size (around 35 bedrooms each), the inns may be smaller – the group includes two of the oldest inns in England.

Hotels of the Cinque Ports operates Short Break and Innbreak packages throughout the year. Rates include two nights accommodation in a room with a private bathroom, early morning tea, full English breakfast, a four course dinner each night, VAT and service. A small supplement is charged for four-poster beds, special and single rooms. A third night is available at the reduced short-break rate. Contact the group for further details of discounts and facilities available.

Activity weekends can be arranged at some of the hotels, including golf, bowls, '1066 Country' battle tour with lectures, cricket and sea-fishing.

Bargain breaks and short-stay holidays

Inghams Travel Eurobreak
10–18 Putney Hill, London SW15 6AX
Tel: 081 780 0909

As well as a huge range of European destinations (including a Euro-Disney Resort), Eurobreaks may also be made to New York and Boston. A new city-break programme, offering flexibility and choice, to over 40 top European destinations, with all combinations of twin and multi-centre holidays is available. The Eurobreak philosophy is to offer good quality, competitively priced and totally flexible short-break holidays, allowing personal selection of hotels and flights.

Bookings can be confirmed immediately and can be made right up to the day of departure, for small parties or large groups. There is a selection of 200 hotels, all with private facilities, flights from 20 UK airports, alternative travel by rail or coach (or self-drive) and a totally flexible choice of departure day and duration.

A range of pre-bookable excursions in every city is offered and customers receive a free Berlitz Travel Guide and Eurobreak Discount Card.

Prices Child prices start at just £89 (by air).

Minotels
37 Springfield Road, Blackpool FY1 1P2
Tel: 0253 292000
Telex: 67596 MARCON-G
Fax: 0253 291111

Minotels is a consortium of mainly proprietor-owned hotels situated throughout England, Scotland and Wales. Minotels offers various different packages including one-night stays, a 'go as you please' voucher and short-stay breaks (see separate Budget Breaks brochure).

At all Minotels, one child under 12 years old will be accommodated free when they share a room with two adults. Payment for children's rooms should be made directly to the hotel concerned.

This group also operates the Minotel Voucher scheme, with approximately 700 affiliated hotels across Europe. The Minotel Voucher offers accommodation in a single, double or triple room with private facilities, continental breakfast and all taxes and services for a single price (subject to slight local adjustment charges). There is free accommodation for one child (up to 12 years) when sharing their parents' room – breakfast payable directly to hotel. Contact Minotels for further details of this scheme.

Short Breaks and Holidays
All prices displayed in the brochure are per person, based on accommodation in a twin or double room with private bath or shower. Single room supplements are payable at most hotels. The holidays are primarily designed for clients using their own cars, but Minotels will arrange transport by train or National Express coach should you wish – simply tick the appropriate box on the booking form and they will contact you to finalise travel details. Almost all holiday prices in the brochure include dinner, bed and full breakfast. Dinner in most cases will be the hotel's Table d'hote menu, but occasionally an equivalent allowance will be given against the A la Carte menu.

Super Savers
Sunday Nights – when you pay the regular price for Friday and Saturday nights, you can enjoy Sunday night for half the normal price.

Winter Savers – there are special discounts for weekend breaks – check hotels displaying the relevant symbol in advance for details.

Queens Moat Houses
9–17 Eastern Road, Romford, Essex
RM1 3NG
Tel: 0800 289331 (freefone)
Telex: 67596 MARCON-G
Fax: 0253 866251

Queens Moat Houses Hotels operates several different short-stay packages in their extensive selection of properties throughout Britain, Belgium, Germany, Holland and Switzerland and the UK. Many hotels have luxurious leisure facilities, such as golf courses, tennis and squash courts, swimming pools and even croquet lawns.

Town & Country Classics Weekend Breaks
The package offers a selection of 99 hotels in 80 locations, throughout England,

Scotland and Wales. There is a minimum of two consecutive nights' accommodation, one of which must be a Saturday. However, you are welcome to spend each of these two nights in a different Queens Moat Houses hotel. Rates for extra nights are available on request and may be higher or lower than the standard break rate. Two hotels in the Town & Country Classics brochure offer a 50% discount on Sunday night's accommodation provided that you have stayed at the hotel for the previous Friday and Saturday night.

Prices quoted in the brochure are per person per night with no single supplements and include accommodation in a room with private bathroom, colour TV, radio and telephone, early morning tea or coffee, full English breakfast and a daily newspaper of your choice each morning, dinner from the table d'hote menu where available, or the equivalent sum placed towards an a la carte meal, VAT and service. Many rooms have garment presses, hairdryers, tea and coffee making facilities and an In-house movie channel on the television – check for availability on reservation. There is often an inclusive dinner dance on the Friday or Saturday night.

Up to two children under 16 may stay (and have full English breakfast) free of charge when sharing a room with two adults. All other meals are charged as taken.

Each additional child will be charged at 75% of the adult rate, as will those children who are accommodated in a separate room. There are three- and four-bedded rooms in some hotels – availability is shown in the relevant brochure entries, but do indicate your wish for such accommodation when making your reservation.

Greatstay Selection
15 hotels in the brochure have been classified as part of the Greatstay Selection. These properties offer superior cuisine, stylish accommodation and first-class service.

Luxury Classic Weekends
A variety of hotels offer weekend breaks with a champagne welcome and various luxury extras. Prices quoted are for two people sharing a double or twin bedded room and apply for a two-night consecutive stay, one of which must be a Saturday. These breaks are subject to availability.

Summer and Winter Savers
Queens Moat Houses offer various special deals for Summer and Winter breaks – check the brochure or reservation staff for further details. These deals often offer substantial savings.

Speciality Weekends
Many Queens Moat Houses hotels arrange (or will arrange on request) special activity breaks including such themes as murder-mystery, antiques and their appreciation, hot-air balooning, etc.

The Town & Country Classics weekend rates do not necessarily apply to Speciality Weekends – contact the hotel concerned for further details.

Disabled Guests
There are special facilities and adapted rooms for the disabled at many of the Queens Moat Houses hotels. Contact the central reservation staff for further details and advice.

Rank Hotels

Rank Hotels Reservations Centre,
1 Thameside Centre, Kew Bridge Road,
Brentford, Middlesex TW8 0HF
Tel: 081 569 7120/7211
Telex: 915888
Fax: 081 569 7109

Rank Hotels Limited operate twenty two hotels throughout Britain, with five of their hotels located in central London. Theirs are luxury hotels, offering all the facilities you would expect to be provided for their regular business clientele.

The group operates a mixture of historic buildings and purpose-built hotels, ranging from a 10th century Manor House to a modern country club. Rank offer a range of packages including Town and Country Weekends, rail-inclusive breaks, one-night stays, golfing breaks and even air-exclusive breaks. They operate a Celebration Package and special offers for Christmas, New Year and Easter – contact the group direct for further details.

Town and Country Weekends

Outside London
Prices shown are per person, sharing a double/twin room and include: Friday (dinner & accommodation); Saturday (full English breakfast, dinner & accommodation); Sunday (full English breakfast). By paying a supplement (variable according to location), you can extend your stay and enjoy an extra night's accommodation and a full English breakfast on Monday morning. There may be a single room supplement in some hotels – check in advance for details.

Up to two children (under 15) sharing a room with one or two adults will be accommodated free of charge. This offer is for accommodation only, and all children's meals (including breakfast) should be paid for directly to the hotel. Most restaurants feature money-stretching children's menus.

Children normally share a room with adults, but inter-connecting rooms will be offered at no extra charge when available. Each extra child is charged 50% of the normal adult price, which includes breakfast.

London breaks
Prices shown are per person per night, sharing a double/twin room, and include accommodation and breakfast. The rates shown are for Friday, Saturday and Sunday nights, but will also apply for Monday to Thursday nights if these are booked in conjunction with a Friday, Saturday or Sunday. There are no single-room supplements on Friday, Saturday or Sunday nights. A 15% discount for each day of your stay is available on meals taken in the restaurants of most of the hotels.

All weekend breaks in London include free accommodation for up to two children (under 15) sharing a room with one or two adults. This offer is for accommodation only and all charges for children's meals (including breakfast) should be paid direct to the hotel. Most restaurants feature money-stretching children's menus.

London break packages offer several extra features including a free map, visitor travelcard, and discount vouchers for a variety of nightlife and daytime activities (see brochure for details).

Golfing breaks
Three Rank Hotel properties make a particular feature of the golfing facilities available and organise specific golfing breaks. One is on the south coast of England, while the other two are in Scotland.

However, it is worth remembering that many of the other hotels featured in the Town and Country Weekends brochure are situated near to golf courses – this is indicated in the brochure.

Discounts for children on travel-inclusive packages
When travelling by train, children under 15 travel for 50% of the adult fare and all children under five travel free. If you travel by British Airways, children under two travel free of charge, from two to eleven years at the appropriate child fare, and those over eleven at the full adult fare. Enquire to British Rail and British Airways for current prices.

Resort Hotels PLC

Resort House, Edward Street, Brighton, East Sussex BN2 2HW
Tel: 0345 313213 (Central reservations – local call charge)
Fax: 0273 606675

Resort Hotels offer their weekend and short breaks programme 'Carefree Days' at each of their 42 Hotels and Fine Inns in England and Wales. Within 'Carefree Days' you will find details of special interest breaks, 'Carefree Options', plus other packages such as split holidays, celebration weekends and special savers.

Carefree Days
Prices are per person sharing a twin or double room. You can choose to have half board (bed, breakfast and evening meal) or simply pay for accommodation and breakfast. Weekend stays are from Friday to Sunday and are for a minimum of one night whilst midweek breaks are from Monday to Thursday and are for a minimum of two nights half board at a Resort Hotel or bed and breakfast at a Fine Inn.

Up to two children under 16 sharing a room with their parents will be accommodated free of charge. Cots are free but must be booked in advance. Meals for children are charged as taken and can be chosen from a special Jolly Jester menu.

90 Choosing and planning your holiday

High chairs are available on request at all hotels. Many Resort Hotels enjoy the facilities of a Health and Leisure Club featuring a heated swimming pool, sauna, solarium, spa bath and exercise equipment. Some also offer massage and beauty treatments.

Carefree Options – Special interest and activity breaks

Carefree Options is a range of active or passive hobbies, interests or sports that can be enjoyed in or near a Resort property. Try ballooning in Berkshire or taking it easy during a Healthy Options weekend at a country house-style hotel. Whatever your interest, Options endeavours to arrange a special-interest package for you. Ring 0345 313 213 for a copy of the brochure or to make a booking.

Welcome Lodge

Charnock Richard
Mill Lane, Charnock Richard, Chorley, Lancashire PR7 5LR
Tel: 0257 791746 (reservations)
Fax: 0257 793596

Oxford
Peartree Roundabout, Woodstock Road, Oxford OX2 8JZ
Tel: 0865 54301 (reservations)
Fax: 0865 513474

Newport Pagnell
M1 Motorway, Newport Pagnell, Buckinghamshire MK16 8DS
Tel: 0908 610878 (reservations)
Fax: 0908 617226

London – Scratchwood
M1 Motorway, Hendon, London NW7 3HB
Tel: 081 906 0611 (reservations)
Fax: 081 906 3654

The four Welcome Lodges are run by the Forte Hotels group and are situated in the grounds of Welcome Break service areas. They offer comfortable facilities, including lounges and bars and, as you would expect, have ample parking facilities. All sites (except Charnock Richard) feature family rooms, which will be allocated on request when available. However, all Welcome Lodges offer free accommodation for children under 16 when they share their parents' room – meals will be charged as taken. Each room has its own bathroom and TV.

Welcome Lodge often operate special weekend packages, which have a standard price per room per night and include English or Continental Breakfast – contact the relevant reservations number for further details, or 0800 40 40 40 (central reservations).

Camping, caravan sites, mobile homes

One of the advantages of a self-catering holiday in a tent, caravan or mobile home is that for a relatively low cost you can enjoy a holiday that allows you to really 'please yourselves'. The independence of self-catering holidays is a great advantage to those with children – there are no stuffy and formal hotels or unsympathetic landladies to cope with and your family can eat when and whatever you please.

To put minds at rest, we need to define what you can actually expect from a camping site. However idyllic the setting, it is important to remember that if you chose to pitch your own tent or caravan in a farmer's field, you cannot expect even the most basic of facilities. You are on your own. There may not be anyone around to help you put your tent up in the pouring rain, help you to dig your car out of the mud and, ninety nine times out of a hundred, no washing or toilet facilities. All of these problems seem trivial in the blazing sunshine, but it is best to imagine the worst before plunging into a 'back to nature' mood. Farmers can often be extremely helpful, but they will never tolerate fools. If you *are* planning to take this option, it is important that you make sure you follow the rules of the Country Code. All its rules are common sense, but it really is essential reading. This section of the book includes many organised campsites with toilet and shower blocks and many have shops and entertainment, making them ideal for families. Another advantage of such sites is that there will be other children for your family to play with! Details of campsites in any particular area in Britain can be obtained by contacting the relevant tourist board, the Camping and Caravanning Club or the National Caravan Council (see Section 7).

One of the most civilised forms of camping is a holiday at a site where luxury tents are ready-pitched and fully equipped for your use. If you are expecting khaki-coloured army-surplus tents, think again. Most family tents are large, 'bungalow-style' tents divided into separate rooms by zipped partitions. They all have transparent window panels and the tent sides can be raised to extend your living space when the weather is good. Inside they may have all mod cons, and some even have their own separate toilet tents. All these facilities should be checked before booking.

If you are wondering what the difference is between a caravan and a mobile home, the answer is, not a lot. Some companies make a distinction, apparently based on the caravan's lack of a plumbed in toilet and shower, but they often do not differ in size and they are both completely stationary! It is probably worth the extra expense in having your own toilet and bathroom facilities; traipsing through wet grass with children in the early hours is not much fun. Many caravans have a television which will prove its worth ten times over should the weather be bad. You can often hire linen and pillows and sometimes the site owner may supply basic groceries on request – check with the site for details of facilities available well in advance. With many sites giving much thought to families, this type of holiday can be ideal for those travelling with children.

When comparing what companies have to offer, the cheapest deal does not necessarily represent the best bargain. It is essential to compare the site facilities before deciding on your destination. These vary from the basic to the fantastic and should be a major factor in any decision you make. A self-catering holiday obviously appeals because of the money that can be saved, but do you really want to cook every day? The take-away food outlets and cafes on some sites could be a terrific advantage. How far is the site from the nearest town or beach? Is there a regular bus service? Is there a swimming pool? What can you do if it is wet? Finally, most parents appreciate some time alone together on holiday, so look out for those sites that offer organised children's games and babysitting facilities.

Prices
Unless otherwise stated 1992 prices have been quoted. These are intended only as a guide to the type of holiday offered and travellers should check with the company concerned for 1993 pricing details.

Becks Holidays

Southfields, Shirleys, Ditchling, Hassocks, W. Sussex BN6 8UD
Tel: 0273 842843

France

Caravan holidays on 4 sites in Brittany, Royan and the Vendée. The caravans sleep 6 or 7 and have electricity and gas supplies, mains water, shower, WCs, fully-equipped kitchen and patio furniture. Cots are available free on request. Blankets and pillows are provided. All sites have resident couriers, swimming pools (including children's pools), shops, bar, restaurant, take-away meals, laundry facilities and children's play area. Additional facilities include showers, WCs, guest rooms, TV room, tennis courts and bicycle hire (not all available at all sites).

Prices (per adult per week: 1992 prices) £78–£200

Discounts Children under 4 go free and 4–13s pay £0–£42 each. Each child must pay £11.50 insurance.

Brittany Caravan Hire

15 Winchcombe Road, Frampton Cotterell, Bristol BS17 2AG
Tel: 0454 772410

France

Mobile home holidays in Brittany and on the west coast, inclusive of self-drive travel arrangements. Mobile homes have gas and electricity supplies, running water, portable flush WC and fully-equipped kitchen and include showers. Patio furniture and pillows supplied. Cots are available free in any accommodation on request. Most sites are close to the beach and have shops, bar, restaurant, take-away meals, showers, WCs, laundry facilities, children's play area and sports facilities. All sites carry a 4-star rating and have swimming pools and evening entertainments in high season.

Prices The company has 5 seasons and the prices quoted cover mobile home hire and a short ferry crossing for 2 adults plus a car for 14 days.
Low season (May): £288
High season: £880

Discounts Up to 2 children free on all sailings; £10 before June and from 7 September. Additional children pay £18 each.

Cabervans

Luxury Motor Homes, Caberfeidh, Cloch Road, Gourock, Scotland PA19 1BA
Tel: 0475 38775

Europe, UK

Motorhomes and caravans (2–7 berth) all with shower and fully-equipped kitchen. Extra charge for sleeping bags and sheets, if required. Drivers must be age 23 or over.

Camping, caravan sites, mobile homes

Prices General range: £240–675 per week inclusive of VAT, insurance and unlimited mileage.

Canvas Holidays Ltd

12 Abbey Park Place, Dunfermline
KY12 7PD
Tel: 0383 621000

Austria, France, Germany, Italy, Spain, Switzerland

Experienced and award-winning company specializing in tent and mobile home holidays. Bungalow-style tents have three bedrooms to sleep 6 people, electricity and gas supplies, fully-equipped kitchen, living room and patio furniture. Bedding and pillows are not supplied. Optional extras include your own toilet tent, a children's pack of cot and high chair (£2.50 per night) and a beginner's tent for children who want to try camping on their own (£3.50 per night). Mobile homes have two bedrooms and sleep 6–8 people. All are similarly equipped to tents, but have hot and cold running water, shower and WC. Blankets and pillows are supplied, but not linen. Children's couriers operate in selected camps, organizing games and activities. Special children's clubs everywhere, with treasure hunts, plus children's pack including transfers, crayons, and 'passport'. Babysitting and baby patrols can be arranged for a small fee. Most sites have swimming pools and the company provides free watersports equipment at many sites with free instruction at selected camps. Most have shopping, laundry, shower, WC and games facilities.

Prices (two weeks)
Low season, 2 adults: £155
High season, 2 adults: £699

Discounts Infants 0–3 free; 4–13 free, excluding high season (high season £14 per holiday); teenagers £19–£39 per holiday. All pay insurance of £15.

Carisma Holidays Ltd

Bethel House, Heronsgate Road,
Chorleywood, Herts WD3 5BB
Tel: 0923 284235

France

Mobile home and tent holidays on 6 sites, all with their own private beaches or on the lakeside, in south-west France, Vendée and Brittany. The mobile homes have 6–8 berths, gas and electricity supplies, hot and cold running water, shower, WC, fully-equipped kitchen and patio furniture. Bedding is supplied, linen is not. Bungalow-style tents sleep 6 in 3 bedrooms. They have a fully-equipped kitchen, living room and patio furniture. No bedding or linen. Couriers on-site organize children's games and family entertainment, and will babysit for a small charge. Facilities on-site include shop, bar, restaurant, take-away meals, hot showers, WCs, sports and laundry facilities, children's play area, swimming pool. A doctor is available daily on each site.

Prices please refer to brochure.
Discounts Children under 4 travel free.

Eurocamp Travel Ltd

Canute Court, Toft Road, Knutsford
WA16 0NL
Tel: 056563 3844

Austria, Belgium, France, Germany, Ireland, Italy, Luxembourg, Spain, Switzerland, UK, Yugoslavia

Tent and mobile home holidays on over 200 sites. Bungalow-style tents sleep 6 in 3 bedrooms, have electricity and gas supplies, fully-equipped kitchen, living room and patio furniture. Bedding and pillows are not supplied. Mobile homes sleep 6–8, are similarly equipped to tents, but also have hot and cold running water, shower and WC. Blankets and pillows are provided. Cots available in all accommodation on request. All sites have many of the following facilities: shops, bar, restaurant, take-away food, hot showers, WCs, washing and drying machines, ironing facilities, play areas, sporting facilities and swimming pool. A children's courier service operates on 80 sites, organizing games and activities for children of all ages. A baby-sitting service is available at an hourly rate (where children's couriers are on site) and children under 15 receive their own travel pack of information, games to play, etc. Baby packs of high

chair, playpen and baby bath are available on selected sites for £2 a night per pack. Travel arrangements, including ferries, motorail and flights can be arranged on request.
Prices (per car with 2 adults for 2 weeks, including insurance and short sea ferry crossing). For holidays in tents:
Saver season: £255
Mid: £555
High: £699
Mobile home holidays extra.
Discounts On travel-inclusive holidays, children up to nine are free, but insurance cover of £14.80 must be paid. Children aged ten to thirteen are free apart from 11 July to 28 August when they must pay a supplement.

Haven

Swan Court, Waterhouse Street, Hemel Hempstead, Herts HP1 1DS
Tel: 0442 233111

France, Spain, UK

Self-catering holiday parks and villages, with chalets and caravans at coastal resorts in England and Wales, France and Spain. Parks are small, medium and large, with corresponding levels of entertainment. Most have pools, indoor and outdoor games, children's and teenage clubs, bars, shops, restaurants, launderettes and evening entertainment for both adults and children. Accommodation is in 4–8 berth caravans or 2–3 bedroomed chalets and villas.
Prices (6-berth caravan in Dorset per week, inclusive of gas, electricity, colour TV, linen, crockery and utensils)
£105 (September), £244 (June), £205 (July) or £320 (August).
Discounts Book 14 nights get £25 off. Book second holiday in same year get £25 off. Short breaks at 60% discount.

Haven France and Spain

Northney Marina, Northney Road, Hayling Island, Hants PO11 0NL
Tel: 0705 466111

France, Spain

Mobile homes and tents on established sites, on the French coast, the Dordogne and the Loire, but including 2 on the Costa Brava. Mobile homes have 7–8 berths (some intended for children rather than adults), shower, WC and kitchen. Tents sleep 6. Crockery and utensils are provided, but bring your own linen. Cots provided free. The sites have restaurants and swimming pools, and most provide children's play areas and games. French Motorail connections on request.
Prices (including ferry, gas and electricity)
From £140 for 7 nights including short sea ferry crossing, gas and electricity.
Discounts Children under 14 go free.

Holimarine

171 Ivy House Lane, Bilston,
W. Midlands WV14 9LD
Tel: 0902 880800

UK

Family holiday parks in south-west England and Suffolk. Sites have swimming pools, paddling pools, playparks and indoor sports, tennis, snooker and amusements. There are also free discos, daytime entertainment, and cabaret shows in the evenings for both children and adults. Accommodation is in luxury lodges, maisonettes, villas or caravans half-board or self-catering, according to site. All are equipped with fridge, TV, bath or shower, crockery and utensils. Bring your own linen and towels. Cots available at most sites (small charge at some sites) but there is no organized babysitting. There is a children's Holiphant club.
Prices (per property per week, including gas and electricity)
5-berth caravan: £80–£290
4-person flat: £80–£320

Matthews Holidays

8 Bishopmead Parade, East Horsley,
Surrey KT24 6RP
Tel: 04865 4044

France

Mobile home holidays in south-west France. Accommodation is available in 3 sizes, sleeps 6, has gas and electricity supplies, fully-equipped kitchen and patio furniture. Blankets and pillows are supplied. The largest units have a WC and shower, while the smallest has WC only. Good showers are, however, on all sites for mobiles without these facilities. In addition, most sites have their own shops, bar, take-away meals, laundry, irons and sports facilities, including swimming pool; others use nearby facilities in local towns.

Prices average price per unit in Benodet, July/August)
£692 per fortnight without shower, £836 with, *excluding travel*
£835 per fortnight without shower, £979 with, *including travel* for 2 adults with 3 children under 14 via Dover. Other routes possible.

Discounts On travel-inclusive holidays, three children under 14 travel free on all dates.

The Mount Holiday Park

Par, Cornwall PL24 2BZ
Tel: 072 681 2616

UK (Cornwall)

Caravan and chalet holidays on a site close to the sea, between Fowey and St Austell. All caravans are fully serviced, with mains drainage, fully-equipped kitchen, gas and electricity, colour TV, WC, hot and cold water and a shower. Chalets have 2 bedrooms and sleep 6 people. They are fully serviced and have colour TV. Blankets and pillows are supplied in all accommodation, but own linen must be provided. Cots can be hired for £3 per week. Facilities on site include an adventure playground, shop, launderette and ironing room. There is a restaurant in a central club-house. Sports facilities, including golf and riding are available nearby. Sea fishing is a popular sport and equipment is available locally. It is the park's policy to accept only family bookings.

Prices (per fully-serviced caravan or chalet per week)
6-berth caravan: from £75 to £185
8-berth caravan: from £110 to £297

Romany Caravan Holidays

c/o Northumbria Horse Holidays
(see **Special Interest Holidays**)

UK (Norfolk)

Romany caravan holidays in the Waveney Valley in Norfolk. The horse-drawn caravans sleep 4–5 people, have modern interiors and are fully equipped. Gas is provided free for lighting and cooking. You can meander along quiet country lanes, staying overnight at the Waveney Valley home base, or at one of the selected parking sites. Before starting out tuition is given in hitching up and handling your horse, which has been specially chosen for its friendly temperament.

Prices (per week's rental)
Low: £209 + VAT
High: £329 + VAT

Sandpiper Camping Holidays

Sandpiper House, 19 Fairmile Avenue, Cobham, Surrey KT11 2JA
Tel: 0932 868658

France

Three sites with ready-erected tents, fully equipped with cookers, beds, electric lights and fridges. On-site couriers available providing children's entertainment. One of the sites has a play tent.

Prices Family of 4 from £199, self-drive including ferry.

Discounts Children (up to a maximum of 3) are free.

Seasun Holidays Ltd

Seasun House, 4 East Street, Colchester, Essex CO1 2XW
Tel: 0206 871212
Fax: 0206 869889

France, Italy, Portugal, Russia, Spain

Self-catering holidays (and hotel packages) by coach, car and air to seaside resorts on the Mediterranean and the Algarve. Accommodation is in apartments, mobile homes, chalets, caravans, cabins and tents. All are fully-equipped, although a returnable deposit is payable on some sites from which gas, electricity

96 Choosing and planning your holiday

and laundry costs are deducted. Most of the sites provide day and night entertainment with swimming pools, mini-golf, mini-olympics, inflatable castles, water slides, competitions, laundry rooms, shops and restaurant.

Prices (per adult per fortnight, including coach travel and ferry crossing)
Tents: £69–£250
Cabins and caravans: £79–£300
Mobile homes: £99–£400
Apartments: £99–£450

Discounts Under-2s go free provided they do not occupy a seat. Children aged 3–15 receive discounts of £15–£40 (high season), with many free places in low season.

Solaire International Holidays

1158 Stratford Road, Hall Green, Birmingham B28 8AF
Tel: 021 778 5061

France, Spain

Mobile homes, caravans and tents sleeping 4–8 people at many sites in France and 1 in Spain. Crockery, utensils and pillows are supplied (bring your own bedding), and the homes and caravans have their own shower and WC. Sites include all modern facilities and many provide play areas for children (contact company in advance for details). All have swimming pools. Baby packs can be booked in advance at £2 per night. Mini tents are £3 per night. Travel by ferry and coach/own car service available to some sites.

Prices (per 2 adults, per fortnight, including ferry crossing)
Brittany: £140–£539
An extra adult would pay £30–£50. Check with operator about any extra charge for hiring a mobile home.

Discounts Children under 14 travel free. Children aged 14–16 pay £20–£40 per fortnight. Some sites offer special discounts for long bookings.

Treble B Holiday Centre

Looe, Cornwall PL13 2JS
Tel: 0503 262425

UK (Cornwall)

Family-run site two miles from sea for people wishing to bring their own tents, tourers, caravans and equipment. Chalet and caravan rental also available, as well as luxury superior cottages. All rented accommodation is fully equipped but bed linen and towels are not provided (apart from cottages). Caravans are 6–8 berth with electricity, shower, WC, TV and fridge. Site has shower and WC facilities, water and drainage points, electricity, battery charging and ice-pack hire services. There are shops, a launderette, swimming pool, children's play area, TV lounges and games facilities. Family entertainment is staged nightly during the months May to September. Cots and high chairs may be hired for around £7 per week in rented accommodation only (no more than one of each per family).

Prices £2.50–£4.00 per person; £1.25–£2.25 per child; £1.80 for electrical hookups.

Uley Carriage Hire

Weavers Workshops, The Street, Uley, Glos GL11 5TB
Tel: 0453 860288

UK (Cotswolds)

Seven-night tours of the Cotswolds in horse-drawn Romany caravans, which sleep 4. Calor gas cooker, crockery, cutlery and chemical WC provided (no linen). An instructor will accompany you on the first day, and is on 24-hour call after that. Most stops are at pubs or farms with washing facilities. Cots might be available (enquire when booking).

Prices (per caravan and horse per week)
Low: £260
High: £380
Please check as subject to change.

Welcome Caravan Company

1st Floor, 54 The High Street, Thames Ditton, Surrey KT7 0SA
Tel: 081 398 0355

France

Holidays in mobile homes, 12 at coastal sites and one in Paris. All homes sleep 6 and have hot and cold water, shower, flush WC, fridge, gas cooker and oven, bedding (no linen), crockery and utensils. All sites have shops, restaurants and swimming pools. Camping cots are available free of charge.

Prices (per 2 adults sharing a mobile home for 2 weeks, including ferry via Dover and Calais)
Brittany: £332–£798
(Each extra adult pays £35–£45, and each child aged 4–13 pays £25–£30.)
French Riviera: £452–£844

Discounts All children under 14 go free in low and mid season but *all* children pay £10 travel insurance.

Westents Ltd
88 New North Road, Huddersfield
HD1 5NE
Tel: 0484 510544
France

Camping holidays on 5 sites in France. The bungalow-style tents sleep 6 in 3 bedrooms and have gas supply, electricity, fully-equipped kitchen with fridge, living room and patio furniture. Bedding and pillows are not provided. Cots are available free on request. All sites have shop, take-away food, play area, sports facilities, swimming pools, showers, WCs, washing and ironing facilities. A bar is also available on all of the sites. Beaches are 1–3km away. Couriers organize children's games and evening entertainments for all (the Squirrel Club).

Prices For 2 adults (children under 14 free)
High season: £699 for 14 nights in bungalow tent.
Ring company directly for special offers.

Discounts All children under 14 go free throughout season, but insurance of £13.50 each must be paid.

Canal holidays

Self-catering waterway holidays have always been popular, both in the UK and abroad, but how practical are they for families with young children? One of the difficulties on boats is the confined space which can be a problem with toddlers and energetic youngsters. The craft used by the companies listed in this section have 2–12 berths (beds or bunks), and without exception, all recommend that you hire a larger boat than you actually need. This gives you room to move around comfortably and store all the paraphernalia that babies require. Having pored over the brochures for some time, we conclude that this is sound advice and not just a tactic to get more money out of you.

Most companies feel that carrycots would be fine on a boat, but children of toddling age could be quite a problem unless carefully confined. All companies can provide cotsides which are supposed to stop small children falling or clambering out of bed. However, Mrs S. Austin of Chingford wrote to us with some useful information on this subject after a 'disastrous' holiday on the Norfolk Broads:

'Cotsides do not extend to the full length of the berth. Therefore, if your toddler should wake in the night, she can crawl to the end of the bed and fall to the floor – a drop of some three feet. We were lucky to have hired a bigger boat as we folded the extra blankets to form a makeshift barrier at the foot of the bed, thus preventing a possible visit to the local hospital.'

Having hired an older style craft arranged on two levels, Mrs Austin also discovered the advisability of having all the cabins on one level so there would be no steps for children to trip over or fall down.

On the question of safety, all the companies supply buoyancy aids (except for babies) and children are advised to wear them at all times. Ideally, at least two members of the family group should be able to swim. It is essential to have a minimum of two adults on the boat as steering and mooring is a four-handed job. For this reason, a single parent with young children would not be advised to take a boating holiday.

Boats are priced according to the number of berths, and the fee is a fixed one. There are no discounts for children, except sometimes where the boat has a variable number of berths, say 4–6. This means that the basic hire fee is calculated on four berths, and the extra two can be used by children (normally under 12 years) at no extra charge. To take advantage of this you must take your own linen or sleeping-bags; an extra person charge will be levied if the hire company supplies linen.

Always enquire before booking what the price includes. You may have to pay extra for fuel, moorings, linen or bicycle hire. It is important to remember when loading your car or struggling on to the train that television reception from canal or river level is notoriously poor!

Prices

Unless otherwise stated 1992 prices have been quoted. These are intended only as a guide to the type of holiday offered and travellers should check with the company concerned for 1993 pricing details.

Canal holidays 99

Adventure Cruisers
Catforth, nr Preston, Lancs PR4 0HE
Tel: 0772 690232
UK (Lake District)

A small company specializing in one-day cruises on the Lancaster Canal, which runs between Preston and Twitfield in the Lake District – a distance of 43 lock-free miles. Fuel is included in the hire cost. Cotsides can be provided, but not cots. Buoyancy aids are provided free. Small dogs may also travel for no extra charge, but the company must be notified when booking (UK residents only).
Prices £40 per day (all inclusive) for up to 6 adults or, if small children in the party, up to 8 are allowed

Blakes Holidays
Wroxham, Norwich, Norfolk NR12 8DH
Tel: 0603 782911 *(all British waterways, Norfolk Broads, Thames, Cambridgeshire, English and Welsh canals and rivers, Scotland)*
0603 784131 *(Ireland and France)*
Fax: 0603 782141
France, Ireland, UK

Established over 80 years, this company arranges the hire of cruisers, narrowboats and yachts of 2–12 berths. Sailing boats tend to be smaller – usually 2–6 berths. All craft have hot and cold running water, shower, WC, fully-equipped kitchen and some form of heating. Bed-linen is supplied, but there are usually extra charges for fuel and pets. Brief instruction is provided free on steering cruisers. Sailing tuition is available free of charge. Cotsides are often free on request.

On European holidays, ferries, rail and/or air connections can be arranged. The holidays cover canals, rivers, lakes and fjords. On many trips bicycle and canoe hire is available from certain locations for a nominal weekly charge.
Prices (per boat per week on UK canals/rivers)
The company has 8 price bands, A–H according to season. Prices shown cover smallest to largest boats.
Low (A): £177–£678
Mid (D): £236–£904
High (H): £295–£1130

For holidays in Ireland and abroad craft are individually priced.
Discounts Children in carrycots travel free. On UK holidays any extra people aged 12 and under may occupy any extra berths on the boat free of charge.

Blisworth Tunnel Boats Ltd
The Wharf, Gayton Road, Blisworth, Northants NN7 3BN
Tel: 0604 858868
UK (Midlands)

Narrowboat holidays and dayboat hire on a variety of canals in the midland system. Narrowboats have 2–12 berths and include shower, WC, cooker, refrigerator, hot and cold water, heating, crockery and utensils. Linen is not provided. Some boats have cotsides that can be bolted on to a bunk (enquire when booking).
Prices (per narrowboat, per week, including fuel, gas, car parking and VAT)
4-berth boat: £343–£572
6-berth boat: £468–£785
day boat, peak season:
Full day (9.30–18.30): £65
Half day (10.00–14.00/14.30–18.30): £35

Bridgewater Boats
Castle Wharf, Berkhamsted, Herts
Tel: 0442 86 3615
UK (England)

A caring family boatyard offering 2–8 berth narrowboats. All holidays start from the medieval market town of Berkhamstead but you can go as far as London, Stratford and Oxford, depending on duration of holiday. Long weekend holidays are also available. All boats have electricity, hot and cold running water, shower, fresh-water flush toilets, fully-equipped kitchen and heating. Bedding is supplied, but there is an extra charge for bedlinen. Cotsides are available if requested when booking. Lower bunks have been specially designed to fold down to make a playpen during the day. Some boats are also equipped with a 'running line' – a rope attached to 'eyes' on the deck – to which toddlers can be harnessed so they can move around without fear of falling in. It is recommended that you take an

extra berth for a baby to allow room for all their paraphernalia. Dogs, but not cats, may travel for a small extra charge (UK residents only).
Prices (per 4-berth boat, approximately)
Low: from £300
High: from £580
Discounts Children in carrycots travel free.

Claymoore Navigation Ltd

The Wharf, Preston Brook, Warrington, Cheshire
Tel: 0928 717273

UK (Cheshire, Lancahire, Yorkshire, North Wales)

Narrowboats (3–10 berths) for hire on many canals and rivers. All boats have electricity, hot and cold running water, shower, WC, fully-equipped kitchen, stereo cassette player and central heating. Bed-linen supplied. Cotsides and buoyancy aids provided free if requested in advance. Basic hire fee also includes all fuel and cancellation insurance. Pets may travel free (UK residents only). TV and hairdryer may be hired for a small extra charge.
Prices (per 4-berth boat, per week)
Low: £290
High: £610 (July–August)
Discounts Children under 2 are free, provided you supply their bedding.

Corsair Cruisers Ltd

Upton Marina, Upton-upon-Severn, Worcs WR8 0PB
Tel: 081 763 1647

UK (Worcestershire, Gloucestershire)

Cabin cruising on the Severn and Avon rivers, around Worcester, Gloucester and the Vale of Evesham. Cruisers are 4–7 berth and equipped to a high standard with cooker, fridge, shower, WC, hot and cold water, gas fire, crockery, utensils, bedding, TV, hair-dryer and shaver-socket. Linen and gas are included in basic hire fee.

Prices Please contact Corsair for current price list.
A £20 Damage Waiver is payable. Pets are charged £20, and there is a car-parking fee of £8.

French Country Cruises

Andrew Brock Travel Ltd, Barley Mow Workspace, 10 Barley Mow Passage, London W4 4PH
Tel: 081 995 3642

France

Pénichettes (5–12 berth French design cruisers) for hire in most regions of France. All boats have electricity, hot and cold running water, shower, WC, fully-equipped kitchen and some form of heating. Bed-linen supplied. Buoyancy aids and cotsides available free on request. Fuel is not included in the basic hire fee.
Prices Available on request.
Discounts None for children, except on holidays inclusive of Channel crossing: approximately £12–£15 off.

Horning Pleasurecraft Ltd

Ferry View Estate, Horning, Norwich, Norfolk NR12 8PT
Tel: 0692 630128

UK (Norfolk)

Self-drive cruisers for hire on the Norfolk Broads. All boats, from 2–10 berths, have electricity, hot and cold running water, shower, WC, fully-equipped kitchen and heating. Bed-linen supplied. Colour TV included. Fuel is not included in the hire fee.
Prices (4-berth cruiser per week)
Standard: £260 (low); £430 (high)
Luxury: £390 (low); £650 (high)

Hoseasons Holidays Ltd

Sunway House, Lowestoft, Suffolk NR32 3LT
Tel: 0502 501010 *(UK)*
Tel: 0502 500555 *(France and Holland)*
France, Holland, UK

Narrowboats, cruisers, sailing boats and stationary houseboats (2–12 berths). All boats have electricity, hot and cold running water, shower, WC, fully-equipped kitchen and some form of heating. Bedding and linen supplied. Buoyancy aids and cotsides available free on request. Fuel is not included in basic hire fee. Pets may travel for an extra cost (£18) (UK residents only). Bicycle and/or dinghy hire may be arranged if requested when booking. All travel arrangements to your destination can also be arranged on request.

Prices (per boat per week in UK; per person per week abroad, including return ferry crossing)
4–6 berth boat in UK: £275 (low), £450–£600 (high)
Luxury cruiser in UK: £600–£650
4–6 berth boat abroad: £100–£250

Simolda Ltd

Basin End, Nantwich, Cheshire
CW5 8LA
Tel: 0270 624075

UK (North, Central)

Narrowboat holidays on the Cheshire/Wales canal system. Choice of 4–6 and 6–8 berth boats, all with shower, hot and cold water, WC, cooker, central heating, bedding, linen, crockery and utensils. Bolt-on cot sides available (no charge). Gas, fuel and car parking are included in the basic hire fee.

Prices (per 4–6 berth boat per week, with 4 people sharing)
Low: £330 approx.
High: £595 approx.
Extra people are charged £18.50 per week.
TV £5.88 per week extra.

Discounts Children under 4 go free; 4–14s pay £9.50 per week. There is no charge for pets. 5% discount is offered for 2 consecutive weeks.

Viking Afloat

Lowesmoor Wharf, Worcester WR1 2RX
Tel: 0905 28667

UK (Shropshire, Worcestershire)

Narrowboats (2–10 berths) for hire from bases in Worcester and Whitchurch. All have electricity, hot and cold running water, shower, WC, fully-equipped kitchen, central heating and radio/cassette player. TV may be hired for a small extra charge. Linen supplied. Fuel is included in the basic hire fee. No extra charge for cotsides, buoyancy aids and pets.

Prices (per boat per week)
4-berth: £369 (low); £670 (high)
6-berth: £462 (low); £840 (high)

Car rental

There is really no question about it – car hire is an expensive business, but it allows such freedom on holiday that the benefits may outweigh any doubts about 'splashing out'.

On top of the basic hire charge, which can be calculated on a daily, weekly or monthly rate, you have to pay local taxes and insurance. Do not be tempted to take out minimum insurance. In Greece and Spain, for example, the minimum does not cover you against claims made by passengers. Most companies recommend taking out additional insurance against personal injury (PAI), and Collision Damage Waiver (CDW) which rules out your having to pay for any damage to the car. Be sure to read the small print carefully. Several companies have rules of hire that may seem peculiar on first reading.

Rates vary a great deal from company to company and country to country. Note too that it is usually cheaper to book your hire car from the UK before departure than on arrival at your destination.

Should you require them, baby seats and roof racks can be fitted in most hire cars if you request them when booking. There is usually a charge for this.

It is unwise to drive a car in a foreign country unless you are conversant with the local rules of the road. The AA and RAC have useful leaflets outlining the rules applicable in various countries. As in the UK, ignorance of the law is no defence, so do find out the basics before you go. Many countries require by law that you carry certain emergency equipment. Check that this is provided by the hire company before departure. (*See also* **Fly/drive holidays**.)

Prices
Unless otherwise stated 1992 prices have been quoted. These are intended only as a guide to the type of holiday offered and travellers should check with the company concerned for 1993 pricing details.

Avis
Trident House, Station Road, Hayes, Middx UB3 4DJ
Tel: 081 848 8733

Worldwide
Avis has 3900 offices in 135 countries.

Worldwide Super Value is the best current Avis deal. European car rental is inclusive of local tax, third-party insurance, CDW and mileage. For US car rental, Super Value rates are available either with or without CDW and include unlimited mileage. Other optional insurances, tax and petrol are payable locally. Prices vary depending on the size of the vehicle and the pick-up point.

Worldwide One-way Rental The car may be picked up in one city or country and dropped off in another, but there may be a fee, depending on the locations involved.

In Touch Services This includes two services which are free to customers renting an Avis car in Europe. *Avis On Call Europe:* Available from 9 European countries, an information service in English on hotels and restaurants; places of interest and special events; general information such as banks, dentists, doctors etc. *Avis Message Centre:* To keep you in touch with what's happening at home, you can leave or retrieve messages at a central contact point for or from anywhere in the world. Ask for details of these services at time of reservation.

Baby seats must be requested when booking. Availability and cost vary from country to country.

Booking for Super Value 2–7 days in advance for a minimum of 3–7 days.

Payment Major credit cards, Avis charge card or Avis travel vouchers.

Eurodollar Rent A Car Ltd
3 Warwick Place, Uxbridge, UB8 1DE
Tel: 0895 233300

Worldwide
Partnership between Swan National in the UK and Dollar Rent-a-Car in the USA.

Eurodrive Leisure rates are available in 14 European countries, but apply only if pre-booked at least 3 days prior to commencement of rental.

One-way rental (picking the car up in one city or country and dropping it off in another) is available free of charge between EuroDollar locations in some countries. Ask when booking.

Baby seats Must be requested at time of booking. availability and cost vary from country to country.

Booking 3 days prior to rental. Minimum rental is 3 days.

Payment With most major credit cards or cash.

Deposit Petrol deposit refundable on return of car.

Hertz (UK) Ltd
Radnor House, 1272 London Road, London SW16 4XW
Tel: 081 679 1799

Worldwide
Hertz has 7500 offices in 130 countries, including 2000 airport locations.

Europe on Wheels is the best current Hertz deal. It covers 35 European countries, including Spain, Greece and Turkey, and rates include CDW/LDW, local taxes and unlimited mileage.

One-way rental In most countries cars can be rented in one town and left in another free of charge. It is usually possible to start rental in one country and end it in another; details should be checked when you book.

Baby seats must be requested at time of booking. Availability and cost vary from country to country, but sometimes they are free.

Booking Usually 7 days in advance with the exception of rentals in selected resorts where the advance booking requirement is 24 hours. Europe on Wheels programme should be asked for at the time of booking.

Payment Details given on request.

Deposit These conditions vary from deal to deal. Except where payment is to be made by Hertz Charge Card or any acceptable charge/credit card, a cash deposit to cover refuelling service charges or any other extra charges will be required.

Holiday Saver If you wish to book and pay 14 days in advance, you can save money with Hertz Holiday Saver rates, which are available to many of the major holiday destinations. Please ring the Reservations Centre and ask for details.

Coach transport and coach tours

Coaches are probably unbeatable as a fast, cheap method of getting from A to B. They cost considerably less than planes and trains and allow the whole family, including the usual driver, to enjoy the scenery and arrive at the destination relatively refreshed. Camping and skiing companies most frequently offer coach travel as a cheap alternative on their holidays, but think hard before embarking with young children on what may be an eighteen-hour journey in cramped conditions.

In researching this section, we found no coach operator offering any special on-coach facilities for children. The most you can hope for are WCs and video films, with some long-distance companies having a drinks and snacks service. Most large operators will have cushions and covers for hire, but it can be a good saving to take your own; a sleeping bag rolls up very small and can easily take two children. Read Sections 4 and 5 on avoiding boredom and think about investing in a personal stereo with connections for two pairs of headphones and some story tapes.

However, many of the companies we list advise against taking very young children on coach trips, and several don't allow them anyway. Coach tours, by their very nature, involve a lot of travel and are not ideal for youngsters with high energy levels and short concentration spans.

If you plan to do a lot of independent travelling by coach, it is worth investigating the special passes and discount tickets that are available. Generally speaking, the longer your stay, the greater the discount. Long-haul operators, such as Greyhound, change their offers from year to year.

Prices
Unless otherwise stated 1992 prices have been quoted. These are intended only as a guide to the type of holiday offered and travellers should check with the company concerned for 1993 pricing details.

Clansman Monarch Holidays

2a Manor Place, Edinburgh EH3 7DD
Tel: 031 226 4220
Fax: 031 226 42481
UK (Scotland)
A range of 14 coach tours lasting 2–7 days, including seasonal theme breaks all year round. The coaches have reclining seats, arm rests, air suspension and forced ventilation. Accommodation is in hotels and the availability of such things as cots and high chairs varies.
Prices (per person, including full board and entrance fees) start from £80.
Discounts Discounts are given at the discretion of individual hotels, and will be quoted on request.

Cosmosair PLC

Tourama House, 17 Homesdale Road, Bromley, Kent BR2 9LX
Tel: 081 464 3444 or 0272 227799 (Bristol), 061 480 5799 (Manchester)
Europe
Sea/coach and air/coach holidays to most European destinations. All departures are

from London or a Channel port. Free coach connection to and from Dover from 90 locations in the UK.
Prices (per adult, half board; 1992 prices)
Short break (4 days): £119 to Paris
Discounts Children under 8 are not allowed on any of these tours. A 10% discount for children aged 8–12 is available if they share a room with 2 full fare-paying passengers.

Evan Evans Tours

26–28 Paradise Road, Richmond, Surrey TW9 1SE
Tel: 081 332 2222
UK
Evan Evans Tours have been in business for 50 years and specialise in a wide range of full and half day tours in and out of London. Their tours visit all the popular tourist destinations e.g. Stratford upon Avon, Stonehenge and Bath, Canterbury and many more. All tours are led by qualified Blue Badge Guides.
Discounts Special prices are offered to children under 17.

Frames Rickards

11 Herbrand Street, London WC1N 1EX
Tel: 071 837 3111 (sightseeing)
071 637 4171 (tours)
UK
Tours to most parts of the UK lasting from half a day to 11 days. Coach tours of more than one day are not recommended for children under 12. The company will advise on suitable tours for children, which are usually day trips in and around London to such places as Oxford, Warwick Castle and Stratford-upon-Avon.
Prices (per adult)
Half-day panoramic tour (all year round): £8 (child), £11 (adult)
4-day 'Quick Look at Britain' tour: £225 (low season) – £250 (high season)
7-day tour of Britain and Ireland: £445 (low season) – £495 (high season)
Please note: 1st April 1993 prices change.
Discounts 20% discount on some longer tours for children aged 5–14.

Special discounts may be available on day trips (details on request).

Greyhound World Travel

Sussex House, London Road, East Grinstead, West Sussex RH19 1LD
Tel: 0342 317317
Canada, US
The principal coach operator in North America. Some excellent deals are available, but must be purchased in the UK rather than at destination. Organized tours are available in conjunction with Brewster, Collette and Tauck Tours. The coaches all have air-conditioning, WCs and reclining seats. No special facilities for babies.
Prices Telephone for the latest rates.
Discounts On bus travel only, under-2s are free when sitting on an adult's lap. Any child taking up a seat (aged 5–11 approx) must pay half the full adult fare. (This does *not* apply to the four-day ticket.)

National Express Ltd

Ensign Court, 4 Vicarage Road, Edgbaston, Birmingham B15 3ES
Tel: 021 622 4373 *or*
071 730 0202 *(London)*
061 228 3881 *(Manchester)*
0329 230023 *(Fareham)*
UK
The largest coach operator in the UK serving hundreds of places daily. For travel information telephone your local enquiry centre; the number will be in the phone book.
Prices There is a huge range, impossible to cover here. Your local enquiry centre will supply details.
Discounts On travel within the UK children under 5 are free provided they do not occupy a separate seat. Children aged 5–15 get about 30% off standard adult fares. People aged 16–23 can buy a discount coach card, which is valid for 12 months and gives the same discounts.

Sonata Travel

227 Umberside Road, Selly Oak,
Birmingham B29 7SG
Tel: 021 472 8636

Belgium, France, Germany, Holland, Italy, Spain, UK

Continental coach tours arranged, as well as weekend city breaks to Paris, Bruges, Amsterdam and many UK cities. Cots and high chairs may be available in some accommodation on request.

Discounts Small children not occupying a seat or separate accommodation go free. Children up to 12 receive 10% discount on the adult fare.

Timescape Holiday Ltd

581 Roman Road, Bow, London E3 5EL
Tel: 081 980 7244

Austria, Belgium, Channel Islands, Italy, Spain

Economy coach holidays, and fly/sail to the Channel Islands. The coaches are executive class and contain video, washroom, toilets, reclining seats, air-jet ventilation and tinted windows. A hostess is on hand, and Timescape pride themselves on their friendly coach crew. Hotel accommodation available in Austria, Belgium, Italy and Spain, and self-catering in Italy and Spain. Suitability for children varies, but can be judged from the comprehensive brochure.

Prices 10-day holidays from £99.

Wallace Arnold

315–317, 4th floor, 109 Hope Street,
Glasgow G2 6LL *and*
Gelderd Road, Leeds LS12 6DH
Tel: 0532 636456 *(Leeds)*or
081 686 2378 *(London)*
041 221 7767 *(Scotland)*

Europe, UK

Long-established coach company offering tours throughout the UK and to most European destinations. Special requirements, such as cots, must be requested when booking. Wallace Arnold also offer air holidays to European destinations and some long-haul packages (Canada, USA and South Africa). All coaches are luxury modern models, are entirely non-smoking and, for continental trips, have an on-board toilet. On all tours, prices include half board (dinner, bed and breakfast) in hotel accommodation.

Prices Prices vary greatly and details should be confirmed direct.

Discounts Details of child discounts are available on request and vary from holiday to holiday.

Cottages, gîtes and farmhouses

The cottage industry, as the length of this section suggests, is a large one. We list companies, large and small, who offer various grades of accommodation mainly in the UK, Ireland, France and Italy. Properties vary enormously – from period cottages to modern bungalows. Our descriptions are necessarily brief, but we hope they will help you to make an initial selection and to find a company that caters for your needs. To save space and needless repetition we state here that all properties have gas and/or electricity supplies, running water, fully-equipped kitchen and bathroom and at least an adequate standard of furnishing.

When choosing your holiday home, do read the small print carefully and compare what is included in the price. In some properties gas and electricity are metered, or a charge may be levied at the end of your stay for the amount consumed. Note too that linen is not usually supplied, although it may be available for hire. Where cots, high chairs and any other children's facilities are provided we say so and quote the fee where appropriate.

If you choose to holiday in France, chances are that your cottage will be called a gîte. This is usually a building that has been renovated with the help of a government grant. Gîtes are usually found in the more rural parts of France and have to meet standards specified by the government. Do not expect luxury, however. They offer simple, clean accommodation but do not necessarily have such things as armchairs. Several companies, Brittany Ferries being one of them, offer gîtes which are not part of this government scheme. The name has simply come to mean a cottage with a basic level of comfort.

If you wish to rent a gîte in a particular part of France, you can obtain lists from the *département* in question. Addresses are available from the French Tourist Office. Note that the best properties are snapped up by Christmas, so book early to avoid disappointment.

Farmhouse holidays tend to fall into two categories: either you rent the property as you would a gîte or cottage, or you can take a chambre d'hôte holiday, which means that you have bed and breakfast in the farmhouse and get to know the family. In some places an evening meal may also be provided. We have had enthusiastic reports from families who have taken chambre d'hôte holidays; they consider them the ideal family holiday as real friendships can form and the French tend to make a great fuss of the children, perhaps leaving you free to do some exploring on your own.

Before you book a property you should read Kathy Rooney's article in *Section 4* that gives advice on the kinds of questions to ask a holiday operator about their properties.

Prices

Unless otherwise stated 1992 prices have been quoted. These are intended only as a guide to the type of holiday offered and travellers should check with the company concerned for 1993 pricing details.

Allez France

27 West Street, Storrington, West Sussex
RH20 4DZ
Tel: 0903 745793

France

This company says it is experienced in dealing with family groups. It offers a selection of properties ranging from tiny country cottages to large seaside villas in most areas of France. Service charges vary from property to property. These include water, gas, electricity, agent's fees and taxes. Cots are available in many places, either free or for a small charge. High chairs are not usual in France. Child-minding can sometimes be arranged locally, depending where you stay. Also a selection of family-run hotels of character, many ideal for family groups.

Prices (per week for 4–9 people)
Low: from £305
High: up to £2640
Discounts Infants free (or nominal charge).

Aultbea Highland Lodges

Torliath, Drumchork, Aultbea,
Achnasheen, Ross-shire IV22 2HU
Tel: 0445 731233/731268

UK (Scotland)

Purpose-built holiday homes on a 23 acre site overlooking the fishing hamlet of Aultbea on the north-west coast of Scotland. Cots and high chairs may be hired for £8 per week and childminding can be arranged locally. STB class and grading – 4 crown facility – highly commended. No pets are allowed.

Prices In 1992 prices ranged from £196 for a couple in low season to £429 for four people in high season. Lodges sleeping six people cost £489 approx. £30 is payable for each extra guest. Apply to company for specific prices.

Discounts Children under 2 are free.

Bath Holiday Homes

3 Frankley Buildings, Bath BA1 6EG
Tel: 0225 332221

UK (Bath)

Historic houses, cottages and flats for rent in and around the city. Some properties specify minimum ages for children, and some do not allow children at all. A few offer free use of cots and high chairs, but at other properties such equipment can be hired by arrangement. Detailed brochure available.

Prices (per property per week)
Low: from £50 to £400
High: from £300 to £600
Fully inclusive with bedlinen, tea towels and towels provided.

B+I Line UK Ltd

East Princes Dock, Liverpool L3 0AA
Tel: 051 236 8325

Southern Ireland

Self-catering holidays with a choice of accommodation in cottages, caravans, river cruisers or holiday villages. 'Village' amenities include shops, launderette, children's playroom, enclosed playground, TV room, showers and sports. Hire of touring caravans can be arranged on request. Cots and high chairs are usually available in all locations.

Prices Available on application.
Discounts Children travel free with all self-catering holidays.

Bell-Ingram Self-Catering Holidays

Durn, Isla Road, Perth PH2 7HF
Tel: 0738 21121

UK (Scotland)

Selection of cottages and country houses ranging from rural simplicity (sleeping 2) to various lodges on their own estates (sleeping up to 15). Some accommodation is classified by 'crowns'; all are illustrated. Cots and high chairs are free, where available.

Prices (per property per week)
Low: from £100
High: £110–£600

Blakes Country Cottages

Tel: 0603 783221 (UK)
(see also **Canal holidays**)

UK

Wide selection of cottages, farmhouses, bungalows and apartments. Some are individual properties of great character or historical interest, while others may be purpose-built holiday villages. Around 80% of the cottages have cots, and some can provide high chairs and baby-sitting. Many cottages have farm animals nearby, which might interest families with young children. Some groups of cottages have facilities, such as swimming pools and jacuzzis. All year round availability of many cottages. Short breaks available from November to March for a minimum of 3 or 4 nights on many cottages.
Prices (average prices for 4 people)
Low: £168
High: £298

Bowhill's

Mayhill Farm, Swanmore, Southampton SO3 2RD
Tel: 0489 877627/878612 or 0329 833093

France

A family-run company established 20 years ago, offering more than 300 villas and farmhouses all over France, many with pools. Chosen for their individual character, situation and amenities, many properties offer additional services such as cleaning, babysitting, etc.
Prices (per 4 people per fortnight)
Average cottage: £629 (low), £951 (high). Villa in East Provence: £1001 (low), £2047 (high).
Discounts Under-4s are free. The prices quoted are approximate. Contact Bowhills for further details and a colour brochure.

Brittany Direct Holidays

362–364 Sutton Common Road, Sutton, Surrey SM3 9PL
Tel: 081 641 6060

France (Brittany)

Illustrated brochure of forty gîtes, cottages and houses. The company also offer holidays in hotels and chambre d'hôte. Colour interior/exterior photos supplied when booking. Cots, where available, carry a small charge. Most places offer baby-sitting which is arranged locally. All offer a starter pack of groceries. In addition to the packages described above, Brittany Direct also organize fly-drive and golfing holidays. On the latter, junior participants are eligible for discounted green fees on *some* golf courses (details available on request).
Prices Available on request.
Discounts Children receive standard ferry discounts.

Brittany Ferries Gîtes Holidays

Information Bureau, 1 Battersea Church Road, London SW11 3LY
Tel: 071 836 5885

France

Nearly 1500 holiday homes comprising gîtes, villas and apartments in west, south-west and the south of France, Normandy and Loire Valley, Burgundy, Franche Comte, Limousin, the Auvergne, Rhone Valley and the Alps. The properties represent the finest selection of France's rural and seaside accommodation and are graded Q–F (very comfortable to basic) by Brittany Ferries. Cots and/or children's beds are available in many places.
Prices (per adult per week, including return ferry crossing with car)
Grade F gîte: £68–£165
Grade Q gîte: £109–£298
Discounts Under-4s are free all year. Children 4–13 pay £15 a week in low season and £24 in high season

Cant Farm

Cant Farm, St Minver, Wadebridge, Cornwall PL27 6RL
Tel: (0932) 247617

UK (Cornwall)

6 luxury cottages overlooking the Camel Estuary in North Cornwall. Three cottages have been developed from original farmhouse and outbuildings and the others built in matching local stone. The farmhouses sleep from five to eight with log

fires, jacuzzi baths, luxury fitted kitchens and utility rooms and some have saunas and four-poster beds. Cant Farm has its own tennis court and 70 acres of farmland and the weekly cost includes gas, electricity, linen and use of tennis court.
Prices Mini breaks available from £116 (low).

Casas Cantabricas

31 Arbury Road, Cambridge CB4 2JB
Tel: 0223 328721

Spain

Family-run company offering a selection of privately-owned cottages in Cantabria, Galicia, and Asturias, northern Spain. Most properties are in villages or rural situations, and most are within easy reach of the sea. The availability of cots varies. Details on request. Baby-sitting may be arranged locally.
Prices (per property per week)
Low: £135–£455
Mid: £155–£595
High: £225–£850

Cerbid's Quality Cottages

Cerbid, Solva, Haverfordwest, Pembrokeshire SA62 6YE
Tel: 0348 837874

France, UK (Wales)

Family company offering a wide variety of privately-owned cottages on the Welsh coast, and the Côte d'Azur in France. Cots available on request. Baby-sitting can be arranged in some locations. Linen and heating is usually inclusive. Pets welcome.
Prices (per property per week)
Wales: £99–£780
France: from £390 (accommodation only)
Discounts In summer small families receive a £10–£40 discount, depending on property chosen.

Character Cottages Ltd

34 Fore Street, Sidmouth, Devon
EX10 8AG
Tel: 03955 77001

UK

Large selection of cottages throughout England, and a smaller selection in Wales, Scotland and Ireland. Cots and high chairs are supplied in some locations. Baby-sitting can sometimes be arranged locally for a reasonable charge.
Prices (per week)
Low: £79–£300
Mid: £130–£500
High: £151–£850

Coast and Country Holidays

15 Town Green, Wymondham, Norfolk NR18 0PN
Tel: 0953 604480

UK (Norfolk, Suffolk)

Cottages and houses of all periods for rent. Around 90% of properties have cots and the vast majority supply them free of charge. Some also have high chairs.
Prices (per family of 4 per week)
Low: from £110
High: from £249
(per group of 9 per week)
Low: £220
High: £384

Coastal Cottages of Pembrokeshire

Abercastle, nr Haverfordwest, Pembrokeshire SA62 5HJ
Tel: 0348 837742

UK (Wales)

Large selection of cottages in Pembrokeshire ranging from modern bungalows to old farmhouses, all in coastal locations. The properties are mainly upmarket, and can accommodate from 2–12 people. All cottages have free cots and high chairs. Babysitting can be arranged in less remote places. Pets are welcome at most properties.
Prices
Short breaks: from £100 per property per week
Low: £100–£190
High: £200–£780

Cornish Traditional Cottages Ltd

Lostwithiel, Cornwall PL22 0HT
Tel: 0208 872559

UK (Cornwall)

Extensive selection of properties ranging from traditional cottages to converted boat-sheds and chapels. Cots are available free of charge in most properties, high chairs in some; check details before booking. Baby-sitting can usually be arranged locally through the property owner or housekeeper.
Prices (per family of 4 per week)
Low: £125–£550
High: £250–£800

A Cottage in the Country

Forest Gate, Frog Lane, Milton under Wychwood, Oxford OX7 6JZ
Tel: 0993 831495
Fax: 0993 831095

UK (Oxfordshire, Cotswolds, Thames Valley and Chilterns)

A varied selection of properties in town and country locations. Tourist Board inspected. Many with an inclusive price, i.e. linen, heating and electricity. Some offering tennis court or swimming pool.
Prices (per property per week)
Low: £70–£250
Mid: £80–£450
High: £90–£650

Countryside Cottages

Vale House, Berkhamstead, Herts
Tel: 0442 870400

Malta, UK

A selection of traditional and modern cottages in the Cotswolds and throughout southern England except Cornwall, plus a few apartments in Malta. Brochures are available and all properties are illustrated with floor plans. Cots and high chairs are provided free where available, and baby-sitting can be arranged at most cottages.
Prices (per property per week)
Low: £97–£229
Mid: £124–£418
High: £162–£483

Dales Holiday Cottages

Otley Street, Skipton, N. Yorks
BD23 1DY
Tel: 0756 799821/790919

UK (Yorkshire and Northumberland)

Mostly traditional, locally-owned cottages, sleeping from 2–11 people. Cots and high chairs are free where available, but linen hire may be extra.
Prices (per property per week: approx)
Low: £135–£210
Mid: £195–£250
High: £220–£350

Discover Britain Holidays

Shaw Mews, Shaw Street, Worcester
WR1 3QQ
Tel: 0905 613746

UK

Personally inspected farms and cottages in all parts of England, Scotland and Wales. Cots, high chairs and folding beds are free where available. Special facilities like helping on the farm, swimming and riding are by arrangement with the owner, as is baby-sitting. Properties are graded from 'homely' to 'luxurious' family accommodation. Both bed and breakfast and self-catering accommodation is available.
Prices (per self-catering property per week)
Low: £120–£300
Mid: £160–£400
High: £200–£500
Bed and breakfast (per night)
Adults: £18.50
Children: £10.00

Dunaird Cabins

St Mary's Road, Birnam, Dunkeld, Perthshire
Tel: 0350 727262

UK (Scotland)

Ten octagonal pine cabins in a two-acre woodland setting. Each cabin sleeps 4–6 people. Cots, high chairs and safety rails for bunk beds are supplied free. Baby-sitting can sometimes be arranged with the owners. A playpark is available nearby.

Prices (per week per 4 people)
Low: From £120
Mid: From £190
High: From £270

English Country Cottages Ltd

Grove Farm Barns, Fakenham, Norfolk
NR21 9NB
Tel: 0328 864041 *(English cottages)*
0328 851341 *(Welsh cottages)*
0328 864011 *(Scottish cottages)*

UK

Extensive selection of over 2500 carefully inspected properties, ranging from cottages and farmhouses to country houses and mansions, sleeping 2–20 people. The company offers countryside, waterside and seaside locations. Some properties in the grounds of country hotels offer hotel-style facilities and a selection of sports. Properties are categorized A–Z, from small to large, with all grades being well represented. Cots are available (free) in most cottages. Many offer baby-sitting by arrangement, high chairs, children's games rooms and equipped play areas. A few properties do not allow children. Short breaks (3 nights) available September to May.
Prices (per property per week)
Off-season: £138–£556
Peak season: £254–£1250

Farm and Cottage Holidays

5 Fore Street, Bideford, Devon
EX39 3PW
Tel: 0237 479146

UK (Devon, Somerset, Cornwall)

Self-catering or half board family holidays. Properties are graded from simply to plushly furnished. Nearly all properties have cots, some offer microwaves, washing machines, videos, etc. Some have linen for hire and many offer baby-sitting by arrangement.
Prices (self-catering property per week)
Low: £84–£275
Mid: £105–£534
High: £115–£649

(half-board accommodation per adult per week)
Low: £93–£168
Mid: £104–£178
High: £115–£200
En suite facilities are available for £7 extra charge per person.
Discounts (available on half board only)
Under-5s pay 50% of adult rate, under 12's pay 70% of adult rate. 12's and over pay adult rate. Children under 2 pay £10 for cot.

Felindre

51 Church Street, Stoke-on-Trent, Staffs
ST4 1DQ
Tel: 0782 744865

UK (Wales)

Nine very comfortable cedarwood bungalows and three cottages set in nine acres of Pembrokeshire National Park near St Davids. The following children's facilities are all supplied free: cots, high chairs, pushchairs, baby baths, potties, nappy buckets and toilet trainer seats.
Prices on request.

Fermanagh Lakeland

Lakeland Visitor Centre, Eniskillen, Co. Fermanagh
Tel: 0365 323110/325050

UK (Northern Ireland)

The company specializes in fishing, boating and cruising holidays. Wide choice of lakeside hotels, guest houses and self catering chalets, many of which offer special facilities for children. Activity Centres for groups and individuals offering instruction in canoeing, caving, windsurfing, pony trekking, etc.
Packages special package holidays can be arranged for those travelling by ferry.
Prices Available on request.
Discounts Some available for OAPs and children and vary according to establishment.

Finnchalet Holidays Ltd

Dunira, Comrie, Perthshire PH6 2JZ
Tel: 0764 70020

Denmark, Finland, Germany, Norway, Sweden

Chalet, farm and hotel holidays. Farm holidays include full board and the brochure lists animals kept on the farm. Most properties have family bedrooms and cots are available on request. The owners mostly speak English. Baby-sitting is generally available. Free cots on ferry crossing.

Prices (including full board, ferry crossing plus berth and twice-weekly sauna)
Finland: £418
Denmark: £275

Discounts Babies go free. Under-11s receive a discount of £11 per day. Under-16s receive a discount of £70 for Finland and £60 for Denmark.

Forest Holidays

Forestry Commission, 231 Corstorphine Road, Edinburgh EH12 7AT
Tel: 031 334 2576/0303

UK (Cornwall, Yorkshire, Scotland)

Selection of cabins and cottages located in open countryside, in woods or near water. Bed-linen included. Pets welcome. Short breaks also available. Send for free colour brochure.

Prices from £75 (3 nights) to £125 (7 nights)

Freedom Holidays Ltd

40 New Street, St Helier, Jersey, Channel Islands
Tel: 0534 25259

UK (Jersey and Guernsey)

Selection of cottages and flats in country and suburban locations. Where available, cots are provided free. Elsewhere the hire of cots, high chairs and all other baby equipment can be arranged. Baby-sitting may be offered in some properties.

Prices Studio apt. (couple and child per week)
Starts at £115

French Affair

34 Lillie Road, Fulham, London
SW6 1TN
Tel: 071 381 8519

France

Hand-picked cottages and houses with detailed descriptions of accommodation. Some properties offer free cots, and baby-sitting is available by arrangement.

Prices Vary widely, depending on property.

Discounts Children receive standard travel discounts (see **Ferries**).

French Country Cottages

Anglia House, Marina, Lowestoft, Suffolk NR32 1PZ
Tel: 0502 517271

France

Self-catering holidays in cottages, gîtes and villas. A selection of family-run hotels and motels are available for stops en route. Cots are provided in some locations.

Prices These are calculated on the number of adults in the party. Details available on request.

Discounts On self-catering and farmhouse holidays under-4s are free. Children aged 4–13 each receive a £20 discount and travel free outside July and August depending on ferry company and date.

Gordon Holiday Cottages

118 Kidmore End Road, Emmer Green, Reading, Berks RG4 8SL
Tel: 0734 472524

France, UK

Small selection of self-catering properties in Scotland, Brittany and some parts of England, ranging from cottages to large farmhouses sleeping 8. Cots may be hired. Some properties have facilities, such as mini-golf, for children, and some have been adapted for the disabled.

Prices (per property per week)
Low: £150–£200
Medium: £180–£300
High: £230–£350

Haywood-Amaro Holidays

Lansdowne Place, 17 Holdenhurst Road,
Bournemouth, BH8 8EH
Tel: 0202 555545 *or* 0202 295006
(24-hour brochures)

UK

Wide selection of coastal and country properties (ranging from converted stables to country mansions) in many picturesque areas of England, Scotland and Wales. All are categorized according to the degree of comfort. Cots are supplied free in many locations, with high chairs and baby-sitting available at some cottages.
Prices (per property per week)
Low: from £81 to £330
High: from £160 to £870
Discounts A 20% lower deposit is necessary for bookings made before the end of January.

Heatherwood Park

Dornoch, Sutherland, Scotland
Tel: 0862 810596

UK (Scotland)

Luxurious Norwegian bungalows set in private Scottish parkland close to local village shops and amenities. Cots are provided on request. Golf and tennis are available nearby.
Prices (per week per 5 people)
From £150 to £250

Holiday Cottages

Water Street, Skipton, North Yorkshire
BD23 1PB
Tel: 0756 700510

UK (Cumbria, Derbyshire, Yorkshire)

Cottages and houses ranging from converted barns to country farmhouses in the Yorkshire Dales, Moors, Peak and Lake District. Cots and high chairs are available at most properties free of charge.
Prices (per property per week)
Cottages: £112–£210 (low); £170–£460 (high).

Holiday Houses Dumfries and Galloway

G.M. Thomson & Co, 27 King Street,
Castle Douglas DG7 1AB
Tel: 0556 2701/2973

UK (Scotland)

Collection of personally inspected houses, sleeping 2–16 people. Many properties are classified with 1–5 crowns (basic to outstanding) by the Scottish Tourist Board. In addition some are graded approved – highly commended. Cots are free where available and there may be some baby-sitting by arrangement.
Prices (per property per week)
Low: £100–£720
Mid: £130–£900
High: £165–£1200

Holidays in Lakeland

Stock Park Mansion, Newby Bridge,
Ulverston, Cumbria LA12 8AY
Tel: 05395 31549

UK (Cumbria)

A selection of 200 houses, flats, cottages and caravans located throughout Cumbria. Some properties are AA-listed and registered with the English Tourist Board. The brochure gives detailed instructions on how to reach each property. Some places do not allow children. In those that do, cots and high chairs may be available free, or may be hired at some properties for £6–£9 per week.
Prices (per property per week)
Low: £80–£370
High: £135–£630

Holiday Scandinavia Ltd

28 Hillcrest Road, Orpington, Kent
BR6 9AW
Tel: 0689 824958
Fax: 0689 835807

Denmark, Finland, Norway, Sweden

This company specialises in tailormade holidays and offers a selection of purpose-built chalets in holiday villages as well as private holiday homes and hotel package holidays in all areas of Sweden, Denmark, Finland and Norway. They also offer farmhouse holidays in Denmark. Hotel cheques, ideal for touring holidays, can be purchased in advance for moderate to first-class accommodation. The chalets are comfortably furnished and all have a fully-equipped kitchen. All 'villages' offer a variety of sports facilities and children's play areas or swimming pool. This company also offers activity holidays (*see* **Special Interest Holidays**).

Prices (per chalet per week for 4 people, including ferry and transportation of car)
Winter: from £163 per person
Summer: from £235 per person

Discounts Under-4s travel free. Children aged 4–15 receive a reduction of up to £86 in summer, £25 in winter.

Home From Home

2a Queens Road, Mumbles, Swansea
SA3 4AW
Tel: 0792 360624/368078

UK (South Wales)

Large selection of cottages, houses and flats in Swansea Bay, Mumbles and Gower. The properties are near safe, clean beaches. Cots included in properties or can be hired with all other baby equipment.

Prices (per property per week)
Low: £75–£200
High: £150–£400

Hoseasons Holidays Ltd

(*see also* **Canal holidays**)
Tel: 0502 500500
0502 500555 (*Continent*)

Belgium, France, Germany, Holland, Italy, Ireland, Spain, UK

Lodges, chalets, cottages and caravans throughout the UK, with gîtes and cottages available in mainland Europe. Many properties are sited in purpose-built self-catering holiday villages which have ample shopping, leisure and sports facilities. There are often special entertainment programmes for adults and children. At some holiday parks you have the option of being self-catering or taking half board. Cots and high chairs are generally available for around £7 each per week. Baby patrols or baby-sitting can also be arranged on request in many locations.

Prices (per property per week in UK; per person per week abroad, including ferry crossing)
UK cottages: from £75 (low); £300–£600 (high).
Continental cottages: from £90 (low).

Discounts UK: special two-week offers, senior citizens' and children's discounts, plus Easter and Spring bank holiday offers. Under-4s may travel free on Continental holidays, but are charged for accommodation if they occupy a bed. Babies in their own cots are free.

Just France

1 Belmont, Lansdown Road, Bath
BA1 5DZ
Tel: 0225 446328

France

Large brochure offering self-catering holidays throughout France in gîtes or cottages. Some have cots available free, some hire them for £5–£10 per week, paid locally.

Prices (per property per week, including ferry crossing for 2 people)
Low: from £235
Mid: from £249
High: from £279

Discounts Children receive standard travel discounts (*see* **Ferries**). If you book a ferry crossing as well as one week's accommodation through this company (before 11 July or after 5 Sept), parties of 8 or more adults or children over 4 receive free ferry travel.

Mackay's Agency

30 Frederick Street, Edinburgh EH2 2JR
Tel: 031 225 3539 *or*
031 226 4364 (brochures, 24 hours)

UK

A wide selection of self-catering cottages throughout Scotland and North England. Many have enclosed gardens and can provide cots and high chairs. Babysitting can be arranged locally at some properties, although not in very remote areas.

Prices
(average per week, per family of 4)
Low: £100–£250
High: £180–£600

Mann's Holidays
20a Gaol Street, Pwllheli, Gwynedd
LL53 5DB
Tel: 0758 613666
UK (North Wales)

Detailed brochure of 500 cottages, farmhouses and caravans. All properties are personally inspected by the company to ensure that they meet the standards described in the brochure. Where available, cots are free. Children especially welcome at all properties.

Prices (per property per week)
A caravan or small cottage several miles from the sea ranges from £60–£140. A beautifully-furnished house overlooking the sea ranges from £120–£390.

Milkbere Holidays
Milkbere House, 14 Fore Street, Seaton, Devon EX12 2LA
15 Tel: 0297 22925
UK (Devon, Dorset)

A selection of 80 cottages, flats and caravans ranging from the simple to the luxurious. All are equipped to English Tourist Board standards. Cots and high chairs are free in some properties, but may be hired in others. Baby-sitting can be arranged in some locations.

Prices (per property per week)
Low: from £95 for a small property
High: £350 for a large property

National Trust for Scotland
5 Charlotte Square, Edinburgh EH2 4DU
Tel: 031 226 5922
UK (Scotland)

Selection of 34 holiday flats and cottages. Each property contains a sheet giving useful information on the locality plus leaflets advising what to see. Cots free where available, but if you are taking small children you are asked to provide rubber undersheets for the beds. Properties sleep up to 10 people. 'Guide To Properties' brochure available on request from the Publications dept. (price £1.35 inc. P&P). Membership details also available.

Prices From £130 to £480

David Newman's European Collection
Box 733, 40 Upperton Road, Eastbourne
BN21 4AW
Tel: 0323 410347
France

Gîte and apartment holidays in many parts of France. Cots may be available in some locations on request. One-week holidays are available in low season.

Prices (per property per fortnight, including ferry for car with 2 adults and 2 children)
Low: £450–£590
High: £750–£850

One-week holidays are available in low season and range from £420–£450 per property. A third child is charged £17 for the ferry.

North Norfolk Holiday Homes
Lee Warner Avenue, Fakenham, Norfolk
NR21 8ER
Tel: 0328 855322
UK (Norfolk)

Personally inspected properties ranging from fishermen's cottages to converted stables, barns and even a hall. The brochure gives helpful descriptions of properties, including information about games provided, lawns to play on and other leisure facilities. Children are obviously welcome at a great number of the properties. Cots and high chairs are

Cottages, gîtes and farmhouses

free where available, but cot linen is not provided. Many houses offer babysitting by arrangement.
Prices (per property per week)
Low: £100–£300
Mid: £150–£400
High: £200–£510

North Wales Holiday Cottages and Farmhouses
Station Road, Deganwy, Conwy, Gwynedd LL31 9DF
Tel: 0492 582492
UK (North Wales)

Large selection of cottages and farmhouses, ranging from converted chapels to a gamekeeper's lodge. Most are privately owned and offer free use of cots. Some are owned by the National Trust and a travelling cot is supplied in most properties. High chairs and babysitting can be arranged in some locations.

Owners Holiday Letting Consortium
Moore House, Moore Road, Bourton-on-the-Water, Cheltenham, Glos GL54 2AZ
Tel: 0451 20927
UK (Cotswolds)

Small selection of characterful cottages and houses. Most properties have washing machines and can organize cots, and baby-sitting. All have colour TV, and some allow dogs.
Prices (per family per week)
Low: £110–£223
Mid: £159–£249
High: £219–£306

Peak and Moorland Farm Holidays
Self-catering: Miss J. Salt, Stonesteads Farm, Waterhouses, Stoke-on-Trent, Staffs ST10 3HN
Tel: 0538 308331
Bed & breakfast: Mrs J. Lomas, Lydgate Farm, Aldwark, Grange Mill, Wirksworth, Derbyshire DE4 4HW
Tel: 062 985250
UK (Peak District, Staffordshire)

A group of farmers and their wives offering bed and breakfast or self catering accommodation in their farmhouses. All those involved are registered with the English Tourist Board. Guests are welcome to get involved with the farm work where practicable. Where available, cots are free.
Prices
B+B per person per night: £11–£25
Self-catering property per week: £90–£300
Discounts On B+B only several farms offer varying discounts for children. For example, one offers free accommodation for under-3s and half price for 3–16s sharing parents' room.

Powell's Cottage Holidays
Dolphin House, High Street, Saundersfoot, Pembrokeshire
Tel: 0834 812791
UK (Wales, South West England)

Wide range of personally inspected cottages, farms, bungalows, flats and houses. Cots are free and available in nearly all the properties, but you must supply your own cot linen.
Prices (per property per week)
From £90 to £850

Les Proprietaires de l'Ouest
34 Middle Street, Portsmouth, Hants
PO5 4BP
Tel: 0705 755715
France

Small, helpful company offering a selection of personally inspected gîtes and apartments in Brittany, the Dordogne, the Pyrenees, Vendée, Normandy and Provence. Descriptions are usefully detailed. Cots are available in many places.
Prices (per person per week, including long Channel crossing; e.g. in party of 4)
Low: from £66
High: up to £249
Discounts Children under 4 are free; 4–13s pay £14–£20 each, depending on season and length of stay.

Recommended Cottage Holidays

8–9 Birdgate, Pickering, North Yorkshire
YO18 7AL
Tel: 0751 75555

UK

Personally inspected cottages all over Britain. Cots and high chairs are available in most properties, but may be supplied by arrangement for an extra charge, if not. Pets welcome at most cottages.

Prices (per property per week)
From: £70

Rendezvous France

1 St Albans Road, Hemel Hempstead
HP2 4XR
Tel: 0442 61666

France

Choice of hotels, chateaux, chambres d'hôtes, gîtes, cottages and villas throughout France. Very few provide specific children's accommodation, but children's prices are available at some hotels. A small charge is made locally for cots. On self-catering holidays, bunks or children's beds are provided free.

Prices (per property per week for 2 people, including ferry crossing)
Low: £290–£335
Mid: £340–£430
High: £375–£455

Discounts Children receive standard travel discounts for sea travel.

S.F.V. Holidays

18–24 Middle Way, Summerstown,
Oxford OX2 7LG
Tel: 0865 311331

France, Italy, Spain

Personally chosen selection of over 400 gîtes and villas with two to ten bedrooms mainly in France. The properties are chosen with children in mind. The brochure has exterior photos of each property and a comprehensive description of facilities, including gardens suitable for children. Most properties overlook, or are within easy walking distance of, the sea. One property offers free use of bikes. Cots are available in all properties if requested when booking at a cost of £15 per week. The company holds a list of baby-sitters and nannies and can make arrangements for you in those areas where it is available. There is also a children's club.

Prices Available on application.
Discounts Children receive standard sea-travel discounts.

Shamrock Cottages

50 High Street, Wellington, Somerset
TA21 8RD
Tel: 0823 660126 *or* 062 76104
(Tipperary)

Ireland

Selection of country cottages throughout Ireland, some registered with the Irish Tourist Board. Cots and baby-sitting are available in some locations by arrangement with the owner.

Prices (examples per property per week, including return ferry crossing for a car and 5 adults)
Simple cottage: Low – £200, Mid – £250–£280, High – £400
Luxurious house: Low – £250–330, Mid – £300–£400, High – £600–£800
Extra adults pay £12–£18 each, depending on the crossing.
Discounts All children are free.

Shaw's Holidays

Y Maes, Pwllheli, Gwynedd LL53 5HA
Tel: 0758 612854/614422

UK (North Wales)

Mainly unillustrated brochure of approximately 750 inspected cottages, flats, houses and caravans. Cots, high chairs and baby-sitting are available in some locations by arrangement. Where permitted, a child in a cot does not count towards the maximum occupancy of the property. Some properties have children's playgrounds.

Prices (per week per family of 4)
Low: from £80
High: from £300

Southern Voyages

Weymouth Quay, Weymouth, Dorset
DT4 8DY
Tel: 0305 777444

France, Spain

Holidays available in gîtes (government approved), hotels, villas, caravans and on camp sites in several parts of France. Villas and apartments available in Northern Spain. The brochure gives detailed descriptions of sleeping arrangements; many properties have children's beds but not cots.
Prices (per adult per week, including ferry crossing)
Low: £104–£126
Mid: £106–£173
High: £137–£235
Discounts Under-14s go free; over-14s pay adult prices.

Stratton Creber

32 Causewayhead, Penzance, Cornwall
TR18 2SP
Tel: 0736 60070

UK (Cornwall)

Choice of a wide selection of properties, traditional and modern. Baby equipment is available for hire on a daily or weekly basis and includes baths, bouncers, buggies, cots, high chairs, papooses, slings, prams and pushchairs, ranging from £2–£10 per week.
Prices (per property per week)
Low: from £120
High: up to £400

Summer Cottages Ltd

1 West Walks, Dorchester DT1 1RE
Tel: 0305 266877

UK

Comprehensive brochure with 770 personally inspected properties catering to all tastes. The brochure shows photos of each property and gives useful information, such as the location of the nearest swimming pool or stables. Cots and high chairs are free where available. Two houses also offer pushchairs. Some places offer baby-sitting by arrangement.
Prices (per property per week)
Low: £56–£425
High: £129–£580

Sun Esprit

Oaklands, Reading Road North, Fleet, Hants GU13 8AA
Tel: 0252 816004
Fax: 0252 811243

France

Sun Esprit devote their holidays entirely to the family market. They offer hotel and self-catering options in France, based in Aquitaine (south west coast) and the Alps. Their resort staff include qualified British nannies, who provide babysitting on 6 nights each week. Freedom for parents and excitement for the youngsters are the hallmarks of a holiday with Sun Esprit. No surcharges. Sells direct to the public and through ABTA agents.
Prices and discounts Available on request.

Sunvista Holidays Ltd

5a George Street, Warminster, Wilts
BA12 8QA
Tel: 0985 217373

France

Wide variety of individually selected villas in many areas of France. Also hotel touring holidays.
Prices (based on a house for four, including short ferry crossings and insurance)
Low: £400–£1600
High: £650–£3000

Toad Hall Cottages

Union Road, Kingsbridge, Devon
TQ7 1EF
Tel: 0548 853089

UK (Devon)

Cottages in waterside or rural settings sleeping from 2 to 14 people. Cots, high chairs and linen are available for hire. Baby-sitting can be arranged at some cottages. Many cottages accept pets.
Prices (per property per week)
Low: £80–£230
Mid: £90–£350
High: £175–£1000

Vacances en Campagne/Vacanze in Italia

Bignor, Pulborough, West Sussex
RH20 1QD
Tel: 07987 433 *(France, Corsica and Spain)*
07987 426 *(Italy)*

Corsica, France, Italy, Spain

Large selection of 'up-market' country cottages, houses and farmhouses in many regions of France and Italy, including Corsica. The company stresses that in mainland France none is close to the coast, so these properties are strictly for country-lovers. All are fully equipped and well furnished. Linen is included in Italy but is usually extra in France; heating is always extra. Where available, cots are provided free. Baby-sitting is sometimes available locally for a small charge, and a maid service is offered in some locations. Travel arrangements at advantageous prices can be made on request.
Prices (examples per property per week, including short Channel crossing for 2 adults and 2 children with a car)
Basic farmhouse in Normandy: £290–£470
Mid-range farmhouse in Normandy: £384–£642
A large château in the Auvergne for 4 adults and 4 children ranges from £542–£851.
Luxurious farmhouse in Provence, with pool, sleeping 15: £2300.

Farmhouse in Tuscany sleeping 4 from £226–£478
Discounts Children under 4 travel free on ferry; reduced prices for children under 14.

VFB Holidays Ltd

Normandy House, High Street, Cheltenham, Glos GL50 3HW
Tel: 0242 526338

France

A wide selection of regularly inspected gîtes, country cottages and houses throughout France. Although the brochure is unillustrated, the company has a two-stage booking process which means that bookings are made on the basis of having seen 4–6 colour photographs of the property, so that you know exactly what you are getting. Cots are available in a few places – children's beds are more common. At some properties baby-sitting can be arranged.
Prices (per person per fortnight, including ferry crossing with car)
4-person gîte: £99–£110 (low); £132–£145 (high).
4-person villa on Côte d'Azur: £170 (low); £260 (high).
Discounts No charge for children under 4 for ferry-inclusive holidays. In some seasons, children under 14 travel free.

Welsh Holidays

Snowdonia Tourist Services, High Street, Porthmadog, Gwynedd LL49 9PG
Tel: 0766 513837/513829

UK (Snowdonia)

Selection of cottages in sea and mountain locations in and around Snowdonia National Park. Cots and baby-sitting are frequently available.
Prices (mid-range cottage, per family of 4 per week)
Low: £105–£250
High: £195–£500

Windermere Lake District Holidays

The Chalet, Bank Terrace, Bowness on Windermere, Cumbria
Tel: 05394 43627

UK (Lake District)

Illustrated brochure of cottages and apartments around Lake Windermere. Cots and high chairs can be hired (no price specified).
Prices The average price for a cottage for 4 during high season is £240 per week. From March–July cottages start from £100.

Cruises

Cruises have the image of being leisurely holidays for the privileged and elderly, but this is by no means universally so. Needless to say, voyages of long distance and duration can be very expensive. However, many companies offer short breaks and special offers.

Cruise ships are frequently lively and action-packed, attracting adults and children of all ages. There are many theme-based cruises, with daily lectures and film shows preparing you for shore trips to sites of special interest. These may seem highly attractive to an adult enthusiast, but check the provision made for children's entertainment before booking tickets for the whole family. Many ships have special children's facilities – clubs, playrooms, qualified attendants and babysitters. You will usually find medical staff aboard too. The sports and leisure amenities are frequently impressive, and the price, of course, includes free access to everything, including evening entertainments. Compare this to a similar land-based holiday, and it all adds up to very good value.

However, there are some possible drawbacks to taking a cruise with children. Do consider very closely the length of time you'll be on board and the facilities available. If babysitting is not part of the service, you may find it impossible to have any time to yourself and mealtimes could be something of a problem. If possible, try to find out if children are a regular feature of the cruise you plan to take. If they're not, you might find your child companionless and that you have to provide entertainment and supervision every waking moment.

If large cruise ships do not appeal, there are many small companies offering cruises, usually on luxury yachts sleeping up to twelve people. Obviously they do not have such a comprehensive range of entertainments and activities, but they do offer a more intimate atmosphere and greater flexibility.

Prices

Unless otherwise stated 1992 prices have been quoted. These are intended only as a guide to the type of holiday offered and travellers should check with the company concerned for 1993 pricing details.

Chandris Ltd

5 St Helen's Place, London EC3A 6BJ
Tel: 071 588 2598

Baltic, Bermuda, Black Sea, Canary Islands, Caribbean, East Mediterranean, Greek Islands, North Cape, Norway, South America, Turkey

Chandris Cruises operate 4 ships. One cruises around the Greek Islands and Turkey, a second cruises to the Baltic, Norwegian Fjords and Spitzbergen as well as the Eastern Mediterranean. The third ship operates out of Miami to the Gulf of Mexico and the fourth from San Juan to the Caribbean.

The Celebrity Cruises fleet comprises three ships operating cruises to the East and West Caribbean and between New York and Bermuda.

Children are welcome aboard all the ships in the Chandris and Celebrity Fleets and on most of them there is a supervised playroom for part of the day. Depending on the number of children on board a programme of entertainment is specifically organised for them.
Prices Fly/Cruise Programmes from London last from 5 days to 53 days and range from an all inclusive price of £538 per person. The prices for all the cruises include flights from UK, transfers between airport and ship (except at New York or Miami) plus full board and entertainment throughout the cruise. For some programmes hotel accommodation for the night before the cruise is also included. There are opportunities to add a Florida Land Programme including a self-drive car to your fly/cruise holiday for an extra charge ranging from £25 per person per day.

CTC Lines

1 Regent Street, London SW1Y 4NN
Tel: 071 930 5833

Africa, Australia, Black Sea, Canary Islands, Caribbean, CIS, East Mediterranean, Far East, Mediterranean, North Cape, New Zealand, Scandinavia, South Pacific

British company with a Ukrainian crew offering worldwide cruises from 2 to 59 nights, departing from Tilbury, London and Liverpool. Cabins have 2, 3 or 4 berths available. Cots can be supplied for children of 2 and under. Any special food requirements must be requested when booking. Cinema, leisure, restaurant and bar facilities are available on board, plus a children's swimming pool, playroom and babysitting service. Special activities are organized for children on cruises during school holidays.
Prices (per adult per fortnight to Canaries in 4-berth cabin)
From £742. Money Saver Schemes also available.
Discounts These are available for children in all sizes of cabin, provided there are as many adults as children in the cabin. Children under 3 receive 90% discount, 3–12s receive 50%, and 13–17s receive 25%. Special children's prices are available on shore excursions.

Cunard

South Western House, Canute Road, Southampton SO9 1ZA
Tel: 0703 634166 or 071 491 3930

Black Sea, Caribbean, Iceland, Mediterranean, Mexico, Scandinavia, South America, UK, US

Cruises lasting 12–40 days on 7 luxury ships, including the QE2. On-board facilities range from shops to saunas, swimming pools, gymnasium, theatre and library, plus gourmet restaurants and an extensive (free) programme of entertainments. All cabins have private shower and WC; some also have baths. The QE2 is the only ship of the 7 on offer to provide nursery facilities and an organized children's programme.
Prices (per double room)
QE2: £2995 for 21-day cruise in lowest grade cabin; £11,580 for 21-day cruise in highest grade cabin
Vistafjord: £3590 for 28 days (lower grade cabin); £8350 for 28 days (highest grade cabin);
Princess: £2190 for 12-day holiday; £1495 for 12-day holiday
Discounts Some discounts available for children under 2.

Equity Cruises

77–79 Great Eastern Street, London EC2A 3HU
Tel: 071 729 1929

Alaska, Canada, Caribbean, Egypt, Mediterranean, Mexico, North Africa, South America, South Pacific, UK, US

Large company offering cruises from a number of cruise lines: Carnival, Cycladic, Holland America Line, International Nile Cruises, La Palma (Intercruise), Regency, Renaissance Cruises, Siosa and Star Lauro. Fly/cruise and cruise/stay holidays are available. The ships offer various degrees of comfort with some sports and recreation facilities. Some are fully air-conditioned.

124 Choosing and planning your holiday

Prices Each company represented has several different ships and each calculates costs differently. Some quote separate rates for low and high season, others add a supplement to the basic rate for different seasons. The best family deal is a 4-berth cabin, but not all the ships appear to have them. The pricing system is rather complex and the destinations and durations extremely varied. Prices start at £590 per adult for a 9-day Mediterranean cruise, but thereafter the possibilities are endless. Our impression is that there are some reasonably priced cruises available, but it's worth phoning the company for advice.

Discounts Like prices, each company calculates differently. Discounts on children's fares can be up to 60%.

P & O Cruises

(*part of* P&O *q.v.*)
Tel: 071 831 1331

North and West Africa, Australia, Canada, CIS, Egypt, France, Gibraltar, Greece, Hong Kong, Holland, Iceland, Italy, Mexico, Portugal, Scandinavia, Spain, Turkey, US

Nine- to 34-night cruises aboard either the *Canberra* or the *Sea Princess*. Free entertainments programme and use of extensive sports facilities. The Junior Club for children aged 2–12 is generally open from 9.00 a.m.–8.00 p.m. (except lunchtimes). A special children's tea at 5.15 leaves parents free to enjoy dinner alone. Baby foods and a night nursery (*Canberra* only) are provided for children under 2, but no childminding facilities are available for this age group during the day. Babies under 6 months are not allowed on these cruises. 'Family cabins' sleep 4 in bunk beds and are only available on the *Canberra*.

Prices (per adult)
From: £699 (10 nights); £1099 (14 nights); £1260 (18 nights); £1699 (26 nights)
These prices apply to 4-berth family cabins which have handbasins only. Larger cabins with private facilities are more expensive.

Discounts Depending on the length of cruise and choice of accommodation, children receive the following discounts. 6 months to under 2 receive up to 70% discount. 2–11s up to 60% discount. 12–16s up to 40% discount.

Fly/drive

Most major airlines and many travel agents can arrange fly-drive packages for you on request, but those we list here are established operators, mainly offering deals to the USA and Canada.

If you plan to stay in a resort or a city with good facilities on the doorstep, or a reliable system of public transport, it would be an unnecessary expense to hire a car and have the additional problems of parking. However, for independent holidaymakers who do not want to be tied down and want to see the most of a huge place like America, a car is more an essential than a luxury.

Cars come in all sizes, from Economy (Fiesta etc.), to Luxury (Granada and estate cars). In the US, most are automatic and have air-conditioning. Child seats are compulsory in many states for children under 5. These are usually available for a small charge.

The way fly-drive holidays are put together varies from company to company. Watch out for deals that offer lots of free gifts, such as flight bags, duty free vouchers and cameras. They do not necessarily offer better value and may have complicated exclusion clauses. Car rental offers no discounts for children or families, but some states, particularly California and Florida frequently offer one or two weeks' very low or free rental. This does not include any insurance, tax or collision damage waiver (*see* **Car rental** introduction for further information). Specific children's discounts are often available and you should contact the companies concerned for further details.

Accommodation vouchers may be purchased in advance in the UK and do offer a genuine saving on what you might pay on the spot. However, you can make a saving yourself if you're prepared to stay in independent motels and trust to luck that rooms will be available. Remember – America is a mobile nation and places to stay are thick on the ground. In European destinations it is advisable to reserve accommodation in advance, particularly if you holiday in peak season.

Prices
Unless otherwise stated 1992 prices have been quoted. These are intended only as a guide to the type of holiday offered and travellers should check with the company concerned for 1993 pricing details.

British Airways Fly/Drive

Atlantic House, Hazelwick Avenue,
Three Bridges, Crawley, West Sussex
RH10 1NP
Tel: 0293 518060 *or* 061 493 3344

Canada, US

Scheduled flights with British Airways to up to 47 destinations in the USA and Canada. Some flights also depart from Manchester. Fully inclusive pre-planned fly/drive tours and independent travel arrangements including Hertz car hire, hotel accommodation and go-as-you-please hotel vouchers. Pre-paid accommodation vouchers worth from £28–£41 per room, are offered by 4 major hotel chains.
Prices Start from £339 for a fly/drive package to Florida.
Discounts Child discounts range from 10% to 50% (US) and up to 50% (Canada).

Getaway America
34 The Mall, Bromley, Kent BR1 1TS
Tel: 081 313 0550

US
Scheduled flights to more than 250 cities across America with Alamo car deals available for up to 7 days free. Accommodation vouchers may be purchased in advance for use in 3 hotel chains. In addition there are tours which include national parks. Escorted coach tours, adventure holidays and cruises are also available.

Prices (per adult, per mid-week flight) New York, from £229; Los Angeles, from £309; Florida, from £279. Prices for Florida, Los Angeles, California and Hawaii include one week's car hire, with further week available from £30 in Hawaii and Florida, and from £55 in Los Angeles and California. No car hire is included for packages to New York, Illinois and New Jersey. Special rates apply to the rest of the USA.

Discounts Children sharing hotel accommodation with parents usually stay free. For children up to age 17 special fares apply.

Jetsave
Sussex House, London Road, East Grinstead, West Sussex RH19 1LD
Tel: 0342 312033 *(London depts.)*
0345 045599 *(Manchester depts.)*

Canada, South Africa, US
Fly/drive holidays, with the option of purchasing accommodation vouchers. The vouchers are valid in over 3000 hotels and may be purchased in advance for £27–£55 per night with supplements ranging from $1–$95 per night. Cots and high chairs are not supplied, but can be requested locally. Car hire is provided through Alamo Rent-a-Car.

Prices Flights from around £325 (charter price) per adult – with car hire starting from £35 per week.

Discounts On many flights under-2s travel free (no seat). On other flights a charge may be made. This may be 10% for South Africa and Canada, and varies for the US. Children aged 2–12 receive up to 33% discount on scheduled flights to South Africa and Canada, whilst discounts for the US vary.

Magic of Italy
227 Shepherds Bush Road, London W6 7AS
Tel: 081 748 7575
Brochures: 081 741 1349 (24 hours)

Italy
Motoring and fly/drive holidays. If you take your own car, the company organizes ferries. Fly/drive holidays offer flights to 9 Italian cities and car hire, which includes local taxes, third-party insurance, collision damage waiver and theft waiver, and unlimited mileage. A variety of companies are used for car hire.

Prices (7 nights for 4 people, per person)
from £153 to £203

Discounts These are based on children up to 12 sharing accommodation with parents.

Mundi Color Holidays
276 Vauxhall Bridge Road, London SW1V 1BE
Tel: 071 834 3492

Spain
Specialists in fly/drive, tailormade and city break holidays including the Paradores and other quality hotels/self-catering establishments, particularly in lesser known beautiful regions of Spain. Multi centre city holidays can be arranged on request. Many hotels offer swimming pools, children's play areas and sports facilities. Baby-sitting can be arranged on request in some hotels.

Prices (per adult, B+B including flight)
City hotel: from £187
Fly/drive (12 days): from £815 on half board and with Avis car.

Discounts Under-2s pay £25 for travel (no seat/no baggage allowance); cots and meals payable locally. Children aged 2–11 receive reductions up to £100 in hotels, and £50 in self-catering accommodation.

Northwest Fly/Drive USA

PO Box 45, Bexhill-on-Sea, East Sussex TN40 1PY
Tel: 0345 747800 (toll free)
0424 224400 *(Bexhill)*
061 499 2471 *(Manchester)*
041 226 4175 *(Glasgow)*

US

Scheduled flights from Gatwick and Glasgow to Minneapolis/St Paul, and Boston and onwards to over 200 destinations in the US. Accommodation vouchers may be purchased from £26 per night for 1–4 people, giving access to a wide range of motel/hotel chains. Car hire can be arranged through Alamo or Hertz.

Prices (per adult for a mid-week return flight)
Boston: from £333
Orlando, Miami, Florida: from £330
Additional weeks for car hire: Boston £58; Florida £27.

Discounts Under-2s pay 10% of adult fare. Children aged 2–12 pay up to 67%.

Peregor Travel

146 High Street, Ruislip, Middx HA4 8LJ
Tel: 08956 39900

Caribbean, Mexico, US

Fly/drive and tailor-made holidays to a wide variety of destinations. There is also a choice of airline, car hire and hotels.

Prices on request

Discounts These depend on which airline you travel with and the holiday package you choose. Quotations are given on request.

Unijet

'Sandrocks', Rocky Lane, Haywards Heath, West Sussex RH16 4RH
Tel: 0444 458611 *(International flights)*
0444 458181 *(European flights)*
0444 459191 *(American holidays)*

Worldwide

Flights to 17 destinations in the US, plus Canada, Australia, Mexico, New Zealand, Far East, South Africa and South America. Accommodation vouchers may be purchased in advance for 4 major hotel chains. Car rental is provided, and sometimes up to 2 weeks free rental may be on special offer. Connecting flights within the UK can be arranged on request.

Unijet America offers a comprehensive programme of holidays, featuring self catering, one, two and three centre holidays, coach tours and fly cruises.

The company also operates fly/drive deals to Gibraltar, Greece, Malta, Portugal, Spain and Turkey (some destinations available in summer only). Accommodation vouchers are not available. In addition to their main fly/drive business, Unijet has holidays in Europe in villas, apartments and hotels.

Prices available on request.

Discounts Under-2s are charged £15 to cover airport taxes in Europe, but pay 10% on intercontinental flights. Children aged 2–16 receive £10–£50 discount on European charter flights, and 25–33% discount on scheduled intercontinental flights.

Valley USA

2 Bradwood Court, St Crispin Way, Rossendale, Lancashire BB4 4PW
Tel: 0706 212333

US

Daily flights from Gatwick to Charlotte, with internal connections to 49 cities in the states of Florida, N. Carolina, S. Carolina and Virginia. Accommodation vouchers may be purchased in advance and are valid in over 2000 hotels. All rooms sleep up to 4 people. There may be an additional charge for cots. Car hire is arranged through Alamo and Dolla and offers unlimited mileage. Collision damage waiver is included.

Prices available on request from Valley USA

Discounts Discounts for infants (under-2s) and children (2–11s) available on request.

Holiday centres

The growth of holiday centres has been a rapid one. Where once the name brought to mind only large organizations, such as Butlins and Warners, there are now literally hundreds of establishments, from caravan sites to luxury hotels, which describe themselves as holiday centres.

What do they offer? They vary a great deal but the list is usually impressive. Most gear themselves to the family market and try to cater for all interests and preferences. Firstly, accommodation is available in anything from rooms to suites, from traditional chalets to high-tech 'lodges'. Most offer self-catering facilities, but many also offer full and half board so you can have a rest from cleaning and cooking.

The great advantage of holiday centres is that you are free to do what you like when you like and you don't have to go far or pay for it. Everything from archery to trampolining is free – often with free qualified instruction. The exceptions to this are sports such as riding and windsurfing, where special equipment is necessary and must be hired at an hourly rate. A great boon for parents are the funfair rides and massive swimming pools with impressive water slides which children can spend the entire holiday on, if they choose, at no extra cost. In the evening there are often professional shows with 'big name' entertainers.

Teams of hosts and hostesses (Redcoats, Bluecoats, or whatever) are constantly around in the larger centres to organize activities, keep an eye on children and generally help things go with a swing. And although most centres offer entertainments for virtually every hour of the day, there is no obligation to join in, and there are plenty of places to sit quietly and relax.

If you dread hearing your children claim, 'I'm bored', take them to a holiday centre and those words won't have a chance to pass their lips!

Prices

Unless otherwise stated 1992 prices have been quoted. These are intended only as a guide to the type of holiday offered and travellers should check with the company concerned for 1993 pricing details.

Barrowfield Hotel

Hilgrove Road, Newquay, Cornwall
TR7 2QY
Tel: 0637 878878

UK (Cornwall)

An AA '3 star' graded hotel, with a 'commended' quality grading from the English Tourist Board. Accommodation is in luxury en-suite rooms, penthouse suites and family suites. Facilities include indoor and outdoor swimming pools, spa baths, saunas, gymnasium, solarium, games room, coffee shop, snooker room, launderette and hairdressing salon. Nightly entertainment in peak season. Short breaks also available. Open for Christmas and the New Year.

Prices (per person half board)
3 days from £98
7 days from £163

Beverley Park Holidays

Goodrington Road, Paignton,
South Devon TQ4 7JE
Tel: 0803 843887

UK (Devon)

A large number of 4-, 5- and 6-berth caravans set in 20 acres overlooking the sea at Goodrington Sands near Paignton. All caravans have a WC, hot and cold water, bath or shower, fridge and colour TV. Blankets, china and cutlery are provided, but bring your own linen and towels (and rubber sheets for children). Linen may be hired on arrival, if you wish. Cots are available for £6 per week. Amenities include a launderette with drying and ironing facilities, heated indoor and outdoor pools, sauna, children's pool, spa bath, crazy golf, children's playground, games room with table tennis and pool and tennis court. There are two shops, a takeaway food bar and restaurant. Free evening entertainment is provided, but no baby-sitting services are available. There is also a touring site where campers have full use of all the facilities.

Butlins Holiday Worlds and Hotels

Bognor Regis, West Sussex PO21 1JJ
Tel: 081 809 4649 *(Brochures)*

UK

Catered and self-catering holidays in five Holiday Worlds and five hotels. A variety of accommodation is available. 'County suites' have 2 or 3 bedrooms and sleep up to 6 people; 'Country Rooms' have 1, 2 or 3 bedrooms. Standard accommodation is in comfortable bedrooms sleeping up to 3 or 5 (minimum occupancy 2 persons, half board). All catered accommodation has private bath or shower and a maid service. Full board is available.

Self-catering accommodation is in chalets and caravans sleeping up to 7 people. They have a bathroom and fully-equipped kitchen. All centres have good shopping facilities plus snack bars, restaurants, bars and free entertainments. There is also free qualified instruction in a wide variety of sports. Redcoats organize special clubs for 6–9s (Wizz Kids) and 9–12s (Teamsters),

with one centre offering special activities for 13–16s. Nurseries are available for children under 6 where they can be left for short stays in the care of qualified staff. A child-listening service is also offered at night. Infants up to 9 months may be left in a free nursery at night with qualified nurses. Special facilities include nappy-changing and mothers' room, launderette and ironing room. Pushchairs and cots may be hired. Butlins hotels offer many of the same facilities.

Prices Phone 0345 700700 for price details.

Discounts Children under 2 go free on all catered holidays; 2–14s pay 50% at all centres. On hotel holidays children under 4 are not accepted. The Metropole, Blackpool, Grand Scarborough and Grand Llandudno are adults-only hotels.

Center Parcs

Center Parcs Ltd, Head Office, Rufford,
Nr Newark, Notts NG22 9DP
Tel: 0623 411411 *(reservations)*
0272 244744 *(brochures)*

UK (Suffolk, Nottingham)

Two all-season holiday centres at Sherwood Forest and Elvedon Forest. Accommodation is in 1-, 2-, 3- or 4-bedroomed villas and 1-bedroomed apartments for Elvedon Forest. The villas are fully equipped for self-catering, whilst the apartments have tea and coffee making facilities only. All accommodation has private patio area, radio, bathroom, WC, colour TV with satellite and video channels. Linen is provided and there is no charge for gas or electricity. In each of the 2-, 3- and 4-bedroomed villas, a travel cot suitable for babies aged up to 9 months, playpen and high chair are provided free of charge. These can be provided in 1-bedroomed accommodation for a small charge as can a wooden cot for a larger baby. Cot linen is not provided.

Included in the villa rental price is access to the Subtropical Swimming Paradise, featuring waves, water chutes, whirlpool baths, wild water rapids and baby and toddler pool. A kindergarten is available and baby-sitting can be arranged, both at extra cost. There is a wide range of activities outside the dome available at both villages, including BMX tracks, children's play area, nature trails and jogging trails. On most outside activities you pay for the hire of equipment such as windsurfers, BMX bikes, and tennis courts. There is an international selection of restaurants, most of which serve children's portions.

Weekends and midweek breaks are available in addition to weeks.

Corton Beach Holiday Village

The Street, Corton, Suffolk NR32 5HS
Tel: 0502 730200

UK (Suffolk)

A family-run holiday village overlooking the beach at Corton. Accommodation is in villas which sleep 4–6 or 6–8; most have a sun patio. All are fully equipped with kitchen and bathroom and colour TV with a free video channel. There are some luxury villas with three-piece suites. The village has heated swimming pools, a toddlers' pool, a full-sized Krypton Factor assault course, an adventure playground, organized children's activities and a resident children's entertainer. There is a cafe and take-away food. Scuba-diving and horse-riding can be arranged, and jetskiing and clay pigeon shoots are run by the owners. Linen is provided, but not towels. A cot and high chair may be hired for £7 per week. A baby-sitting service is available at £2 per hour plus a £1 booking fee. Electricity is metered.

Prices (per villa per week)
£85–£300, depending on time of year and size and standard of villa.

Flamingo Land Holiday Village

Kirby Misperton, Malton, North Yorkshire YO17 0UX
Tel: 065 386300

UK (Yorkshire)

Fully-equipped 6-berth caravans a short walk from Flamingo Land Zoo and Family Fun Park, whose rides and shows are free to guests. All caravans have a shower room and WC, kitchen with fridge and a TV. On-site facilities include a heated indoor swimming pool, fitness centre, sauna and jacuzzi, children's play area, splash pool, shop, games room, launderette and take-away food. Cutlery, crockery, blankets and pillows are provided, but not linen. Gas and electricity are free. In the Fun Park there are children's roundabouts, a circus, a Formula One car track, high-flying chairs, a looping roller coaster, a log flume and many other attractions. The zoo is the largest privately-owned zoo in Europe and has regular dolphin, seal and parrot shows. In the village there are video shows and a disco every evening to which children are welcome. A cotside can be attached to the bottom bunk free of charge.

Prices (per caravan per week)
£230–£395, depending on season and size of caravan.

Holiday Club Pontins

Pontins Ltd., P.O. Box 100, Sagar House, The Green, Eccleston, Chorley, Lancashire PR7 5QQ
Tel: 0257 452452

UK

Holiday Centres, Villages and Chalet Hotels in 21 locations throughout England, Wales, Southern Ireland and the Channel Islands, providing accommodation for up to 7 persons in either 1 or 2 bedroom chalets. Each chalet has washing facilities, toilet and lounge with self catering chalets having a fully-equipped kitchen. Colour TVs are provided in most chalets. Full board, half board and self catering options are available, full board providing three full meals each day with either self service or table service. Facilities include shops, restaurants, take-away food, swimming pools and many exciting sports activities. There is also a special Childrens Club which has supervized activities, and a special Captain Croc character who helps provide fun and games. Each location provides nightly entertainment programmes.

Prices and discounts All prices and discounts for 1993 can be found in the 1993 Summer Brochure. Apply to number above for further details.

Pontins

See **Holiday Club Pontins**

Radfords Country Hotel

Lower Dawlish Water, Dawlish, South Devon EX7 0QN
Tel: 0626 863322
UK (Devon)

Set in 6 acres of gardens and fields, Radfords can accommodate up to 37 families in large family bedrooms and 2-bedroom family suites. The bedrooms have baby-listening intercoms and a bathroom complete with airing cupboard and drip-drying area. There is an early tea for children, although they are also welcome to eat in the dining room in the evening; and parents can prepare baby foods at any time. Facilities include an indoor heated pool and a children's pool, supervised by a qualified lifeguard, a jacuzzi and spa bath. There is a colour TV lounge, a playroom and a games room with skittle alley, pool, darts and table tennis. Three mornings a week there is a free playgroup. Reliable babysitters are always on duty (each night) at no extra charge. Children's activities include swimming galas, parties, discos, entertainers and horse-riding. The Radford Club for 5–12 year-olds meets one morning per week for indoor and outdoor activities.
Prices on application
Discounts Discounts depend on age of child and time of year.

The Saunton Sands Hotel

near Braunton, North Devon EX33 1LQ
Tel: 0271 890212
UK (Devon)

A 4-star hotel with sweeping views over Barnstaple Bay and the 5 miles of Saunton Sands. Accommodation is in twin or double rooms, most of which have room for a third bed or cot. There are also some rooms with a single or twin-bedded room en-suite sharing a private bathroom. All bedrooms have bathroom, satellite TV and baby-listening intercom. Children's teas are served between 5 and 6 p.m. Facilities include a large indoor heated pool with a paddling pool, new outdoor terrace pool, sauna, spa bath and solarium, a fitness area, squash, tennis, putting green, table tennis, billiard room, mini cinema, a children's games room and a nursery staffed by an experienced nanny from 9 a.m.–5 p.m. where children can be left for an hour or two free of charge. There is a wide programme of entertainment especially for children. The hotel also has fully furnished self-catering apartments and suites on a separate tariff.
Prices (per adult per day, half board)
Low: from £60–£72
High: from £64–£77
Special seasonal break rates also apply.
Discounts In low season when children either share parents' room, or 2 children share a single room, under-2s are free, 2–5s pay 40%, and 6–11s pay 60%. Where there are more than 2 children and another room is needed, the first additional child is charged the adult rate and the second one, if 2–5 pays 60%, if 6–11, pays 75%.
In high season when children either share parents' room, or where 2 children share a single room, under-2s pay £3.50, 2–5s pay 50%, 6–11s pay 70%, 12–13s pay 90%. The first additional child in a separate room is charged the adult rate, the second additional child, if 2–5 pays 66%, if 6–11 pays 75%, if 12–13 pays 90%.

Sussex Beach Holiday Village

Bracklesham Bay, West Sussex
PO20 7JP
Tel: 0243 671213
UK (Sussex)

Self-catering and catered holidays by Bracklesham Bay, a secluded and safe beach near Chichester. Accommodation is in 2-, 4- and 6-berth bungalows with fully-equipped kitchen, bathroom and WC, carpets, electric fire, radiator and colour satellite TV. Self-catering bungalows are not provided with linen and cots may be hired per week. Electricity is metered. The village has a shop, a restaurant, cabaret club, village pub and a launderette. Younger children can join the Effelump Club, which has a busy programme of films, cartoon shows, fancy dress parties, talent contests and games. There is also an adventure club for 10–15-year olds. Membership of these clubs and all the facilities in the village is free. Visitors have a wide choice of sporting activities such as darts, football, archery, billiards, snooker, crazy golf, table tennis and tennis. There is a heated pool, sauna, solarium and gym, hairdresser and beauty parlour. The village runs a special football week with coaching by players from West Ham Football Club. There is nightly entertainment with visiting guest artists.

Prices available on request.

Discounts Children under 5 staying in full-board accommodation are free.

Vauxhall Holiday Park

Acle Road, Great Yarmouth, Norfolk
NR30 1TB
Tel: 0493 857231

UK (Norfolk)

Self-catering caravans and chalets in Great Yarmouth. Each holiday home is fitted with a fully-equipped kitchen and all have a bathroom with hot and cold water, shower, WC, basin and shaver point. Towels and linen are only supplied in holiday suites. There is a colour TV with a video and satellite link for films. Details of cot and linen hire are sent on receipt of your booking. There is a launderette, supermarket, hair-stylist, restaurant and take-away meals. The Jolly Roger Club, supervised by experienced staff, organizes sports days, Punch and Judy shows, cartoons and many other activities. For children under 5 there is the Jack and Jill nursery and play area, and a paddling pool. Other free facilities include two adventure playgrounds, games room with pool and snooker and a heated swimming pool supervised by a lifeguard. In the evening there are magicians, jugglers, fancy dress parties, discos and competitions in the Family Club House. Most weeks there is a celebrity cabaret in the Holiday Inn (adults only). Vauxhall Holiday Park also has a touring and camping site.

Prices (per holiday home per week)
4-berth caravan: £90–£312
6-berth caravan or chalet: £96–£429
8-berth caravan: from £168

Home swapping

If you want a cheap way to take the whole family on holiday to a comfortable base in a faraway spot without breaking the bank, home-swapping must be the answer. The idea is that you supply details of your home and family to a company who list you in a yearly directory which is then circulated to all the people listed. You can browse through it at your leisure, choosing a country and a property that appeal to you. You then exchange photos and letters with as many people as you like before making your final decision. References may be taken up and a holiday agreement exchanged for added security.

The company listed here undertakes all the administration involved in setting up a home exchange, but they charge heavily for this service. However, it does remove the element of risk by verifying details of all properties, and it saves you a lot of work.

The advantages of home exchange are many. You have someone looking after your home, even feeding your pets, while you're away, so you have peace of mind. But most of all you can afford to visit faraway countries, living in comfortable accommodation, having the free use of a car if you want, and getting to know the area like a native. Families of all sizes, including one-parent families, can exchange homes and the savings to be made are enormous.

What about dishonesty and damage to your property? We are assured that complaints about these things are very rare. Your best security is to get to know your exchange family well by corresponding in some detail.

Prices

Unless otherwise stated 1992 prices have been quoted. These are intended only as a guide to the type of holiday offered and travellers should check with the company for 1993 pricing details.

Intervac International House Exchange Service

6 Siddals Lane, Allestree, Derby
DE3 2DY
Tel: 0332 558 931

Worldwide

Long-established company which publishes 3 directories a year in January, March and May, listing thousands of home for exchange and rent in more than 50 countries. The registration fee covers the cost of your entry and copies of the 3 directories. You may include a photo of your property for a small extra charge. The company is happy to advise on any aspect of the exchange procedure. They also offer a 'last minute' phone service to members still looking for a holiday after publication of the third directory.

Cost Listing your home for exchange costs £43 a year (£52 to overseas members). Including a photo costs £6 extra. A second home listing costs £10.

Hotel holidays

When you think about taking the family away for a hotel holiday, you will probably be thinking of a traditional package-style trip. This section contains many examples of companies offering holidays in Spain, Greece, Florida and other popular locations, but also includes companies that organise holidays in Britain and other spots far from the traditional package resorts.

We do not pretend to be comprehensive in this listing, but all the companies included either cater particularly for families or provide an interesting or unusual service. Many offer more than just hotel accommodation and you will also find examples of self-catering deals, themed holidays (such as golfing trips) and self-drive packages within this listing.

There are many fantastic deals for families on the market and, as the amount of disposable income per family diminishes, travel companies try harder and harder to make us part with our hard-earned cash. Keep your eyes open for special offers in the Sunday newspapers and take note of weekly updates on television and radio travel shows for advice and particular bargains.

Many brochures advertise free places for children but be sure that you read the small print carefully before becoming too enthusiastic. In general, if you don't mind sharing a room with your children (usually up to two) you are eligible for a substantial saving. It would be worth your while to check the approximate size of the room you will all be sharing as a hotel's idea of how many beds can be fitted into a double room may differ considerably from yours.

One of the problems with free-holiday deals is that they may be offered outside normal school holidays. British parents are legally entitled to take their children away from school for two weeks holiday per year, but exam and revision times should be avoided at all cost when thinking about such an option.

One parent families are often eligible for discounts. It is always worth contacting the company concerned to enquire about special deals and group holidays. The head office of Gingerbread (for address, *see* **Section 7**) can provide you with further information about holidays arranged specifically for lone parents and their children both by them and other companies (eg Virgin) produce separate brochures listing the facilities available at each resort in detail. Before finalising your trip, you may also want to assess which airport is best suited to your travel requirements. Full details of the facilities available at British airports is given in our survey in **Section 3: Holiday transport**.

See also **Bargain breaks and short-stay holidays**.

Prices

Unless otherwise stated 1992 prices have been quoted. These are intended only as a guide to the type of holiday offered and travellers should check with the company concerned for 1993 pricing details.

Aer Lingus Holidays

83 Staines Road, Hounslow Middx
TW3 3JB
Tel: 081 569 4001

Ireland

Holidays in southern Ireland from 2–7 nights. Accommodation ranges from guest houses to luxurious hotels. Most have cots, and some provide children's menus. Fly/drive holidays, cottages, golfing holidays and coach tours are also available.

Prices Two nights, B+B in Dublin costs from £115 (guest house) or £140 (comfortable hotel).
A 7-night motoring holiday (including car hire) costs £240–£270 (guest houses) or £295–£360 (hotels).

Discounts Children under 2 are charged 10% of air fare (cots and meals payable locally); 2–11s receive up to 50% discount if sharing with parents.

Air France Holidays

69 Boston Manor Road, Brentford,
Middx TW8 9JQ
Tel: 081 568 6981

France

Specialists in late bookings, Air France Holidays offers breaks and short holidays in Paris, the French Riviera including Monaco, and fly-and-drive packages from 11 gateways in the UK to 13 destinations in France.

Prices packages include car hire, Air France Rail from Paris and accommodation in hotels and self-catering.
£120 for 2 nights accommodation in a 3-star hotel in Paris.
£197 for a weekend fly-and-drive in Nice for 2 people.

Discounts Please contact Air France Holidays for details of special offers.

Airlink Holidays Ltd

9 Wilton Road, London SW1V 1LL
Tel: 071 828 7682

Greece, Portugal, Spain

Hotel and self-catering holidays on Greek islands, Cyprus, Lanzarote, Gran Canaria, Tenerife, Menorca and the Algarve. Accommodation is in small hotels and pensions.

Prices (per adult per fortnight, B+B)
Corfu: £174–£350
Rhodes: £219–£490
Algarve: £215–£339

Discounts Children under 2: £15 (no seat, meals extra); children (2–17): 1 week, reduction £20; 2 weeks: reduction £40.

Airtours PLC

Wavell House, Holcombe Rd,
Helmshore, Rossendale, Lancashire
BB4 4NB
Tel: 0706 260000 (*Reservations*)
Tel: 0706 240033 (*General*)

Canary Islands, Cyprus, Egypt, Greece, Madeira, Malta, Menorca, Portugal, Spain, Tunisia, US

Hotel and self-catering accommodation in many locations, with special emphasis on the Greek and Spanish islands. Most hotels have swimming pools, including a children's pool. Cots and baby-sitting can usually be arranged on request, payable locally. Airtours Getaway Gang – membership free to all 3–12 year olds – at certain properties: 1 hour per day, 6 days a week, May to October. Departures are from UK airports.

Prices (per adult per week, self-catering, including flights)
From £99

Discounts Under-2s travel free (no seat). Some hotels offer free places to the first child per booking aged 2–11 (sometimes 2–15), provided they share parents' room. Additional children sharing receive additional discount on holidays. This applies to all destinations (except the Canaries) throughout the season. On apartment holidays the first child per booking aged 2–19 receives a free holiday subject to availability.

Albany Tours

Royal London House, 196 Deansgate,
Manchester M3 3NF
Tel: 061 833 0202
Fax: 061 834 4890

Australia, Canada, New Zealand, US

Wide range of hotel holidays, fly/drive and tours. Accommodation in 3–5 star hotels, most with swimming pools, some with play areas and baby-sitting services. Most provide cots.
Prices (per adult for duration specified, 1992 prices)
Orlando: hotel, 7 nights, £439–£1072 (land only, including car hire)
Canada: 14-day tour (including flight) £2329–£2405
Discounts Refer to brochure for details.

Aquasun Holidays

41 Crawford Street, London W1H 1HA
Tel:071 258 3555

Malta

Small company offering holidays in penthouses, apartments and hotels in Malta and Gozo. Most have swimming pools. Cots are available either free of charge or for £1 a day payable locally.
Prices (per adult per week, including flight and transfers)
Self-catering: from £165 (low), £255 (mid), £315 (high). Children from £145 to £260.
3-star hotel, half board: from £210 (low), £300 (mid), £390 (high)
Discounts Under-2s pay £30 (no seat on plane); meals payable locally. Children under 12 receive a 33% discount on scheduled flights when booking direct.

Arctic Experience Ltd

29 Nork Way, Banstead, Surrey SM7 1PB
Tel: 07373 62321
also at: The Flatt Lodge, Bewcastle, Cumbria CA6 6PH
Tel: 06978 356

Alaska, Canada, Faroes, Finland, Greenland, Iceland, Norway, Sweden

Hotel holidays and tours (both independent and escorted) throughout Scandinavia, Iceland and Greenland. Accommodation ranges from tents and guest houses to high grade hotels. Some have swimming pools, saunas and gymnasiums. Sporting activities and 'adventure excursions' are available on most holidays. Most hotels provide cots. Travel by air or (on some holidays) by sea. Many activity holidays are also available, as well as fly/drive, self-catering, camping and cruises.
Prices (per adult)
Eight days' B+B in a guest home, by air: Iceland: from £389
Discounts Available on most holidays. Details on request.

Balkan Holidays

Sofia House, 19 Conduit Street, London W1R 9TD
Tel: 071 491 4499

Bulgaria, Romania, Turkey

Hotel, villa and apartment holidays. Turkey may only be visited on two-centre holidays. Cots should be requested when booking. High chairs are occasionally available. Baby-sitting is available in 3 Bulgarian resorts.
Prices (per person, per week, B+B, summer 1992)
From £153–£403
Discounts Under-2s pay £20 only (no seat on plane). Cots and meals are payable locally. Children aged 2–16 may receive a substantial discount (details available on request).

BelleAir Holidays

314–6 Upper Richmond Road, Putney, London SW15 6TU
Tel: 081 785 3266

Gozo, Malta

Hotel and apartment holidays on Malta, Gozo and Comino. Nearly all hotels have swimming pools and sports facilities. Some have children's pools, play areas and baby-minding services. Cots may be supplied free or cost up to £1.60 per night.
Prices On application.
Discounts Children under 2 pay £15 (no seat on plane); cots and meals are payable locally. Most hotels offer a reduction of about £5 per night for 2–11s, if sharing with parents.

Best Travel Ltd

31 Topsfield Parade, London N8 8PT
Tel: 081 444 3333 *(reservations)*

Greece, Cyprus

Specialist in Greek and Cypriot holidays (under the brand names of Grecian Holidays and Cypriana respectively). Most of the larger hotels can supply cots, high chairs and baby-sitting. Children's swimming pools are available in some locations.
Prices Please refer to brochures for late availability.
Discounts Children under 2 pay £19 charter, £29 scheduled.

Caprice Holidays Ltd

32A Queensway, Stevenage, Herts
SG1 1BS
Tel: 0438 316622
Fax: 0438 740239

Austria, Belgium, France, Germany, Holland, Italy

Hotel holidays to most European cities. Choice of method of travel (own car, coach, air) and hotel category (from budget to luxury). Multi-centre arrangements also available (travel by air/hovercraft/self-drive). The company can arrange restaurant bookings suitable for the type of group travelling (families, couples, etc.). Cots may be hired. The emphasis is on personal choice of the type and budget of holiday desired.
Prices (per adult, B+B, inc. flight)
Wide variety, depending on method of transport, length of stay and choice of hotel.
Paris: from £61 for 3 days
Italian cities, e.g. Florence: from £199 for 3 days
Amsterdam: from £61 for 3 days
Brussels and Bruges: from £61 for 3 days
The maximum price increase anticipated for 1993 is 10%.
Discounts Cots and meals payable locally. There are some limited reductions for children aged 2–11, depending on choice of holiday.

Caribbean Connection

Concorde House, Forest Street, Chester
CH1 1QR
Tel: 0244 341464

Bahamas, Bermuda, Caribbean, US

The widest choice of islands and top quality hotels throughout the Caribbean. Tailor-made holidays combining more than one island are all part of an emphasis on personal service.
Children and infants are welcomed at most hotels. Especially suitable for infants, the luxury villas brochure provides all the facilities you could wish for including housekeeper, cook, nanny, etc.
For children the 'All Inclusive' selection gives parents total freedom while the children are entertained by nannies all provided free.
Prices per adult
From £780 in summer, £1006 in winter for 7 nights on a room-only basis.
Fly Concorde, First Class or Club World.
Quotation requests welcomed for tailor-made holidays.
Discounts Children 2–12 from £299.
Infants under 2 pay only £100.
Full details on request.

Caribbean Villas & Hideaway Hotels

4a William Street, London W1 9HL
Tel: 071 245 1235
Fax: 071 245 1250

Caribbean

Traditionally the company has offered luxury staffed villas, most with private pools, in several Caribbean islands. Many are beach front and the local staff will cook, clean, shop and garden for you. They are perfect for a family (or two) to enjoy a pampered tropical holiday.
In addition they offer a range of accommodation including self-catering and a selection of hotels from the moderate to luxury categories. Two/three-centre holidays are a speciality.
Prices Please call for a brochure.

Castaways

2–10 Cross Road, Tadworth, Surrey
KT20 5JU
Tel: 0737 812255

Spain

One- and two-week hotel holidays in the quieter areas of Majorca. Some hotels offer baby-sitting, special meals, organized games and excursions for older children.
Prices Please refer to brochure.
Discounts Under-2s £25 (no seat); cots and meals payable locally. One or two children (2–12) sharing parents' room receive 40% discount depending on departure date. (Many hotels have 4-bedded rooms.)

Celebrity Holidays & Travel

18 Frith Street, London W1V 5TS
Tel: 071 734 4386

Cyprus, Turkey

Hotel, villa and bungalow holidays. The hotels are of good quality and all rooms have private facilities. The self-catering accommodation is very comfortably furnished and fully equipped. Cots may be hired for £2 per day. Fly/drive and 2-centre holidays can be arranged on request. Breakfast is provided for all accommodation.
Prices (per person per week, B+B hotel accommodation, including flight and transfers)
From £389
Discounts Under-2s travel free, but may pay around £30 for the flight. Children aged 2–12 pay 15% of adult fare.

Celtik Holidays

Celtic Line Travel Ltd, 94 King Street, Maidstone, Kent ME14 1BH
Tel: 0622 690009

Spain

Hotel and villa holidays in Menorca. Accommodation ranges from rustic tavernas to luxurious beach houses. Cots, high chairs, pushchairs and babysitting can be arranged on request.
Prices (per adult per week, including B+B and flight)
Low: from £143
High: from £200
Discounts Children receive a 30–40% discount, except in peak season when they receive a standard £15 reduction. (Children do not have to share parents' room in order to qualify for these discounts.)

C.I.E. Tours International

185 London Road, Croydon CR0 2RJ
Tel: 081 667 0011

Ireland

Motoring holidays and weekend breaks in Southern Ireland. Accommodation ranges from guest houses to quality hotels. Most provide cots. Self-catering cottages are also available, as well as coach tours (no children under 12). Travel may be by air or sea.
Prices (per adult for 6 nights in a B+B, including flight and car-hire)
Guest houses: £263–£291
Hotels: £326–£377
Three nights in Dublin (by air) costs £195–£299.
Discounts Under-2s usually go free; meals and cots payable locally. Reductions of between £33 and £82 are available for children aged 2–11 if sharing with parents.

Citalia

Marco Polo House, 3–5 Lansdowne Road, Croydon CR9 1LL
Tel: 081 686 0677

Italy

Hotel, villa and apartment holidays in Tuscany and Umbria beach resorts, plus cities and lakes throughout Italy. Hotels are generally 2–5 star, the grander ones offering an impressive variety of sports facilities, plus air-conditioned rooms with colour TV and fridge as standard. The simpler hotels are still very pleasant, but offer fewer facilities. Some have children's pools, playgrounds and games rooms. Cots may be available on request. Self-catering holidays in villas and apartments offer a variety of accommodation, all comfortably furnished. Some offer a limited maid service and cots are supplied free.
Prices available on request, or phone 081 686 5533 for a brochure.

Hotel holidays

Discounts Under-2s are free in hotels, villas and apartments when there is maximum occupancy, otherwise they receive ordinary child reductions. Children aged 2–16 receive £27 discount throughout season.
In hotels, when sharing a room with 2 adults, under-2s pay £90 (no seat on plane). 2–11s receive 20% discount if sharing with 2 adults, 10% if sharing with one adult or another child.

Club Cantabrica

Holiday House, 146–148 London Road, St Albans, Herts AL1 1PQ
Tel: 0727 833141/866177

France, Greece, Italy, Majorca, Spain

Hotel, apartment, camping and caravan holidays, with a choice of travel by coach/ferry, air or self-drive. Some hotels have swimming pools and rooms with private balconies. At all resorts a wide variety of equipment may be hired, including cots, pushchairs, sleeping bags, pillows, hairdryers, irons and parasols. At selected sites a special children's club is available 6 days a week for 3–12 year-olds, offering organized games, competitions and entertainments. Childminding can be arranged in the evenings for a small charge.

Prices Variety of offers – please refer to brochure.

Discounts Children under 2 travel free (no seat) on all holidays, all summer; cots and meals payable locally. With 3 children 1 child travels free for every 2 full-paying adults on the coach/self-drive holidays up to 18th June and from 7th September. Children are half price (coach and self-drive) from 19th June to 10th July and 24th August to 6th September.

Club Méditerranée

106–110 Brompton Road, London SW3 1JJ
Tel: 071 581 1161

Worldwide

Wide choice of holiday villages, mostly in warm climates. Accommodation in hotels, bungalows or straw huts. Most resorts have swimming pools, bars, restaurants, nightclubs and many sporting activities. Some clubs do not accept children under 12, others have minimum age limits ranging from 4 months to 8 years. There are children's clubs (free), and the few which accept babies also provide crèches, cots, potties, sterilizers, bottle-warmers and pushchairs.

Prices (per adult per week, with flights, full board, including entertainment, wine, a range of sports and tuition)
Spain: £548 (low), £756 (high)
Thailand: £1192–£1628
Tunisia: £415–£691
Greece: £428–£535
Club membership fee: £8 per adult, £4 per child over 4.

Discounts Some give discounts of 90% for children under 2, 60% for 2–4s and up to 30% for 5–11s. Reductions are lower in high season.

Color Line Holidays

Tyne Commission Quay, Albert Edward Dock, North Shields NE29 6EA
Tel: 091 296 1313

Norway

Touring and 1- or 2-centre holidays with accommodation in hotels, pensions, farmhouses or family hostels. Self-catering chalets are also available for self-drive holidays. Some hotels offer swimming pools, playrooms and entertainments.

Prices
Low: £38 adult single fare
High: from £85
Children receive a 50% discount.
For a return crossing with cabin accommodation, 7 nights self-catering chalet (including car on ferry) is from £190 per person.
Child rates range between £80 and £100.

Discounts Under-4s always travel free unless they require a separate berth.

Cresta Holidays

Cresta House, 32 Victoria Street, Altrincham, Cheshire WA14 1ET
Tel: 061 926 9999 *or* 0345 626263

France

Hotel, apartment, and cottage/gîte holidays with a choice of fly/drive, self-drive

or air travel. The company also offers villas with pools. Several hotels have rooms with air-conditioning, telephone and colour TV. Swimming pools and sports facilities are generally available. Children are often free when sharing parents' room. Cots are subject to availability, and are mostly free of charge.
Prices available on request.
Discounts
BY CAR:
Hotels Children under 2 are free unless cots are charged for. Cot hire and meals are payable direct to the hotel. Children aged 2–13 receive reduction of £30 but are often free if sharing parents' room.
Apartments Under-4s are free. 4–13s pay approx. £20.
BY AIR
Hotels Infants under 2 pay a nominal flight charge of £35. Cot hire and meals payable direct to hotel. Children 2–11 years receive a reduction of £50.
Apartments Reduction approximately £10 per child.
(These reductions are 1992 reductions given as a guideline. 1993 reductions available on request.)

Cyprair Holidays

23 Hampstead Road, Euston Centre, London NW1 3JA
Tel: 071 388 7515

Cyprus

Hotel and apartment holidays offering accommodation usually with air-conditioning, central heating, private bathroom, balcony, telephone and radio. Swimming pools and tennis courts are available in many locations. Cots and highchairs are free and babysitting is available in larger hotels. At some hotels activities are organized for children, and most provide special food for children and babies.
Prices (per adult per fortnight including flight)
From under £300.
Discounts These vary. It is best to consult the brochure or ring the company for details.

Davies and Newman Travel Ltd

Norway House, 21–24 Cockspur Street, London SW1Y 5BN
Tel: 071 839 1192

Austria, France, Germany, Holland, Norway, Portugal, Spain, Switzerland

Hotel holidays (mostly short-breaks) in Europe. Accommodation ranges from 3-star hotels to luxury French châteaux. Some have swimming pools, tennis courts, games rooms and other sports facilities. Fly/drive holidays are also available. Cot hire is free in some hotels (ask when booking).
Prices Available on application.
Discounts Children under 2 pay a standard rate, depending on destination; 2–11s receive up to 60% discount if sharing with parents.

Embassy Leisure Breaks

PO Box 671, London SW7 5JQ
Linkline 0345 581811 *(brochure requests and bookings)*

UK

Over 150 hotel holidays in London and throughout Britain. Every hotel has a bar, restaurant, bedrooms with private bathrooms, colour TV, telephone and tea/coffee-making facilities. Cots can normally be supplied on request. Children's menus available.
Prices (per adult per week)
Bottom of range: £175–£190
Towards top of range: £350–£450
Discounts Children under 16 stay free when sharing a room with 2 adults. This offer is valid for 1 child per adult up to a maximum of 2 children. Two children under 16 travel free on rail-inclusive packages if travelling with 2 adults. Children occupying their own room receive 50% discount on the full adult price.

Enterprise Summersun

Groundstar House, London Road, Crawley, West Sussex RH10 2TB
Tel: 061 831 7000

Balearic Islands, Bulgaria, Canary Islands, Cyprus, Greece, Italy, Madeira, Malta, Portugal, Spain, Tunisia

Part of the Owners Abroad group, Enterprise offer a vast range of holidays in their Summer 93 brochure, ranging from small family-run hotels and *pensions* in unspoilt resorts to large lively hotels with many facilities.

Every property in the Enterprise Summer Sun 1993 brochure has been rated on its suitability for children. Certain properties offer such facilities as a children's pool or pool section, a playground, early suppers and special menus. Selected Sol hotels have a 'Family Floor' which is exclusively reserved for families with young children. There is a centre for children up to 4 years old, special facilities for mothers and a babysitting service. Certain properties in Tunisia are ideally suited for families.

Enterprise Sunbeam Clubs cater for children aged 3–11 years. These clubs are free and generally operate six days a week. Each of the clubs has specially trained children's representatives. Activities vary from resort to resort and include painting, beach games, parties and competitions. Sunbeam Clubs are operated at selected properties and resorts throughout the Enterprise Summersun and Enterprise Tunisia brochures. Enterprise Summersun also features an expanded range of single-parent family offers. These are available all season at hotels in Bulgaria and Romania.

Prices Lead-in price for a family of 2 adults and 2 children based on a Gatwick flnight is around £448 for 7 nights holiday, self-catering in Ibiza.

Discounts Children under 2 on the day of their return flight travel for £19, provided that they sit on an adult's lap during their journey. Meals and cots are available, payable locally. Discounts are available to first and second children, within the age limits specified at each individual property, sharing a hotel room or apartment with 2 full-paying passengers. The first child pays the child price in an apartment or studio and receives £50 off the adult fare for all holiday durations when sharing a room in a hotel.

Falcon Family Holidays

Groundstar House, London Road, Crawley, W. Sussex RH10 2TB
Tel: Central reservations 061 831 7000

Balearic Islands, Canary Islands, Cyprus, Greece, Portugal, Spain, Tunisia

Hotel and apartment holidays in a wide variety of Mediterranean resorts. Falcon have their own free children's clubs for three different age groups: 3–5 years; 6–11 years; 12–16 years (at selected hotels only). Creche available at one hotel, 6 days a week for toddlers from 15 months to 2½ years. Cots and pushchairs may be hired. Low-cost child insurance. Hotels offer supervised early suppers, free room patrols and parents' utility rooms.

Prices Self-catering (including flight): family of 3 from £480; family of 4 from £560. Hotel (half-board, including flight): family of 3 from £579; family of 4 from £678.

Discounts Children under 2 pay £19 for flight (no seat); cots and meals are payable locally. Thousands of free child places all season for early bookers; all first and second children qualify for the same low prices all season.

Finlandia Travel Agency Ltd

3rd Floor, 227 Regent Street, London W1R 7DB
Tel: 071 409 7334
Fax: 071 409 7733

CIS, Denmark, Far East, Finland, Norway, Sweden

As their name would suggest, Finlandia is particularly known for its knowledge of Finland, but the programme also includes holidays throughout Scandinavia, Russia and the Baltic states. The company offers flexible tour programmes including city breaks, multi-centre or combination tours from Helsinki to Stockholm, St Petersburg and Moscow by air, rail, or sea. Tailor-made tours and activity holidays are also arranged. Special 3- and 4-day pre-Christmas programmes to 'Santa's Lapland' are particularly popular; these include reindeer driving, snow-mobile safaris and a meeting with Santa Claus on the Arctic Circle. There is also a 6-day 'Christmas in Lapland' programme. All necessary thermal clothing is provided. The company offers other unusual tours, including a special New Year icebreaker experience in winter and, in summer, a 4-wheel-drive jeep safari and other activity programmes in 'The Land of the Midnight Sun'. There is a new programme to the Far East including Beijing, Bangkok, Singapore and Tokyo.

Prices City Breaks (per person for 3 days including flights, accommodation and breakfast) from: £283
Lapland Santa Claus Tours
From: £585

Discounts Low-cost flights to Helsinki are available from £205. Special rates for families are available on Santa Claus tours. Full details are available on request.

Golden Gateways

Kent Crusader Ltd, London Road, Southborough, Tunbridge Wells, Kent TN4 0PX
Tel: 0892 511808

Belgium, France, UK (South East)

Hotel (in England) and self-catering (in Belgium) holidays at seaside and inland locations in good quality accommodation. Many hotels have babyminding services, and a few have children's playgrounds and entertainment staff.

Prices An average price is £50 for 2 nights' bed and breakfast. Overseas hotel prices include ferry travel.

Discounts All hotels have reductions for children under 14 sharing parents' room, ranging from no charge (except meals) to half price.

Great British Holidays

61 Seamoor Road, Bournemouth
BH4 9AE
Tel: 0202 751700

UK

Specialists in holidays and short breaks throughout the UK. All holidays can be booked in advance by payment of deposit. 198 hotels and self-catering apartments and lodges in 108 destinations. Low priced rail and air travel available. Many hotels provide baby-listening services.

Prices (per adult on half board)
Hotel only from £25 per night to £160 per week

Discounts 0–15 years accommodated free (pay meals as taken direct to hotel) or 0–4 years accommodated free (pay meals as taken), 5–15 years 50% of adult rate inc. meals. Check individual hotel entries.

Highlife London and UK Breaks

PO Box 139, Leeds LS2 7TE
Tel: 0800 700 400

UK

Short break holidays throughout the UK in comfortable hotels, some with swimming pools and games rooms. Cots are widely available. Travel is not included but rail/air travel can be arranged to most cities.

Prices (per adult per night, B+B in 3-star hotels)
Windermere from £36
Plymouth from £33
London from £40
Edinburgh from £40

Discounts Children under 16 stay free in parents' room subject to availability. Meals extra. 75% of full adult price if occupying own room, meals included.

Horizon Holidays

Broadway, Five Ways, Edgbaston, Birmingham B15 1BB
Tel: 021 632 6282, 081 200 8733, 061 236 3828

Mediterranean

OSL Villa Holidays offer villas accommodating up to ten people in the Mediterranean. Most have swimming pools, and all include a free car. *HCI Club Holidays* offer a choice of hotels, apartments and villas, from full board to self-catering. Holidays include free sports and entertainment, and children's and teenager's

clubs. *Horizon Summer Selection* covers Europe, Africa and Florida; includes long haul, villas and apartments, and lakes and mountain holidays. Many hotels offer playgrounds, kindergartens, baby-sitting service, and the Hippo Club for 3–11 year olds, which includes games, activities, bedtime stories and baby patrolling.
Prices available on request.
Discounts For various child reductions please see brochure.

(2) and La Gomera (1) in good-quality accommodation ranging from 3-star to 5-star hotels plus self-catering apartments.
Prices (Winter 1992/93) from: £171 per person per week by air, self catering in Lanzarote; from: £232 per person per week, by air, B+B in Lanzarote.
Discounts Children 2–11 receive up to 20% discount. Adult third bed up to 10%. Insurance is free for children up to 16 accompanied by one adult.

Inghams Travel (Lakes and Mountains)
10–18 Putney Hill, London SW15 6AX
Tel: 081 785 7777

Austria, Bulgaria, France, Germany, Italy, Switzerland

Hotel holidays in the lakes and mountain regions, plus Austrian cities and tours throughout Austria, Bulgaria, French and Italian Alps, and Switzerland. Travel can be by air, coach, self-drive or rail. Some self-catering apartments are also available. Many holidays include a free excursion in their price. Some of the hotels offer baby-sitting and cots are prebookable. Many of the resort villages have sports complexes which cater superbly for children and toddlers. Most sporting/activity-related events organized by the resort take place in the school holidays from early July to end August.
Prices (per adult per week, half board in Austrian mountains, including flight) From £274 (summer 1993)
Discounts Children under 2 travel free on charter flights (no seat), and pay £18 on scheduled flights. Cots and meals payable locally. Where hotels offer children's discounts, 2–11s receive up to 40% off. This applies to one or two children sharing a room with two full-fare paying adults.

Inntravel
The Old Station, Helmsley, York
YO6 5BZ
Tel: 0439 71111

Belgium, France

Small, friendly company specializing in short break holidays, mostly out of season, in rural Belgium and France. Staff will offer advice on which hotels particularly suitable for families. Accommodation is in small, family-run hotels. Cots are generally available. Travel is by ferry and own car, or fly-drive to the South of France. The company have produced a new brochure of self-catering and hotel holidays that have been selected as being especially good for families. These properties are near to areas with children's activities (e.g. at a Swiss Chalet where children are encouraged to help on the farm, swim, cycle, etc.).
Prices (per adult for 3 nights' half board) Picardy: from £142
Normandy: £136
Discounts Children under 4 go free (meals extra). Reductions of £15 per holiday are offered to children aged 4–14. For Self-drive and Fly/drive holidays, children under 2 go free, 2–11 have a reduction of £50.

Inghams Travel (Canary Travel)
10–18 Putney Hill, London SW15 6AX
Tel: 081 785 7777

Canary Islands

Winter and Summer programmes in Gran Canaria (4 resorts), Tenerife (3), Lanzarote

Intourist Travel Ltd
Intourist House, 219 Marsh Wall, London E14 9FJ
Tel: 071 538 8600

CIS

Hotel holidays, including journeys on the Trans-Siberian Railway. No cots available,

but wherever possible, children may use folding beds in parents' room.
Prices (per adult, including all travel and meals)
Moscow + St Petersburg, 1 week: from £385
3-night break in Moscow: from £285
Discounts Children under 2 pay £10 for flight (no seat, and food costs to be paid locally); meals payable locally. 2–11s sharing parents' room receive 25% discount (1 child per adult).

Island Holidays

Maxwell House, South Esplanade, St Peter Port, Guernsey, Channel Islands
Tel: 0481 721897
Fax: 0481 714047

Channel Islands (Guernsey, Herm, Sark)

Hotel, guest house and self-catering holidays. Some hotels offer children's swimming pools, games facilities and babysitting. Cots are usually available for hire.
Prices (per adult per week in Guernsey hotel, including flight from Southampton)
Low: £200
High: £360
Discounts *By air:* children under 2 travel free (no seat); 2–11 receive a 50% discount. *By sea:* similar discounts apply, but age limits vary with each carrier.

Jetsave Holidays

(*see also* **Fly/drive**)
Tel: 0342 312 033 *(inclusive holidays)*
0342 327 711 *(flights only)*

US, Canada, Southern Africa

Hotel and self-catering villa and apartment holidays in Florida, California, South Africa, Hawaii, Botswana and Zimbabwe. All accommodation is geared towards families and most offer swimming pools, playgrounds, games rooms and sports facilities. Cots and baby-sitting can be arranged on request in some locations.

Discounts Details of discounts for children are available on request.

Just France

1 Belmont, Lansdown Road, Bath
BA1 5DZ
Tel: 0225 446328
(*see also* **Cottages, Gites and Farmhouses**)

France

Large range of hotels, from simple Auberges to luxurious châteaux. Some are particularly well-suited to family travel, providing children's menus and baby listening services.
Prices (per adult per week including ferry crossing) from: £192–£667
Discounts Children receive standard travel discounts (*see* **Ferries**) if you book a ferry crossing as well as one week's accommodation through this company (before July 11th or after 5th September), parties of 8 or more adults or children over 4 receive free ferry travel.

Manos Holidays

168–172 Old Street, London EC1V 9BP
Tel: 071 608 1161

Cyprus, Turkey, Greece

Offer 2- to 5-star hotels and mountain resorts in Cyprus, mainly studios and apartments in Greece (but some hotels) and B+B, *pensions* or small hotels in Turkey. They do 'Go Green' holidays for the environmentally-conscious family with discounts for cleaning up the beaches and helping to save the turtles – there is a Tommy Turtle Club for children in various places in Greece. Cots are available in all resorts.

Manos Holidays also do 7-night cruises from Cyprus to Egypt and back, or 7-night cruises to 5 Greek islands, Egypt and the Holy Land.
Prices Please refer to relevant brochure.
Discounts Generally there is a discount for children up to 15 in Greece of 25% and for 2-12s in Cyprus. There is a 5% discount for senior citizens and special offers for singles between April to May and October.

Hotel holidays 145

Meridian Holidays
12–16 Dering Street, London W1R 9AE
Tel: 071 493 2777
Canary Islands, France, Greece, Portugal, Spain
Holidays in comfortable hotels and self-catering apartments, plus golfing holidays in France. Some of the hotels in Greece, Spain, Portugal and the Canary Islands have children's pools, clubs and play areas. Travel in own car to France, and by air to other destinations, but transfer can be arranged.
Prices On request.
Discounts On request.

Metak Holidays
70 Welbeck Street, London W1M 7HA
Tel: 071 935 6961
North Cyprus, Turkey
Hotel holidays, mainly in Turkey. Very few hotels have specific children's facilities, but many have swimming pools and various sports on offer. Coach tours, fly/drive and flight only and a few apartments are also available.
Prices (per adult per week, 1992 prices) Turkish coast B+B: £224 (April) £274 (August)
Istanbul: £250–£270 (budget hotel), £375–£425 (luxury hotel)
North Cyprus: £309–£534 (B+B)
Discounts Children under 2 £40 (no seat, meals extra). Most hotels offer 15% discount for 2–12s sharing with parents.

Multitours
Tel: 071 821 7000
Gozo, Malta
Hotel and apartment holidays mainly in 3- and 4-star hotels. The majority have swimming pools. A few have baby-listening services and play areas. Many have games rooms and water sports facilities.
Prices (per adult per week, B+B)
Malta: £197
Gozo: £214
Discounts Children under 2 pay maximum £25 on all Multitours products, including flights. If flights are not included, children under 2 travel free. Cots and meals payable locally. Children aged 2–12 are sometimes accommodated free (meals extra), otherwise they receive discounts ranging from £20–£100 (breakfast included in some hotels).
One-parent families Some hotels offer special deals. Details on request.

David Newman's European Collection
Box 733, 40 Upperton Road, Eastbourne BN21 4AW
Tel: 0323 410347
Belgium, France, Italy
Chambre d'hôte and château holidays arranged for the independent motorist. Family facilities are frequently excellent, with ample sports and leisure amenities. Cots can often be provided on request, but you may be advised to take your own travel-cot. The proprietor offers lots of helpful, no-nonsense advice. This company also acts as an agent for Cuendet Italia (*see* **International Chapters** in **Villas and apartments**).
Prices These vary greatly. Please telephone for details.
Discounts Children under 4 travel free and are usually accommodated free when sharing parents' room. Cots and meals payable locally. 4–13s receive around 20% discount.

North Sea Ferries Holidays
North Sea Ferries, King George Dock, Hedon Road, Hull HU9 5QA
Tel: 0482 77177 (reservations)
Belgium, France, Germany, Holland, Luxembourg
Hotel and self-catering holidays with self-drive option. Accommodation varies from standard family hotels to de-luxe hotels, offering indoor pools, sauna, gymnasium and TV lounge. Self-catering accommodation is in bungalow parks which usually have ample shopping, sports and leisure facilities. Also Citybreaks, go-as-you-please' driving holidays in the Benelux countries, Germany and France,

146 Choosing and planning your holiday

Disney Holidays, and bungalow parks in Belgium.
Prices Available on request.
Discounts Details of discounts for children are available on request. They vary from holiday to holiday.

Olympic Holidays

Olympic House, 30–32 Cross Street, London N1 2BG
Tel: 071 359 3500 (reservations)

Cyprus, Greece

Hotel, *pension* and villa holidays in accommodation ranging from simple to luxurious. The majority of hotels offer children's pool and cots (often free of charge), and babysitting is available on request. Daytime flight departures to most destinations. Some hotels have facilities for disabled travellers.
Prices (per adult per week, in hotel in Cyprus)
5-star hotel: £495 (low), £599 (high).
Discounts Children under 2 pay £19 on charter flights (no seat). Some free holidays are available for children aged 2–11 in low season. At other times substantial discounts apply. Reductions are available in Cyprus for single parents, and some hotels do not require parents to share rooms with children.

Vikki Osborne

Nelson Road, Newport, Isle of Wight
PO30 1RD
Tel: 0983 524221

UK (Isle of Wight)

Wide choice of hotels ranging from modest to luxurious, self-catering apartments and holiday centres. Many have baby-listening services and some have swimming pools, games rooms, paddling pools, play areas, cots, high chairs and special meals. Travel can be arranged by rail, coach or car.
Prices (per adult per week, travelling by car in 1992 were): from £183
Discounts Children under 3 travel free (meals and cots extra). Children aged 3–4 receive slightly over 50% discount; 4–14s receive 50% discount.

PAB Travel

Central Hall, 3 Ryder Street, Birmingham B4 7NH
Tel: 021 233 1252

Ireland

Go As You Please motoring holidays in Ireland with accommodation in hotels, farmhouses or guest houses. Travel by air or ferry for as long as you like. Also self-catering holiday cottages and caravans in Ireland.
Prices From £124 per adult by ferry. Self-catering holidays from £99.
Discounts Child reductions for children up to 12 years sharing parents' room, discount of 25–50%. Self-catering: on certain dates children may be free or the second week may be free.

Page & Moy Ltd

136–140 London Road, Leicester
LE2 1EN
Tel: 0533 542000

Canada, CIS, Europe, Hawaiian Islands, US

Hotel holidays and coach tours in many locations round the world. Accommodation is in comfortable hotels, many with swimming pools. Cots may usually be hired. A Mediterranean cruise is also available.
Prices (per adult per week)
Spain (bed & breakfast and half-board): from £449
Hawaii (room only): £779
Discounts Children 2–11 years, reduction of 25% when sharing with parents. Cots available.

Panorama Holiday Group Ltd

29 Queens Road, Brighton, East Sussex
BN1 3YN
Tel: 0273 206531

Ibiza, Tunisia

Hotel and apartment holidays, plus holidays in mobile homes and tents in Ibiza. Most hotels offer children's swimming pools. Cots and high chairs can often be hired and baby-sitting is usually available

Hotel holidays 147

on request. Fully-trained children's representatives organise special activities and will take children for 5 hours a day, 5 days a week. This facility is available in Ibiza (the Sandcastle Club) and Tunisia (free of charge). The holiday programmes are called 'Ibiza Experience', 'Tunisia Experience', and Panorama also operate 'Rail Experience' and 'Ski Experience' (further details on request).
Prices Please see brochures for details of current prices.
Discounts Children's discounts are available – please refer to brochure.

Portland Holidays

218 Great Portland Street, London
W1N 5HG
Tel: 071 388 5111

Austria, Cyprus, Greece, Jamaica, Kenya, Madeira, Majorca, Malta, Portugal, Spain, Tunisia, US

This company offers a wide range of hotel, villa and apartment holidays. Many hotels are geared to families, offering children's pools, play areas, cots, high chairs and early suppers. Baby-sitting is sometimes available. A children's club (Portland Pirates) is available for 5–11 year olds.
Prices 1 week half board in Majorca from £215 May, from £278 July.
Discounts Children under 2 pay £15 for flight (no seat); cots and meals payable locally. 2–11s receive 10–70% discount in certain hotels on specified dates. This applies to 1 child sharing with 2 adults. Elsewhere, including villas and apartments, 1 or 2 children sharing a room with 2 adults each receive a 10% discount.

Preston Holiday

4 Dollis Park, London N3 1JU
Tel: 081 349 0091

UK (Channel Islands)

Hotel and guest house holidays, mainly on Jersey. Most places have baby-listening facilities, and some have swimming pools, games rooms, children's pools and early meal-times. A few have play areas and special menus. Cots usually available.

There are also a few apartment holidays (Guernsey) and a campsite (Jersey). Travel can be arranged by air or sea.
Prices (per adult per week, half board with travel by sea)
Guest house: from £165
Hotel (medium grade): from £194
Discounts Many free children's places are on offer at certain hotels. Otherwise, children's prices are from £49 per week.

Pullman Holidays UK Ltd

31 Belgrave Road, London SW1V 1RB
Tel: 071 630 5111

Israel

Hotel, kibbutz, inn and self-catering apartment holidays, with fly/drive option. Some hotels offer children's pools and babysitting on request. Cots may be hired.
Prices (per adult per week, fly/drive kibbutz holiday)
£316 (winter) – £490 (summer)
Discounts Children under 2 travel for £35 (no seat); cots and meals payable locally. 2–12s sharing room with 2 adults receive 20–30% discount.

Scandinavian Seaways

Scandinavia House, Parkeston Quay, Harwich, Essex CO12 4QG
Tel: 0255 241234 *(Harwich)*,
091 296 0101 *(Newcastle)*

Denmark, Germany, Norway, Sweden

Touring or 'stayput' holidays with a choice of accommodation in hotels, inns, farmhouses, self-catering apartments or campsites. Facilities for children vary from place to place, but normally cots can be reserved on request.
Prices A short break to Denmark involving 2 nights on ship and 2 nights in a hotel (B+B) costs from £113 per adult in the low season. A self-catering apartment with 2 nights on ship and 2 nights in apartment costs around £75 per adult (winter).
Discounts Children under 4 travel free on ferry. Cots and meals are payable locally. On the prices quoted above, children aged 4–15 receive a reduction of £2 (winter).

Seymour Hotels and Holidays

15 Mulcaster Street, St Helier, Jersey, Channel Islands
Tel: 0534 73485

United Kingdom (Channel Islands)

Small company offering holidays in medium- to high-grade hotels. Some have swimming pools and games rooms, and one also has a paddling pool, children's club, special menus and early mealtimes. Cots are available in all hotels.
Prices (per adult per week, half board 1992 prices)
Medium-grade hotel: £178.50–£262.50
High-grade hotel: £315–£392
Discounts Children under 2 travel free. Under-12s receive discounts of 10% if in own room, or 50% if sharing with 2 adults.

Simply Simon Holidays

1/45 Nevern Square, London SW5 9PF
Tel: 071 373 1933

Greece, Spain

Small company offering holidays in Greece, the Greek Islands and Spain. Accommodation in small family-run hotels, *pensions* and tavernas. Most hotels are by the beach but there are no specific children's facilities, and cots are not always available.
Prices (per adult per week, B+B)
Greece: £180–£230 (*pension*)
£180–£240 (hotel)
Discounts Children under 2 usually travel free, with any accommodation charge payable locally. Reductions are available for a third person sharing, varying according to destination and season.

Ski Esprit

Oaklands, Reading Road North, Fleet, Hants GU13 8AA
Tel: 0252 616789 (24 hours)
Fax: 0252 811243
Brochure: (24 hours) 0252 625175/616789

France, Switzerland

Ski Esprit specialises in family skiing holidays. Their chalet-based programme covers eight major resorts in France and Switzerland. Each resort has in-chalet creches, operating all day and caring for children from 4 months to 4 or 5 years old. They are staffed by British qualified nannies (NNEB or RGN). Older children attending ski school are also looked after by Ski Esprit staff who give them lunch and take them back to lessons in the afternoons or run an activities programme for children who have morning-only classes. Among other family features are a separate high tea for under 14 year olds (to leave dinner as an adults only meal), free babysitting and very generous children's discounts.
Prices From £248 for a fully catered chalet holiday by air.
Discounts Details on application.

Skytours Holidays

Broadway, Five Ways, Edgbaston, Birmingham B15 1BB
Tel: 021 632 6282, 081 200 8733, 061 236 3828

Canary Islands, Corsica, Corfu, Crete, Cyprus, Greece, Ibiza, Kos, Majorca, Malta, Minorca, Portugal, Rhodes, Spain, Tunisia, Turkey

Large package company (part of Thomsons and sister company of Britannia Airways), offering summer holidays all around the Mediterranean. Many hotels and apartments have facilities for children including separate pool, playgrounds, cots, high chairs, early suppers and a free evening baby patrol up to midnight. There may be a charge per night for a cot (if available), which must be paid to the hotel, together with the cost of baby food. Snap Dragon children's club offers daily activities for 3–11s.
Prices For prices and discounts please refer to brochure.

Sovereign Scanscape Holidays

Astral Towers, Betts Way, Crawley, West Sussex RH10 2GB
Tel: 0293 599922

Denmark, Finland, Iceland, Norway, Sweden

Hotel holidays

Regional departures by air and sea. Wide range of accommodation available from country farmhouses to self-catering farmhouse cottages. Legoland in Denmark is popular, where hotels are geared to family needs, and include baby-sitting and child minding facilities. Cots for infants are included in the holiday price. Single parent families are also catered for, with generous reductions for the children.
Prices
Low: Family of 2 adults and 2 children under 16 staying at a farmhouse near Legoland on a 4-night short break, travelling by sea with car, including breakfast: total £484, including entry to Legoland.
High: Details as above: £630.
Discounts Children under 12 travelling by air to Legoland receive a 40% discount on the adult price. Children under 16 sharing with adults up to 50% off.

Steepwest Holidays Ltd

130–132 Wardour Street, London
W1V 3AU
Tel: 071 629 2879/434 1230 longhaul

North Cyprus, Turkey

Wide range of holidays in hotels and self-catering apartments. Fly/drive and flight-only deals are also available. Property throughout Turkey and North Cyprus. Two-centre holidays are also offered. For details of facilities available for children in 1993 season, contact Steepwest Holidays.
Prices (per person for 14 nights self catering): from £325
Discounts Details of discounts available for children in 1993 season are available on request.

Sun Blessed Holidays

19/21 Southbourne Grove,
Bournemouth, Dorset BH6 3QS
Tel: 0800 373111 (freefone)

UK (Channel Islands)

Tailor-made short- or long-stay holidays in Jersey and Guernsey by air from your local airport, or by sea. Several grades of hotel and guest house are available. Many hotels have swimming pools, indoor leisure facilities, tennis courts or games room, and baby-listening service. Cots are provided on request and cost £5 a day on average.
Prices Please refer to brochure.
Discounts In most cases under-2s are free. Children aged 2–11, 50% discount. All discounts are based on a child sharing with two adults.

Sun Esprit

Oaklands, Reading Road North, Fleet,
Hants GU13 8AA
Tel: 0252 816004
Fax: 0252 811243
Brochure: (24 hours) 0252 816004

France

Sun Esprit devote their holidays entirely to the family market. They offer hotel and self-catering options in France, based in Aquitaine (south west coast) and the Alps. Their resort staff include qualified British nannies, who provide babysitting on 6 nights each week. Freedom for parents and excitement for the youngsters are the hallmarks of a holiday with Sun Esprit. No surcharges. Sells direct to the public and through ABTA agents. For changes to 1993 programme please contact the company.
Prices and discounts Available on request.

Sunspot Tours Ltd

96 Tooley Street, London SE1 2TH
Tel: 071 378 8111

Gozo, Malta, Tunisia

Hotel and apartment holidays in the Mediterranean. Most hotels have swimming pools and some have games rooms, tennis courts and watersport facilities, particularly those in Tunisia. A few provide children's pools, baby-sitting, high chairs and early meals. Cots are usually available free.
Prices (per adult per week, half board in a medium-grade hotel)
Malta: £205
Tunisia: £220–£300
Discounts Children under 2 pay £40 each, which includes cot hire. Meals extra.

Discounts of £11–£140 are available in most hotels for 2–11s (2–5s or 2–8s in a few cases) if sharing with 2 adults. There may be free places for children during the summer.

Sunvil Travel

Sunvil House, Upper Square, Old Isleworth, Middx TW7 7BJ
Tel: 081 568 4499

Cyprus, Greece, Hungary, Italy, Portugal

This company offers holidays in hotels, tavernas, villas and apartments, plus a fly/drive option. Some hotels have children's pools and play areas. Cots may be hired for about £2.50 per day. The company advise that their sites in Cyprus and Greece are the only areas really suitable for families with young children.

Prices (per adult, self-catering in Corfu, including day flight from Gatwick)
1 week: £245–£309
2 weeks: £277–£342 (These are 1992 prices, a 5% rise is estimated for 1993.)

Discounts Under-2s pay either £15 (Greece) or £30 Cyprus (no seat on flight). No discounts on fly/drive holidays. Children aged 2–11 receive up to 30% discount on the basis of 1 child per 2 full fare-paying adults in Cyprus and up to £50 in Greece. The child must occupy a third bed in adults' room.

Taber Holidays

Norway House, 126 Sunbridge Road, Bradford, W. Yorks BD1 2SX
Tel: 0274 393480 *or* 081 441 4010

Germany, Norway

Specialist operator to Norway and Germany offering travel by air or sea, with or without a car. Escorted holidays and self-catering are also available. Choose from bed and breakfast or half board in a wide range of hotels. Cots, high chairs, etc may be available in some hotels. Special food may be arranged for babies if requested in advance. Facilities for disabled travellers are usually available, but you are advised to discuss your requirements with the company in advance.

Prices and discounts are available on request.

Thomson Holidays

Greater London House, Hampstead Road, London NW1 7SD
Tel: 081 200 8733 *(London)*
021 632 6282 *(Birmingham)*
061 236 3828 *(Manchester)*

Worldwide

A very large package company with hotel, villa and *pension* holidays, as well as 'theme' holidays, such as 'Small and Friendly' and 'Lakes and Mountains'. In the large hotel/apartment resorts in Spain there are lots of facilities for children, including a special club with supervised play and games up to 3 hours a day, 6 days a week, children's pools, early meals, cots (payable locally) and a baby-listening service.

Prices and discounts Details of the vast range of prices and discounts available should be obtained direct from the operator or relevant brochure.

Travelscene Ltd

6 Travelscene House, 11/15 St. Ann's Road, Harrow HA1 1AS
Tel: 081 427 8800
Reservations: 081 274445

Europe, North Africa, US, Canada

Short breaks to the cities of Europe. Travel can be arranged by air, train, car or coach and accommodation is offered in 2–5-star hotels. 35 departure points in UK.

Prices (per adult in high season, including flights)
Paris (3 nights B+B): low £149 (Heathrow)
Amsterdam: low season £135 (Gatwick), £159 (Heathrow)
5–10% increase in high season

Discounts Under-2s pay £15; cots and meals payable locally. Children aged 2–11 receive £20 discount each on rail and air, £10 on coach.

UK Express

Third Floor, Whitehall House,
41 Whitehall, London SW1A 2BY
Tel: 071 839 3303

Turkey

Package holidays and coach tours to coastal and city destinations in a wide range of accommodation. Larger hotels may have children's swimming pools. Cots may be available – ask when booking.

Prices (per adult per week, B+B in Istanbul, including flight)
Low: from £250
High: from £295

Discounts Under-2s pay £45; 2–11s sharing a room with 2 full fare-paying adults receive 25% discount on all holidays.

Please note: These are 1992 prices given as a guideline only.

VFB Holidays Ltd

Normandy House, High Street,
Cheltenham, Glos GL50 3HW
Tel: 0242 526338

France

Auberge holidays in rural areas of France, such as Alsace/Jura, Aquitaine, Auvergne, Brittany, Burgundy, Languedoc, Loire Valley, Midi/Pyrenees, Normandy and Provence. The company will suggest suitable hotels for children, which provide recreational facilities, early suppers, special menus or babysitters. They can also organize a tour of France arranging 3–4 night stays in any of their 50–60 'Stay Put' hotels. Another favourite with families is the activity holiday programme in the French Alps, where such outdoor pursuits as tobogganing can be arranged during the months of July and August. Self-catering cottages available in some of the loveliest rural areas of France, no less than 450 in current programme.

Prices (per adult, including return Dover/Calais ferry crossing with car and 3 nights' accommodation with half board) Hotels vary widely, but average £120–£160.

Discounts These depend on the individual hotel. Children sharing parents' room may receive a discount of 20–30%.

Virgin Holidays Ltd

The Galleria, Station Road, Crawley,
West Sussex RH10 1WW
Tel: 0293 562944

Canada, US, West Indies

Hotel holidays in the US (Florida and New York), Jamaica, the Bahamas and the Cayman Islands. Hotels are of a high standard and most have swimming pools, games rooms and sports facilities. Some also provide children's pools, special menus, play areas and cots. (Some hotels may charge for cots.) Apartments, fly/drive, fly/ride, and cruising holidays are also available. Virgin produce an excellent range of holidays that are suitable for disabled travellers. They are listed in a separate brochure, available from the number listed above. They also offer fly-rides (Harley Davidsons) and train journeys coast-to-coast.

Prices (per adult per week, accommodation only)
Florida Flydrive: from £299
Flydrive Ski: from £299
California Flydrive: from £389
Boston Flydrive: from £249

Discounts Under-2s pay £40 on all holidays; 3–11s sharing with 2 adults receive 25–35% discount.

Rail travel

Rail travel has been around a long time and many people tend to think that this must make it cheap. In some instances this can be perfectly true, but in others it can be quite wrong – even horrifyingly expensive. For example, travelling from New York to San Francisco by trains costs more than travelling by air; it also takes two days longer! However, speed isn't everything. Perhaps you want to take things at a more leisurely pace and see the countryside rather than zoom above it hidden in cloud.

Whatever reasons you have for choosing to travel by train (and despite the few practical incentives that train companies offer), there are some good value tickets to be had, and 'family cards' often allow very generous discounts. It is well worth taking a small amount of food and an ample supply of cartoned drinks on such a journey. The buffet car may be non-existent or only serve drinks unsuitable for young children. You are advised to read the articles on avoiding boredom in Sections 4 and 5. A flat-pack of toilet paper is often a godsend!

Certain discount tickets or rail passes may be bought in the UK, usually through tourist offices. If you're committed to train travel, do buy them as the savings can be substantial.

A word of warning about rail travel: in some countries, such as Turkey and Italy, it can be tediously slow, with trains stopping at every small station and taking literally hours to cover a distance that could be bicycled more quickly. Rapid or Express trains do exist, but in our experience they can still be unbelievably slow. In some countries, too, rolling stock can be very old. It might look picturesque, but it can also be pretty uncomfortable, so take something soft to sit on. If you can book seats, do so. It's maddening and frustrating to queue for hours on the morning of your departure to find that the train was booked solid the day before.

If you have a long journey planned, it is always advisable to book a couchette or sleeper. Children find this exciting, but the main reason is to get some comfortable sleep and help keep up the children's normal routine. It also gives you a little peace and quiet. *Note:* It is not unusual for a baggage allowance to be specified by the rail company and for it to vary, depending on what class you travel. Children's baggage allowance is usually half that of adults'. Check with the company concerned or the National Tourist Office of the country concerned before travelling.

Prices

Unless otherwise stated 1992 prices have been quoted. These are intended only as a guide to the type of holiday offered and travellers should check with the company concerned for 1993 pricing details.

The entries in this section are arranged alphabetically by country rather than by railway name.

Railways of Australia

c/o Longhaul Leisurail, GSA
Department, PO Box 113, Peterborough
PE1 1LE
Tel: 0733 51780

Railways of Australia operate a network of air-conditioned rail services linking coast to coast of this vast country. Itineraries of up to 5 or 6 days on trains such as the Ghan or the Prospector allow you to travel in comfort through the sometimes fierce heat, but still get a real insight to Australia. Some states, such as Queensland and New South Wales, operate their own rail tours. Hot and cold showers are standard on most trains, as are dining cars, lounge club cars or buffet cars. Sleeping accommodation is in several categories, ranging from private bedrooms with day lounge to reclining seats. On long-haul journeys, sitting cars are not usually available – you must occupy a sleeping berth, and this often means that meal charges are compulsory. Video TV and recorded music are available on some trains.

Discounts Children under 4 travel free provided they do not occupy a separate seat or berth. Bassinets are available if requested at time of booking. Children aged 4–15 pay half fare, unless occupying a sleeping berth and having meals, when they pay full fare.

The best value for continuous travel is the *Economy Austrailpass*. A 14-day pass costs approximately A$415 and provides seated accommodation. Economy berths are only available on a limited number of services and the economy pass does not allow you to pay additional fare and travel first class.

Note: If you plan to stay several days in various cities and resorts, point-to-point rail travel may be more economical. *CAPER* tickets, which must be booked and paid for at least 7 days in advance, can save you up to 30% on some journeys.

Austrian Federal Railways

c/o Austrian National Tourist Office,
30 St George Street, London W1R 0AL
Tel: 071 629 0461

Austrian Federal Railways cover 5800 km and have direct connections with all European countries. A small extra charge is payable for reserving a seat, and for travel on express trains. Motorail links are available between Vienna, Italy and Yugoslavia.

Discounts Children up to the age of 6 travel free provided they do not occupy a separate seat; 6–15s pay 50% of the adult fare.

Nationwide Area Tickets (valid for 1 month) entitle you to travel on any train in Austria and include a 50% discount on tickets with private rail companies and Lake Constance shipping companies. The *Rabbit Card* (valid for ten days) allows you to travel on four days out of the total ten on any train in Austria.

Province Tickets, for travel in the 8 provinces of Austria, entitle you to similar discounts.

Group tickets for 4–9 people travelling on same route in same class are entitled to 30% reduction, and children travel at half the reduced price.

Belgian National Railways

Premier House, 10 Greycoat Place,
London SW1P 1SB
Tel: 071 233 0360

Belgian trains offer no special facilities either for children or overnight travellers. However, many types of ticket are available and some may be purchased at Victoria in London. On internal tickets journeys cannot be broken, but international tickets are more flexible. Bicycles can be hired from main Belgian railway stations all year round. A list of participating stations is available on request. Children under 6 must not occupy a seat on crowded trains – however, trains are rarely crowded.

Discounts Children under 6 travel free on internal tickets, provided they do not occupy a separate seat on crowded trains; 6–11s pay 50% of adult fare. On international journeys travellers aged 12–25 receive a discount. At weekends a first person receives 40% discount, and up to 5 others (if travelling in a group) receive 60%. The first person does not have to be in a group to qualify for this discount.

B-Tourrail tickets, valid for 5 days, must be used within a 17-day period. There are reductions for people under 26. The *50% Reduction Card*, costing 550 francs and valid for one month on the whole network, allows you to buy an unlimited number of single tickets at half fare.

Details of other reductions for young people and families are available on request.

British Rail (BR)

> Administration Head Office, Euston House, 24 Eversholt Street, London NW1 1DZ
> Tel: 071 928 5151

An extensive and efficient rail system covering all parts of the UK, with an Inter-City link to Ireland. Details of services and fares are best requested from the nearest local station or BR-appointed travel agent. Sleepers may be booked in advance. A buffet/restaurant car is available on many services.

Discounts On ordinary single and return tickets children aged 5–15 pay half the reduced adult fare. Children under 5 travel free. Ask at the ticket office.

The *Family Railcard* costs (1992 price) £20 and is valid for one year. It allows adults a discount of between 25–33% on standard class tickets, and up to 4 children under 16 years to pay £1 each. There are no discounts on sleepers. This card is only available to people aged 18 and over who are resident in the UK.

All-Line or *Regional Rover* tickets offer unlimited rail travel within specified areas, with certain ferry sailings included. The All-Line Rover costs £200 (standard class) or £320 (first class) for 7 days and £320 (standard class) or £500 (first class) for 14 days. Under-5s travel free. Children aged 5–15 receive one-third discount.

Danish State Railways (DSB)

> c/o Scandinavian Seaways, Scandinavia House, Parkeston Quay, Harwich, Essex CO12 4QG.
> Tel: 0255 240240, 0255 241234

DSB and a few private companies cover the country with a dense network of rail services supplemented with buses on quieter stretches. Refreshments are available on most trains.

Discounts Under-4s travel free. Children aged 4–11 pay half fare. Discounts are available for groups of three or more.

Hellenic Railways Organization (OSE)

> c/o National Tourist Organization of Greece, 4 Conduit Street, London W1R 9TG
> Tel: 071 734 5997

Limited and slow railway network on the Greek mainland only. While trains are cheaper than buses, they can be old-fashioned and rather spartan. A popular misconception is that they are much slower than buses. This is untrue, and over long distances may be preferred as less crowded, and equipped with toilets. It is advisable to book seats. There are no special arrangements for children. Sleeper cars are available on long-distance trains and may be booked in advance.

Discounts Under-4s travel free; 4–12s pay half fare.

Season tickets, both first and second class, are available at reduced rates. They allow unlimited travel for the duration of the ticket. An individual Tourist Card can be granted to individual passengers, families and groups of up to five people. These operate within a chosen time limit (10, 20 or 30 days) and allow the holder reduced rates.

Irish Rail

> CIE, 185 London Road, Croydon, Surrey CR0 2RJ
> Tel: 081 686 0994

An extensive service covering all parts of the country, offering Inter-city and suburban connections. As no journey is very long, there are no sleeping cars.

Discounts Children under 5 travel free. Other children under 16 pay half fare up to a maximum of £10 on scheduled services.

The *Family Ticket* allows discounted travel for families (father travelling with up to 4 children, mother travelling with up to 4 children, or husband and wife together without children). Details from CIE.

Israel Railways

c/o Israel Government Tourist Office, 18 Great Marlborough Street, London W1V 1AF
Tel: 071 434 3651

Not an extensive network, but cheaper than buses. The railway is not highly thought of as a means of public transport, but it is good for people who are not in a hurry (especially families and children) because it is slow and scenic. But there are frequent quick and comfortable trains during the day between Tel Aviv and Haifa. Trains run between Nah, Tel Aviv, Haifa and Jerusalem. Seats may be reserved in advance for a small extra charge. Most trains have a buffet car. No trains run on the Sabbath or on Jewish holy days.
Discounts Children under 3 travel free; 3–12s travel at half fare.

Italian State Railways (CIT)

Marco Polo House, 3/5 Lansdown Road, Croydon CR1 1LL
Tel: 081 686 0677
Fax: 081 686 0328

An extensive and efficient network covering most of the country. There are several classifications of train, from very slow 'locale' trains that stop at every station, to first-class 'super-rapidos' which offer a fast service between the main cities. Rail travel is a good (and cheap) way of seeing Italy.

The CIT office in Croydon can sell tickets from England to Italy and rail passes within Italy in addition to tickets to and from specific places in Italy. The staff will also give timetable information.
Discounts Children under 4, not occupying a seat, travel free; 4–12s pay 50%. Children under 12 are eligible for any pass at a discount of 50% on the adult fare.

Biglietto turistico allows non-Italian tourists unlimited travel on any train and may be purchased in the UK or at major railway stations in Italy. They are valid for 8 consecutive days to 30 days and cost from £88–£152 (second class), depending on duration.

Chilometrico tickets are valid for 3000 km and may be used by up to five people at the same time for a maximum of twenty separate journeys. These cost £90 (second class).

Japan Railways (JR)

c/o Japan National Tourist Organization, 167 Regent Street, London W1R 7FD
Tel: 071 734 9638 *or*
4296 2029/4296 0794 *(Paris)*

General information on fares and timetables is available from the tourist office. For more specific information contact the Paris office. JR operates an efficient and complex network of rail services. There are play areas on some of the modern express services.
Discounts 1 adult can take 2 children free (under 6 years old), but must pay for additional children; 6–11s travel at half fare.

A *Japan Rail Pass* is available to foreign tourists and costs 27,800 yen for 7 days. This must be purchased outside Japan. It is available from Japan Airlines, if you are flying with JAL, but there are several other agencies where it may be purchased independently. Details from the tourist office. The rail pass also allows travel on some bus and ferry services.

Malayan Railways (KTM)

c/o Tourism Promotion Board (Malaysia), 57 Trafalgar Square, London WC2N 5DU
Tel: 071 930 7932

The Tourism Promotion Board can only give information, and does not provide a booking service. The train service is very comfortable and of a high standard, but is slow. Two main lines operate: one runs along the west coast to Singapore then turns north to Kuala Lumpur and Butterworth, meeting Thai Railways at the border; the other line runs to the north-east near Kota Bharu. Express or normal services are available. On overnight trains it is possible to book cabins for daytime use as well as night – useful for families with children. Sleepers must be reserved in advance. Some trains are air-conditioned.

Discounts A *KTM Railpass* entitles the holder to unlimited travel in any class and to any destination for a period of 10 or 30 days but is available only from rail stations in Malaysia; not issued by this agency. Sleepers are extra.

Moroccan Railways

c/o Moroccan Tourist Office, 205 Regent Street, London W1R 7DE
Tel: 071 437 0073

Tangier, Rabat, Casablanca, Meknes, Fez, Marrakesh, Taza and Oujda are linked by modern, air-conditioned trains. Timetables are available from the tourist office. Fares are very low.

Discounts Children under 4 travel free; 4–12s receive 50% discount. Groups of six or more receive a 20–50% discount.

Family Railcard: 2 adults with up to 3 children, under 18 years old, receive 20–50%.

New Zealand Railways Corporation

c/o Longhaul Leisurail, GSA Department, PO Box 113, Peterborough PE1 1LE
Tel: 0733 51780

Some long-haul trains have air-conditioning, snack bar and restaurant. Other trains have no refreshment facilities. Seats are guaranteed if booked up to 72 hours in advance. No sleeping accommodation is available. Take your own carrycot for babies. There are often Mother & Baby rooms in stations.

Discounts Children under 4 travel free; 4–14s pay half fare.

The *Travelpass*, valid for 8, 15 or 22 days, offers similar child discounts and allows unlimited travel by train, coach and ferry. The cost of a pass is $396, $492 and £644 for each duration. Prices subject to change.

Norwegian State Railways

21–24 Cockspur Street, London SW1 5DA
Tel: 071 930 6666

A fairly small railway system with no connections to the extreme north of the country. Bus and air routes cover the gaps. Special compartments with washing and nappy-changing facilities are available on long-distance trains.

Discounts Children up to 4 travel free; 4–15s travel at half fare.

A *Nordic Railpass* allows 21 days' unlimited travel in Norway, Denmark, Finland and Sweden. The 1992 cost of this pass is £155 (second class) and children aged 4–12 pay half price, 12–25 £115.

Peruvian Railways

c/o Peruvian Tourist Office,
10 Grosvenor Gardens, London SW1 0BD
Tel: 071 824 8693

Peru has three separate rail lines – the central, southern and Cusco–Machu Picchu lines. Tickets for the central line must be bought from the main station in Lima, tickets for the Cusco–Machu Picchu line bought in Cusco, and tickets for the southern route bought at the point of departure.

Discounts Children from 3 to 12 years pay half fare. There are no special passes.

Philippine Railways

c/o Philippine Embassy – Department of Tourism, 17 Albemarle Street, London W1X 7HA
Tel: 071 499 5443

A railway system exists only on the largest of the Luzon Islands, connecting San Fernando in the north with Manila, and Legazpi to the south. Refreshment cars are usually available, as are sleeper cars. We were advised that most people prefer to travel by bus. Within Manila the Light Rail Transit System (LRT) is an elevated railway covering 16 stations over 15 km. The fare is fixed to and from any one point. Travel time from end to end is 30 minutes.

US Rail (AMTRAK)

c/o Longhaul Leisurail, GSA Department, PO Box 113, Peterborough PE1 1LE
Tel: 0733 51780

America is a vast country, and given the distances, perhaps it's not surprising that train fares are frequently more expensive than air fares. All trains have reclining seats and a food service car. Many are air-conditioned. Sleeping accommodation is available in various grades of bedrooms and 'roomettes', where seats convert to bunks at night. Family rooms are available on request on major routes – normally for 2 adults and 2 children.

Discounts Children under 2 travel free if accompanied by an adult; 2–15s pay 50%.

There are some *Special Fares* which must be purchased outside the US, which could be cheaper on certain routes. Details on request.

Safaris, treks and exotic tours

For holidays with a difference and, it must be admitted, not too much thought for expense, this section is essential reading. If you want to get away from the well-trodden tourist track, but still ensure that you don't miss anything, the following agents/operators can arrange the holiday of a lifetime – a real adventure for you and your children.

First of all we must point out that there are many more safari and trek operators than listed here, but research showed that the majority either refused to take children, or did their utmost to discourage them. The nature of these holidays means that they appeal most to independent travellers, usually aged between 18 and 45. Travel may be in anything from converted lorries to air-conditioned coaches, but it is always on dusty, pot-holed roads and it can be exhausting even for a healthy, active adult. Several agents expressed the fear that children's crying or chattering might frighten away the animals that everyone has come especially to see. Concern is currently being expressed that, in their eagerness to live up to the promises of their brochures, safari operators may encourage their drivers to try to get as close as possible to any available wildlife. It is certainly true that they try to pack as much viewing time into the holidays as possible. Travellers are advised to consult the company concerned if they are at all worried about the character of the holiday they are booking.

As these adventure holidays are not geared to children you must go well equipped. Take your own travelling carrycot and baby-sling plus adequate supplies of nappies, baby food, milk and medicines. Although some treks include a doctor on the team, the majority do not, so it is best to go prepared. It's worth remembering that children are very resilient and adaptable, but you know your own child best, so think hard before making a commitment to this type of holiday.

The holidays to exotic locations – China, India, Malaysia and so on – allow you to travel in more comfort but, unless they're beach holidays, still contrive to pack in as much as possible. Sightseeing in cities can be as arduous as clambering over ancient monuments and a pushchair could be very useful, but will it be practical? All the companies we list will give specific advice on request.

Prices

Unless otherwise stated 1992 prices have been quoted. These are intended only as a guide to the type of holiday offered and travellers should check with the company concerned for 1993 pricing details.

Safaris, treks and exotic tours

Abercrombie & Kent Travel
Sloane Square House, Holbein Place,
London SW1W 8NS
Tel: 071 730 9600

Africa, Burma, China, Egypt, India, Indonesia, Israel, Jordan, Mauritius, Malaysia, Morocco, Nepal, Papua New Guinea, Seychelles, Singapore, Thailand, United Arab Emirates

Up-market and tailor-made holidays in a wide variety of exotic places. The company also offers many safaris and treks, but does not encourage children on these trips as they are generally arduous and offer no special facilities for children. However, specific advice will be given on request. On beach-centred holidays, many hotels are geared towards families and offer children's play areas, swimming pools and cots.

Prices (per adult)
Mauritius (7 nights): from £1061 (including flights from London) (1992 prices)

Discounts As these holidays are expensive and many considered unsuitable for children, the company offers specific advice and information on request.

Africa Exclusive
Hamilton House, 66 Palmerston Road,
Northampton NN1 5EX
Tel: 0604 28979
Fax: 0604 31628

Africa

Personalized itineraries to suit your requirements. Holidays in Tanzania, Kenya, Zimbabwe, Mauritius, Namibia, Botswana and South Africa including some beach locations. Wide variety of accommodation including in hotels, safari lodges, permanent encampments and moveable tents.

Children under 12 can be accommodated in several locations. In certain places game viewing holidays are tailored to suit children's interests and ensure their safety.

Prices 15 days in Kenya, between 1st September and 30th November, from £2180. Full details of prices available on request.

Bales Tours Ltd
Bales House, Junction Road, Dorking,
Surrey RH4 3HB
Tel: 0306 885991 (reservations)

Africa, Asia, Australia, Canada, Far East, Middle East, South America

Escorted tours to many exotic locations. Choice of tours, from budget to 'top market'. All tours allow for free time, but as itineraries can be demanding, travellers are advised against bringing children under 8. Hotel holidays are offered in Hong Kong, Cairo and Bali.

Prices Tours range from £499 (8 days in Egypt) to £4380 (top-market tour to Ecuador and the Galapagos).

(New brochure launched in September 1992. The company hopes to introduce more tours.)

Butterfield's Indian Railway Tours
Burton Fleming, Driffield, East Yorkshire
Tel: 0262 470230

India

Railway tours of central and northern India, living and travelling in specially converted carriages attached to scheduled trains. Hotel accommodation is provided at some stops, and guided tours at many. Much travelling and very limited space, so not recommended for children. Because sleeping space is so precious, children are charged at full rate unless they share parents' beds. The South India Railway Tour, which includes return flights from London, is more leisurely, with comfortable sleeping facilities on the trains. A comprehensive 20-day tour of Pakistan has also been recently introduced with a private railway carriage and attached restaurant car. Further details from the above address.

Prices Available on request. Prices vary from tour to tour.

Discounts May be negotiable for children but are subject to space availability.

China Travel Service (UK) Ltd

24 Cambridge Circus, London
WC2H 8HD
Tel: 071 836 9911

China

Guided tours to various parts of China, lasting from 11 to 25 days. This is the largest operator in China, with 330 branches throughout the country. Options via Korea, Nepal, Russia and the CIS or Thailand. Itineraries can be hectic and are not recommended for children. Accommodation is in twin-bedded hotel rooms, which can be very basic, however, 4- or 5-star hotels are also available now. Children are generally expected to share adults' room (bed provided if necessary).

Prices (per adult)
£585 for 11 days
£1998 for 21 days

Discounts Details of child discounts available on request.

Thomas Cook Faraway Holidays

PO Box 36, Thorpe Wood,
Peterborough PE3 6SB
Tel: 0733 61849 (Mon-Sat 9–5.30)
Freefone: 0800 88 1212

Bermuda, Caribbean, Egypt, Far East, Hawaiian Islands, Indian Ocean, Thailand

Thomas Cook specializes in upmarket long haul destinations.

Discounts Children under 2 pay a flat rate of £85 to all destinations. Air Canada gives 90% discount to children under 2, and 25% discount to children aged 2–11. Children aged 2–11 get up to 50% of adult price to the Caribbean, Indian Ocean and Bermuda if sharing a room with 2 adults. For the Far East, the reduction is up to 50%, maximum of one child sharing.

Creative Tours Ltd

2nd Floor, 1 Tenterden Street, London
W1R 9AH
Tel: 071 495 1775

China, Hong Kong, Japan, Korea, Philippines, Singapore, Thailand

This company is part of Japan Airlines and specializes in independent tours of Japan, with stopovers in Hong Kong, the Philippines, Korea, Singapore and Thailand. Hotel accommodation is not described. Featured hotels range from first class to deluxe or superior deluxe; details sent on request.

Prices from £1008 for 5 nights with return flights.

Discounts Children under 2 pay 10% of the full adult price; 2–12s pay 50%. These discounts may vary, depending on the season.

Jasmin Tours Ltd

High Street, Cookham, Maidenhead,
Berks SL6 9SQ
Tel: 06285 31121
Fax: 06285 29444

Middle East, Far East

Escorted tours, some with guest lecturer. Most itineraries are quite demanding as there is a lot of sightseeing involved. Accommodation is in comfortable, well-appointed hotels. The availability of child-minding varies; it is best to enquire at the hotel. People are advised against bringing children under 5.

Prices Available on request.

Discounts Children up to 12 are entitled to a discount between 10 and 30%, depending on the hotel and provided they share a room with 2 adults.

Passage to South America

41 North End Road, West Kensington,
London W14 8SZ
Tel: 071 602 9889
Fax: 071 602 4251

South America

Features for children: Highest cable car ride in the world (Merida, Venezuela),

Safaris, treks and exotic tours

horse riding, river rafting, skiing (Argentina and Chile), nature tours in the Amazon to learn about the local flora and fauna (Brazil), steam train trips, beaches and cities.
Prices Available on request.
Discounts Up to 50% discounts for children on flights.

Regent Holidays (UK) Ltd
31A High Street, Shanklin, Isle of Wight PO37 6JW
15 John Street, Bristol BS1 2HR
Tel: 098386 4212/4225 or
0272 211711 *(Bristol)*

Albania, China, Cuba, Czechoslovakia, Greenland, Hungary, Iceland, Laos, North Cyprus, North Korea, Turkey, Vietnam

Packages to some familiar, as well as some highly unusual, destinations. Accommodation is in 2–5-star hotels and facilities for children vary enormously. The company is happy to advise.
Prices (per adult in low season, including flight)
Weekend in Turkey, B+B: from £178
1 week in Albania, half board: from £550
Discounts On flights under-12s receive 10–40% discount, depending on destination and carrier. In accommodation under-2s are often free, while 2–12s generally receive 50% discount.

Silk Cut Travel Ltd
Meon House, College Street, Petersfield, Hants GU32 3JN
Tel: 0730 265211

Africa, Borneo, Brazil, China, Caribbean, Far East, Indonesia, Mauritius, Malaysia, Seychelles

Safaris to Kenya, Madagascar and Borneo, beach holidays in the Caribbean, Kenya, Malaysia, Mauritius and the Seychelles, and escorted tours to most destinations above. Accommodation is in moderate to superior hotels. Cots can be provided on request.

Prices (per adult based on 2 sharing a twin room)
Mauritius (beach holiday, 7 nights half board): from £954
Borneo (17-night tour, inc. accommodation and some meals): from £1679
Discounts Please telephone for details.

Tracks Africa/Europe Ltd
12 Abingdon Road W8 6AF
Tel: 071 937 3028

Africa, Europe

Thirty different itineraries offered, although the long-haul trips of 15–18 weeks may not be suitable for children. Over-12s may be accepted on 5, 6 and 10 week trips to Southern and Eastern Africa if accompanied by an adult.
The company also covers most European destinations, including Eastern Europe.
Prices Europe from £129. Africa from £230 (not including flights).

Travelbag PLC
12 High Street, Alton, Hants GU34 1BN
Tel: 0420 88724

Australia, Canada, Far East, Indonesia, Malaysia, New Zealand, South Pacific, Thailand, US

Touring and resort holidays in Australia, New Zealand, North America and Thailand, with the option of stopovers in India, US, Canada, Africa, Malaysia, the Far East and South Pacific. Most accommodation is in 3–4 star hotels. This company also sells flights only.
Prices (flights only)
Perth: from £549
Auckland: £649
Round-the-World: from £799
Hotel prices will be quoted on request.
Discounts On flights children under 2 pay 10%; 2–12s pay 50%. Accommodation discounts will be quoted on request.

Voyages Jules Verne

21 Dorset Square, London NW1 6QG
Tel: 071 723 5066

Africa, Albania, Asia, Canada, China, CIS, Egypt, Far East, Indonesia, Jordan, South America

Guided tours to many destinations, with accommodation in the best available hotels (sometimes basic). Internal travel by bus, train and boat. Itineraries are tiring and not recommended for children under 8, but the company is prepared to negotiate.

Prices Depends on destination – please telephone for details.

Worldwide Journeys and Expeditions

146 Gloucester Road, London SW7 4SZ
Tel: 071 370 5032/3

Worldwide

Special interest tours and safaris to East and Central Africa, India, Tibet, Nepal and South America. The tours take in almost everything from local culture to conservation. The company discourages taking children under 12 on safaris and the more rigorous expeditions, but special arrangements can be made and specific advice is given on request. On easier trips that are hotel-based, children of any age may travel. Tailor-made holidays to anywhere in the world can be arranged on request.

Prices (per adult)

A Kenya safari with one week's full board camping and one week B+B in a coastal hotel starts from £1300 (including flight).

Discounts As these holidays tend to be rather expensive and children something of a rarity, discounts will be calculated on application.

Sailing

Many of the holidays in this section tend to be instructional, offering you the opportunity to learn how to sail a wide variety of craft. This type of holiday is not generally suited to children under 9, and some companies have even more rigorous age limits. However, family needs are catered for by some, and several companies offer junior courses for children as young as 5 (more often 9 or 10) where they can learn to sail dinghies. Crèche and babysitting facilities are offered by a few companies.

Charter boats (skippered or independent) will often accept children of any age, providing safety harnesses for toddlers and buoyancy aids for all children apart from babies. For families flotilla sailing seems to be the best bet. It allows you the freedom to skipper your own yacht, but gives you the security of travelling in a small convoy following the lead boat which has a trained captain and crew to navigate and help you out en route if necessary. For this sort of holiday at least one adult on your boat must have some sailing experience or have taken a short flotilla sailing course prior to, or at the beginning of, the holiday. A second adult is necessary on each boat to help with ropes and mooring. Single parents with young children would need to team up with another group or family in order to manage the yacht safely. It is worth contacting your local Gingerbread group, who may be able to put you in touch with like-minded one-parent groups (*see* **Section 7** for the address of Gingerbread's head office).

If you're an experienced sailor there's nothing to stop you chartering a boat and taking it where you please – but bear in mind that rough seas (particularly in British coastal waters) will necessitate keeping the children below and this can lead to frayed tempers all round. From our research on this section we would recommend that novice sailors start with an easy voyage in safe and calm waters – probably somewhere in the Mediterranean. Proficiency is acquired fairly quickly and you'll soon be able to move on to more adventurous sailing.

Prices
Unless otherwise stated 1992 prices have been quoted. These are intended only as a guide to the type of holiday offered and travellers should check with the company concerned for 1993 pricing details.

Bosham Sailing

Bosham Lane, Bosham, Chichester, West Sussex PO18 8HP
Tel: 0243 572555
UK (South Coast)

Dinghy sailing classes of 5, 7 or 12 days' duration. Two age groups: Junior (9–14 years) and adult (over 14). Optional test for Royal Yacht Association at end of course. Half board accommodation in private homes in Bosham or full board or no accommodation at all. No childminding facilities are provided, but unaccompanied children are catered for.
Prices A 5-day course, including all equipment and instruction: £145–£175 (adults), £135–£156 (children). (Packed

lunches are optional at £2 per day.) 2-day weekends £72–£83. Accommodation per night: B+B £12; half board £16; caravan £15 per night (or £89 per week).
Please note: These are 1992 prices, add about 4–5% for 1993 prices.

Elmsworth Sailing School

(*see also* **Sunsail**) The Port House, Port Solent, Portsmouth, Hants PO6 4TH
Tel: 0705 210510

Elmsworth Sailing School offers the complete range of RYA Dinghy Sailing and Yachting Courses from Chichester Harbour, The Solent and Lake Windermere. Special Camp and Sail activity holidays, Teenage Cruising and Adult all-in Dinghy Sailing holidays are available during high season. Further details and full-colour brochure available upon request.
Prices Start from £70 for a Weekend Dinghy Sailing Course.

Greek Islands Sailing Club

66 High Street, Walton-on-Thames, Surrey KT12 1BU
Tel: 0932 220416

Greece

Shore-based dinghy-sailing and windsurfing holidays on the 'unspoilt' islands of Paxos, Ithaca and Cephalonia, for both beginner and expert. The company offers a wide range of craft plus expert tuition by RYA-qualified instructors, with special instruction for children. Full safety cover and buoyancy aids provided. Accommodation is either in studios, shared apartments or villas (*see* entry in **Villas**). Cots, high chairs and baby-sitters are available, with a Club creche operating on Paxos and Cephalonia.
Prices (per adult per fortnight, including return flight, transfers, airport tax, accommodation, craft and tuition)
Low: £470
Mid: £635
High: £770
Discounts On two-week holidays children aged 2–12 receive 5–10% discount, depending on season.

Jubilee Sailing Trust

Test Road, Eastern Docks, Southampton SO1 1GG
Tel: 0703 631395

UK, Canary Islands

Adventure tall ship sailing voyages in the UK and Canary Islands for able-bodied and physically disabled people aged 16+ aboard the 180 ft purpose-designed *STS Lord Nelson*. Voyages lasting from 3 to 10 days provide a challenging opportunity for crew members to experience the great tradition of sailing a square-rigger at sea. Once on board *STS Lord Nelson* everyone can expect to experience all aspects of tall ship sailing, from handling sails, helming and navigating to keeping watch, cleaning the ship and helping in the galley. Voyages include stop-offs at foreign ports. Experienced yachtsmen, casual sailors and complete novices are welcome.
Prices (per adult voyage including accommodation, meals and wet weather gear)
UK voyages, 3–10 days (Mar to Nov 93): £200–£695
Canary Island voyages, 7–23 days (Dec 92 to Mar 93): £295–£600
Discounts 50% discount for doctors and watchleaders with RYA Yachtmaster Offshore Certificate for UK 1993 voyage season.

Lorne Leader

Don & Gillian Hind, Lorne Leader, Ardfern, By Lochgilphead, Argyll PA31 8QN
Tel: 08525 212

UK (Western Scotland)

Lorne Leader is a refurbished sailing trawler with 12 passenger berths, hot and cold running water, heating, WCs and showers. From April to October, skipper and crew take her on 6- and 11-day cruises around the Hebrides. There are different themes, such as birdwatching, art, and folk music, and special family holidays with childminding and more trips ashore.

Sailing 165

Prices (per berth, including all meals and sailing instruction)
6 days: £285–£395
Discounts 10% for children under 11.

Made to Measure Holidays

Conwell House, 43 East Street,
Chichester, Sussex PO19 1HX
Tel: 0243 533333

Mediterranean, Caribbean

Tailor-made sailing holidays of a very high standard in the Mediterranean and yacht charters in the Caribbean. Details available from the company on request.
Prices Available on request

Minorca Sailing Holidays

58 Kew Road, Richmond, Surrey
TN9 2DT
Tel: 081 948 2106

Spain

Dinghy-sailing and windsurfing holidays in Minorca, with expert tuition from RYA instructors at whatever level is required. Accommodation is in self-catering apartments and villas which sleep 4–8 people, or a small, family-run hotel. Some villas have private gardens and swimming pools. Cots are supplied free on request. Childminding can be arranged at certain times of the year; at the moment May and June are most popular.
Prices Available on request.

Rockley Point Sailing School

Rockley Park, Poole, Dorset BH15 4LZ
Tel: 0202 677272

UK (South Coast)

A range of sailing courses lasting 2–13 days for beginners, intermediate and advanced. Also wind-surfing courses and various watersports, such as canoeing. Children aged 8–12 are specially catered for, and there are also courses for unaccompanied 10–17-year-olds. Accommodation is in guest houses, caravans or campsites (bring your own tent).
Prices (excluding lunches, waterproofs and wetsuits)

5-day sailing course: Adults (over 13) from £130, Juniors (8–12) £110.
Accommodation: 6-berth caravan from £200 per week; guest house (half board) £15 per night.
Discounts 5% off course fee for 3 members of the same family; 10% discount for 4 members.

Sovereign Sailing

Astral Towers, Betts Way, Crawley,
West Sussex RH10 2GB
Tel: 0293 599944

Greece, Sardinia, Turkey

Specialist operator offering windsurfing, dinghy sailing and yachting holidays in Greece, Turkey or Sardinia for novice or expert. 'Surfbusters' programme teaches children, whilst parents sail for the day. Guaranteed money back if you're not satisfied. Fly from Gatwick or Manchester.
Prices (per person for one week 'Surf and Sail' in Greece, low season, based on twin rooms in hotel accommodation on a bed and breakfast basis with flight from Gatwick): from £270.
Discounts Child discounts of £70–£100.

Sundown Marine Yacht Charter Ltd

Sundown House, Rectory Lane,
Woodmansterne, Surrey SM7 3PP
Tel: 07375 51271

Australia, Caribbean, Far East, France, Germany, Italy, Pacific Islands, Turkey

Selection of sailing holidays. Choose from sail-yourself yachts, flotilla holidays, crewed sailing/motor yachts and crewed Turkish caiques. Yachts have 4–10 berths and range from comfortable to luxurious. Children and babies welcome. Cotsides can be arranged, but sometimes you have to provide your own buoyancy aids.
Prices (per yacht/boat, excl. flight)
6-berth yacht (sail-yourself for 2 weeks in Greece): £1090–£1690 (£45 per day extra for captain).
12-berth Turkish caique, including captain, crew and diesel: £200–£450 per day.

Sunsail/Sunsail Clubs

The Port House, Port Solent, Portsmouth, Hants. PO6 4TH
Tel: 0705 210345 *Yachting*
Tel: 0705 219846 *Clubs*

Corsica, Greece, Sardinia, Caribbean, Turkey

Sailing, cruising and Sunsail Club holidays in the Mediterranean with expert tuition.

Sunsail also offers skippered holidays for those who prefer not to 'do-it-yourself'. Accommodation is either on board or in shore-based club rooms. Older children can learn to sail and windsurf. A crèche run by qualified staff is available at some of the Clubs for children aged 1–12 years. Cots, high chairs, potties, specially designated play areas, sandpits, children's dinghies, windsurfing rigs, a baby listening service and children's pools are among the facilities on offer at some of the Clubs.

Prices (per adult per fortnight including flight and accommodation at Sunsail Clubs)
Low from £300
Mid from £449
High from £549

Discounts Seasonal percentage reduction off adult price for children 2–12 years sharing room with parents in cot or Z-bed.

Skiing

Skiing must surely be one of the fastest growing markets for family holidays. While skiing's up-market image still exists in resorts such as Gstaad and St Moritz, a growing demand for winter holidays that don't break the bank has spawned many companies and many new resorts. In fact, fierce competition among tour operators has brought benefit to all skiers. This is reflected in the discounts offered to families.

Children aged 2–12 (some companies say 2–15) are eligible for all sorts of discounts, which are dependent on their sharing a room with two full-price adults. These discounts may be offered as a flat fee, a specific monetary reduction or a percentage discount. It really does pay to compare; some companies use the same accommodation but charge wildly different amounts.

Of course, the expense of a skiing holiday isn't just the travel and accommodation. You have to have equipment and ski passes, and to be honest, the proper (waterproof) clothing is a great advantage. It's not very comfortable skiing in wet jeans. If you don't want to splash out on clothing until you know that you're committed to skiing, it's possible to hire it from many ski shops.

But to get back to families – how can parents go on skiing holidays without taking it in turns to mind the children? It is both amazing and gratifying how many resorts offer excellent childminding facilities. Nurseries often take babies, some as young as one month, more often 6–9 months. They are staffed by qualified nurses or nannies who organize fun and games for toddlers, as well as supervising meals. Older children, usually from 3-12, may attend kindergarten, a playgroup that also offers ski instruction for an hour or so a day. Babysitting may usually be arranged through the company representative or the local tourist office. All these childminding facilities are normally paid for at the resort, and the number of hours per day that they are available varies from 5–14. Some companies offer these services free and have their own English-speaking nannies.

A lot of people believe that starting children on the slippery slope is best done around the age of 3, and the consensus of opinion is that they love it. Skis come in all sizes and there are generous discounts on equipment hire and ski passes in virtually every resort. Do note, however, that some resorts do not supply separate instruction for children; they are expected to join adult groups. As far as we can see, the main drawback to this is that the children usually outshine the adults.

If the company you travel with offers children's ski schools organized in the resort, do make careful enquiries as to exactly what care they provide.

Prices
Unless otherwise stated 1992 prices have been quoted. These are intended only as a guide to the type of holiday offered and travellers should check with the company concerned for 1993 pricing details.

Abercrombie & Kent Travel

Sloane Square House, Holbein Place,
London SW1W 8NS
Tel: 071 730 9600

Skiing hotels programme operated all year, with summer Alpine holiday packages. Your nanny can accompany you for a small fee when you book an eight-person chalet. Children's discounts are substantial and there is daycare available if required.

Prices Guidelines: Adults from: £120–£1009 for 7 nights. Children from: £289–£829 for 7 nights.

Balkan Holidays

Sofia House, 19 Conduit Street, London
W1R 9TD
Tel: 071 491 4499

Bulgaria, Romania, Slovenia

Ski holidays in the resorts of Borovets, Vitosha and Pamporovo from December to April. Accommodation is in hotels, purpose-built wooden chalets or apartments. Ski instruction, equipment hire and passes are available at reduced rates if ordered when booking. A nursery in Borovets takes children from 2 to 4 years and is open from 9.00 a.m.–11.00 p.m. (prices vary). Children's ski schools are available in both resorts and take children from 4 to 7 years.

Prices (per adult per week, half board, including flight): from £171

Discounts Children under 2 travel for £20; cots and meals payable locally. 2–12s sharing parents' room receive a 40–50% discount, depending on location. In both resorts children receive substantial discounts on all excursions, and may sometimes be free.

Inghams Travel

10–18 Putney Hill, London SW15 6AX
Tel: 081 785 7777

Austria, Bulgaria, France, Italy, Romania, Slovenia, Switzerland

77 resorts ranging from rustic alpine villages to modern purpose-built resorts. Accommodation includes hotels up to luxury 5-star, *pensions* and self-catering apartments. Travel is either by air, rail or car, but accommodation only is also available.

Prices From: £144 per person per week by air, self-catering in France. From: £168 per person per week by air, half board in Bulgaria.

Discounts 2–11s receive up to 50%, with free insurance for children up to 16 when accompanied by an adult.

There is a 20% discount for adult third bed and up to 8% discounts for groups of 11 or more.

Interhome

383 Richmond Road, Twickenham,
Middx
Tel: 081 891 1294

Austria, Belgium, Canary Islands, France, Germany, Greece, Holland, Italy, Switzerland, Slovenia

Ski holidays in a wide variety of resorts. No travel is included but ferry arrangements can be made on request. Accommodation is in apartments and hotels. Childminding and children's ski schools are available in some resorts, but the company takes no responsibility for this and their availability should be checked with the tourist office for the relevant country. Arrangements for cot hire can 'possibly' be made – ask when booking.

Prices (per 4-person apartment in Switzerland): from £150 (late January)

Discounts On hotel holidays children aged 3–12 receive 50% discount if sharing parents' room. On short ferry crossings under-3s travel free; 3–13s pay £25 each in winter. No child discounts on self-catering accommodation.

Lotus Supertravel

Hobbs Court, Jacob Street, London
SE1 2BT
Tel: 071 962 9933

France, Switzerland, US

Ski holidays with accommodation in chalets and a variety of hotels. Childminding is available in certain resorts, but the mini-

mum age varies. Where available, cots in chalets are free, but hotels usually charge and the cost is payable locally.
Prices Available on request.
Discounts On chalet holidays, children under 2 pay £25 (cots and meals payable locally). Discounts for 2–15s vary, depending on departure date and whether they have a main bed or an extra bed. Substantial discounts available on ski passes.

Made to Measure Holidays

Cornwell House, 43 East Street, Chichester, Sussex PO19 1HX
Tel: 0243 533333

Austria, France, North America, Switzerland

Made to Measure Holidays offer family skiing holidays – full details are available from the company on request.
Prices Available on request.

Ski Chamois

18 Lawn Road, Doncaster, South Yorkshire DN1 2JF
Tel: 0302 369006

France

Ski holidays in 5 resorts, with a choice of 2 jumbo chalets. Travel options include plane, sleeper-coach and self-drive.
Prices (per adult for 10 days sleeper-coach, half board)
Chalet: from £199 (low) up to £369 (high).
Discounts All under-2s £40 per week and all children under 13 qualify for discounts. Details available on request.

Ski Esprit

Oaklands, Reading Road North, Fleet, Hants GU13 8AA
Tel: 0252 616789 (24 hours)
Fax: 0252 811243
Brochure: (24 hours) 0252 625175/616789

France, Switzerland

Ski Esprit is the leading specialist in family skiing holidays. Their chalet based programme covers eight major resorts in France and Switzerland. Each resort has in-chalet creches, operating all day and caring for children from 4 months to 4 or 5 years old. They are staffed by British qualified nannies (NNEB or RGN). Older children attending ski school are also looked after by Ski Esprit staff who give them lunch and take them back to lessons in the afternoons or run an activities programme for children who have morning-only classes. Among other family features are a separate high tea for under-14-year olds (to leave dinner as an adults-only meal), free babysitting and very generous children's discounts.
Prices from £248 for a fully catered chalet holiday by air.
Discounts Details on request.

Ski Thomson

Greater London House, Hampstead Road, London NW1 7SD
Tel: 081 200 8733 (London)
021 632 6282 (Birmingham)
061 236 3828 (Manchester)

Andorra, Austria, Canada, France, Italy, Spain, Switzerland, Slovenia, US

Ski holidays with travel by air, self-drive or snow train. Accommodation is in hotels, chalets and self-catering apartments, but children under 17 are not allowed on chalet holidays unless you book the whole chalet for your party's exclusive use. Cots and children's meals are available in most resorts. Kindergartens and ski schools are available at most resorts. The minimum age varies, but some do take young babies. New for this season are 13 Family Choice resorts in Austria, France and Andorra which offer pre-bookable ski and day kindergarten. Ski Thomson Children's Clubs are available at St Johann in Tyrol, Austria, Val Thorens and Risoul in the French Alps. 3–8-year-olds are looked after by a specially trained Ski Thomson children's rep. This service is pre-bookable and costs £30 per child per week.
Prices (per adult per week, including return flight and transfers)
Self-catering studio from £128
Half board in hotel accommodation from £193

Choosing and planning your holiday

Discounts On air holidays under-2s pay £15 (no seat); cots and meals are hirable locally. Children aged 2–11 receive a 10–60% discount, depending on departure date. Substantial discounts on ski passes and equipment hire are available in most resorts.

Ski-Val

39A North End Road, West Kensington, London W14 8SZ
Tel: 071 602 7444
Fax: 071 371 4904

Australia, France, US

Ski holidays in Val-d'Isere and Colorado, with travel by air, snow train, coach/ferry or self-drive. Accommodation is in hotels, clubs, chalets and apartments. Nurseries, kindergartens, children's ski schools and clubs are available in some resorts, and baby-sitting can usually be arranged with chalet staff on request.

Prices Refer to brochure

Discounts May be available – please refer to brochure.

Swiss Ski

Swiss Travel Service, Bridge House, Ware, Herts SG12 9DE
Tel: 0920 463971

Switzerland

Ski holidays in hotels or self-catering studios, with the option of air travel or self-drive. Some hotels classified as 'Ideal for Family' have extensive reductions and special facilities, such as a free kindergarten. However, virtually all resorts have a supervised kindergarten, but none appears to take children under 2 years.

Prices (per adult per week, half board, including flight and transfers)
3-star hotel: from £348
4-star hotel: from £455
5-star hotel: from £596

Discounts On air holidays, under-2s £20 (no seat); cots and meals are payable locally. Children aged 2–11 receive a 50% discount all season, provided they share parents' room. When not sharing there is a £100 discount per child. Children aged 12–15 each receive a £50 discount. On self-drive holidays children under 2 are free, 2–16s have 50% reduction.

Special interest and activity holidays

Whether you're a steam train enthusiast, aspiring circus performer, devoted horse-rider or musician, you'll find your special-interest holiday to suit your likes. If an activity holiday is up your street, we list many centres offering a wide variety of physical and cerebral pursuits. You don't have to be a hearty, outdoor type to enjoy one of these holidays, and if you prefer not to participate at all, you're at liberty to do so.

For families who dislike beach holidays and have a sense of adventure, activity holidays are ideal. But what if your children are mad about archery and badminton, while you're keen on photography and abseiling? No problem. Most centres have flexible arrangements to cater for different interests within a family and can create a holiday to encompass all your special interests. And if you have no special interests and don't know what you might like, you can opt for a multi-activity holiday which gives you a taste of everything on offer. Most activities are graded for difficulty, and qualified instructors are always in attendance.

It's worth noting that some centres operate all year round and employ permanent, qualified staff. Others simply take over a school for a couple of weeks during the summer, so the choice of dates is limited, even if the activities aren't.

Children are often not allowed to participate in adult activities, such as rafting or hiking, but the minimum age tends to vary from centre to centre. Around 10 is the general rule, but special programmes are available for children aged 6 upwards. Children under 6 are often allowed on site but facilities vary. Usually they are accepted on the understanding that parents take full responsibility for them.

Perhaps cycling isn't the most obvious choice of holiday if you have young children, but it can be one of the most rewarding for all concerned. The benefits of fresh air and exercise are self-evident, but you also have the advantage of seeing unspoilt countryside at a leisurely pace and going back to your hotel's creature comforts at night. All the tours are graded for difficulty, so if in doubt of your ability, choose something easy to start.

None of the tour operators offering organised hotel-based cycling holidays recommend taking children still in nappies, particularly if they're not dry at night. However, children aged 3, 4 and 5 can have a lovely time sitting in a child seat on the back of your bike. Some companies warn of an in-between time around the ages of 5–7 when children are too big for a child seat, but too young to handle a small bike with gears. After this time they often can handle such a bike, and on the quiet lanes and roads chosen for these tours, have no problem cycling up to forty miles a day – sometimes leaving parents panting in their wake! Obviously you need to have some idea of your child's stamina and road sense, so it's wise to check out both at home before embarking on a cycling holiday abroad.

You'll generally find hoteliers and restaurant owners very helpful about providing children's meals – even if your offspring demand egg and chips rather than boeuf bourgignon. Lunches are left up to you and picnics tend to be the norm.

A word of warning about luggage. Unless the company offers to transport your suitcases from stop to stop or offers panniers to transfer your belongings to, avoid bringing suitcases in the first place; your own panniers or a rucksack are much easier to handle on a bike.

Prices
Unless otherwise stated 1992 prices have been quoted. These are intended only as a guide to the type of holiday offered and travellers should check with the company concerned for 1993 pricing details.

Acorn Activities

7 East Street, Hereford HR1 4RY
Tel: 0432 357335
UK (Herefordshire and Wales)

Over 100 activities listed in the brochure, including riding, sailing, go-carts, microflights, abseiling and caving. Accommodation in farmhouse hotel and an activity centre for families, children and group accommodation.
Prices £30–£80 per day.

Adventure International

Bellevue, Bude, North Cornwall
EX23 8JP
Tel: 0288 355551

Centre with 70 family-size rooms, twin-bedded accommodation. Activities include sailing, surfing, canoeing, rock-climbing, abseiling, archery, shooting, assault course, guided walks (several off site), horse riding, golf, sea fishing and clay pigeon shooting. Swimming pool and games room on premises. Discos and Karaoke nights catering for children.
Prices Unaccompanied child, £189 per week. Option of parent and child activities, child-only activities, or accommodation only, paying for activities on a daily basis. See brochure for full price information.

Anglo Dutch Sports Ltd

30a Foxgrove Road, Beckenham, Kent
BR3 2BD
Tel: 081 650 2347
Fax: 081 663 3371
Austria, Belgium, Holland

Specialise in cycling holidays for the family or the individual. The holidays are based at family hotels on a dinner, bed and breakfast basis. Bicycle hire included, but it is possible to bring one's own. On most packages transport of luggage is included. There is safe cycling in Holland on its cycling paths and in Austria along the Danube on the towpaths. Holidays are from April to October on a daily basis. Duration of 3 to 10 days or longer. City breaks can be added to any of our packages.
Prices from £200 (including ferry crossing with or without car)
Discounts Children under 12 sharing parents' room receive 10–50% discount.

Countrywide Holidays

The Countrywide Holidays Association,
Birch Heys, Cromwell Range,
Manchester M14 6HU
Tel: 061 225 1000

UK (Cornwall, Cumbria, Derbyshire, Gwynedd, Isle of Wight, Norfolk, Perthshire, Somerset, Yorkshire, Grampian region of Scotland)

Activity holidays for children of 6 and over based in the same centre as their parents. Adventure weeks, Family fortnights, Multi-action holidays, Youngsters' Fun weeks, include such activities as sailing, skiing, mountain biking, pony trekking, paragliding, golf, cycling, river rafting, canoeing, rock climbing, archery and many others under the supervision of professional guides and instructors. Play leaders organize games in the evening. Some centres have videos and games rooms. There are no crèche facilities for very young children, but cots are provided free of charge and bottles can be warmed up. Accommodation with full board is in a series of guest houses, including a castle, an Elizabethan manor and many mansions in beautiful grounds. There will be a greatly increased programme for 1993.

Prices (per adult)
From £147–£280, full board including instruction and equipment. (Guidance only). Ski-hire is extra.

Discounts Children under 2 sharing parent's room are free. Non-participating children staying in the centre and sharing a room with each other pay 25% if aged 2–6, 50% if aged 7–13 and 75% if aged 14–17.

Non-participating families staying in the centre receive a discounted rate. Special rates for groups.

Courtlands Centre

Kingsbridge, South Devon TQ7 4BN
Tel: 0548 550227

UK (South Devon)

Activity holidays in a converted manor farm set in 4 acres close to the sea and Dartmoor. Accommodation is in dormitories and twin-bedded rooms. One wing of the house can be rented weekly on a self-catering basis and is ideal for families. A separate bathroom for adults only is available with dormitory accommodation; children use communal showers and washrooms. The price includes 3 meals a day, special equipment for certain sports and evening entertainments, such as discos, games, videos and barbecues. Children must bring sleeping bags and pillow cases; adults are supplied with linen, blankets and duvets. Activities include canoeing, sailing, riding, surfing, waterskiing, archery, orienteering, assault course, hikes, climbing, abseiling, sea fishing and yacht cruising. All cater for different abilities. Children aged 6–8 may participate in young children's adventure holidays. Unaccompanied holidays (Adventure Run) are open to anyone aged 9 years upwards. You can opt for one activity or a multi-activity programme, and the centre will try to accommodate differing interests within families if you give prior notice when booking. Adults pay only for accommodation if they choose not to participate. An indoor sports hall is available for evenings or use in foul weather.

Prices Multi-adventure holidays start from £249 per week.

Discounts Small children sleeping in a cot and sharing their parents' room go free.

Cycling for Softies

2 and 4 Birch Polygon, Rusholme,
Manchester M14 5HX
Tel: 061 2488282
Brochure line: 061 2485134

France

Award-winning company offering between 7- and 14-night cycling holidays in 9 regions of France. The cost includes all scheduled direct flights, accommodation and half board in small hotels, lightweight bicycles and all other necessary equipment. Suitcases are left at the home-base hotel on each tour and belongings are carried in panniers on your bicycle. Various types and sizes of bicycle are available, including small ones for children aged 7 up. Children under 5 may be carried in child seats, provided free on request. All hotels used are happy to supply infants meals if given a little warning. Babysitting is not available.

Prices (per adult per week)
Low: from £458 (two adults sharing a room)

Discounts Infants' discounts available: details on request.

Just Pedalling

9 Church Street, Coltishall, Norfolk
NR12 7DW
Tel: 0603 737201

UK (East Anglia)

Seven-day cycling holidays and short breaks. The cost includes bed and breakfast in guest houses, bicycle hire and all

necessary equipment. Take your own bike if you prefer. Child seats are provided on request. A Deluxe tour offers accommodation in period houses of special interest.
Prices (per adult per week)
£135 to £150 plus £20 bicycle hire

Discounts Children under 2 pay £20 on air holidays; cot hire and meals payable locally. Under-4s are free on self-drive holidays. On other holidays children up to 17 are eligible for various discounts, quoted on request.

Millfield Village of Education

Millfield School, Street, Somerset
Tel: 0458 45823
Fax: 0458 45102

UK (Somerset)

Five-day activity holidays from late July to late August. Choice of 115 different courses, including sports, arts, crafts and academic studies. Children of all ages are welcomed. Free crèche and babysitting facilities. Children over 8 may stay unaccompanied. Accommodation is in the school's boarding houses (room with shared bathroom). Children under 10 share parents' room; older children share together.
Prices (per 5 days' full board)
Adults: approx. £135
Accompanied children: £125 (under 12) or £99 (under 8)
Unaccompanied children: £135
Tuition fees are approximately £61 per course.
Discounts Early booking discount of 10%.

Naturist Holidays

Peng Travel Ltd, 86 Station Road, Gidea Park, Romford, Essex RM2 6DB
Tel: 0708 471832

Canary Islands, Caribbean, France, Spain, US

Nudist holidays in many locations, including France, Corsica, Spain and Florida. Accommodation is situated in nudist villages or nudist areas and campsites in studios, villas, hotels, apartments and bungalows. Naturist cruises may be available in Turkey, but do not take children under 16.
Prices (per family of 4 per fortnight, camping, tent provided)
France, self-drive, including ferry: £306–£412 plus insurance.

Northumbria Horse Holidays

East Castle, Annfield Plain, Stanley, Co. Durham DH9 8PH
Tel: 0207 235354/230555

UK (North East England, Norfolk)

Europe's largest equestrian holiday company catering for all ages, from complete beginners to seasoned riders. Holidays last 7 and 14 days and may be based in one place from which you ride out and back daily, or you can go post trail riding, following a progressive route across country and stopping somewhere different every night. Horse-drawn caravan holidays are available in Norfolk. Accommodation is full board in comfortable twin-bedded rooms in licensed premises; self-catering is also available. There are no age restrictions, but young children must be accompanied and attended to by parents.
Prices (per adult per week, full board)
A 'learn to ride' holiday ranges from £169. A post trail riding holiday ranges from £249.
Discounts Children up to 14 receive a 10% discount if they share a room with 2 adults.

PGL Family Adventure

Alton Court, Penyard Lane, Ross-on-Wye HR9 5NR
Tel: 0989 768768

Austria, France, Holland, Spain, Sweden, UK

Family activity holidays at over a dozen different centres. Accommodation is in hotels, colleges and schools with single/twin/family rooms and dormitories, university apartments, caravan leisure homes or luxury bungalow tents, depending on location. All self-catering accommodation is fully equipped. A few non-residential places are available for families who bring their own caravans and stay in the sites nearby. Activities include sailing, archery, tennis, climbing, swimming, fencing, riding, canoeing, windsurfing, skiing, cycling, squash. Single or multiple activity holidays can be arranged. At catered centres the minimum age is 7, but there are 2 where the minimum age is 4. A Babar Playscheme is available for 4–6s at these 2 centres. There are no age restrictions on self-catering holidays. Minimum ages apply to certain activities.

Prices At a centre in Wales, which has the Babar Playscheme, one week's full board, including activities, prices in 1992 were from £239 per person.

Discounts In the 1992 brochure, children aged up to 11 or 12 years are eligible for discounts ranging from £20 to £100, and in some instances travel free, depending on the holiday destination and rooming arrangements.

Surfrider Activity Holidays

Montague Farm, 6 Watery Lane,
Croyde, North Devon EX33 1NQ
Tel: 0271 890083

The centre in Croyde is very children-orientated, especially in the summer holidays. Children should be over 8 with a very basic swimming ability. The majority are in their teens. Surfing, waterskiing, wave skiing, mountain biking, rock climbing, abseiling, beach volley ball, tennis and coastal trekking are available. The activities are geared to all levels – from gruelling to sedate.

Prices including all food (e.g. a 3-course meal at local restaurant scheme), all coaching and equipment, transport to activities and insurance. £260 single per week in shared rooms, £285 in private rooms, £299 with additional private facilities.

Discounts Negotiable for children. Early bookings receive concessions, and those returning for a repeat visit go free if they bring a small group.

Twickers World of the Red Sea

22 Church Street, Twickenham,
Middlesex TW1 3NW
Tel: 081 892 7606

Egypt, Israel, Jordan

Offer inclusive package holidays and flight only to the above countries. The main Red Sea season is from October through to May – and in general should be viewed as a winter sun destination. Though holiday makers do visit the area in the summer – hot though it is – the heat is made tolerable by the constant summer breeze and low humidity.

One of the main features of the area for families vacationing is, apart from the cultural interest, the Coral Sea – it is the nearest Coral Sea to the UK and in the winter the Red Sea is accessible by direct charters.

The Red Sea offers something for all the family – swimming with dolphins, snorkelling, scuba diving, snuba (enables those too young to learn to scuba dive to get the taste of the beauties of the underwater world), windsurfing and waterskiing. Numerous leading hotel groups, e.g. Hilton, situated throughout this region, enable the company to provide well-priced holidays including self-catering units, family rooms, interconnecting rooms, villas and apartments.

Prices start from: £199

Discounts children get discounts up to 60% depending on location and season – single parent families also catered for.

YMCA National Centre

Lakeside, Ulverston, Cumbria LA12 8BD
Tel: 05395 31758

UK (Lake District)

Week-long programmes in July and August. Activities include rock-climbing, archery, canoeing, sailing, crafts and local visits. Accommodation is in either twin rooms with own facilities, 2–4 bed rooms with shared facilities, or chalets with separate facilities. A daytime playscheme is provided for 4–6-year-olds, but no crèche or cots.

Prices Available on request.

Discounts Various. Details available on request.

Villas and apartments

For many people self-catering accommodation in the form of villas and apartments is the solution to taking a family holiday. You have privacy, comfort and all mod cons, but are rarely far from shops and restaurants, so you can self-cater as much or as little as you like.

What constitutes a villa? Usually they are detached properties standing in their own grounds, but increasingly they may be terraced properties in purpose-built 'villages'. The standard of furnishing and equipment varies, but at the very least it is reasonably comfortable. If the properties are privately owned, as many are, the surroundings can be quite luxurious – but so can the price. If, for example, you'd like to rent Rex Harrison's villa in the south of France, you'll have to pay more than a small fortune!

The accommodation on apartment holidays is equally variable, ranging from multi-storey blocks to maisonettes in historic châteaux. Most claim to be reasonably comfortable and to supply enough facilities to cook a meal. However, from reports we have received, it is evident that some companies may provide facilities that are less than ideal.

If you want a place with character you might have to forgo your own private balcony and the use of a swimming pool, but there may be bonuses in the shape of local farmhouse produce and perhaps a landlady willing to baby-sit. On the other hand several large villa companies have representatives who can organize baby-sitting for you. Always enquire before booking about the availability (and cost!) of cots, bunk beds, high chairs, playpens and so on. You may be pleasantly surprised by how much companies offer for families.

Prices

Unless otherwise stated 1992 prices have been quoted. These are intended only as a guide to the type of holiday offered and travellers should check with the company concerned for 1993 pricing details.

Auto Plan Holidays Ltd

Energy House, Lombard Street,
Lichfield, Staffs WS13 6DP
Tel: 0543 257777

Austria, France, Germany, Italy, Norway, Switzerland

Self-drive holidays with accommodation in apartments, villas, chalets, bungalows and small hotels. All self-catering accommodation is fully equipped, but no linen is supplied. On travel-inclusive holidays the prices include ferries, 2 hotel stops en route, AA 5-star insurance and your accommodation.

Prices (per adult per week)
A travel-inclusive holiday in a self-catering chalet is from £164 (low), £225 (high).

Discounts On inclusive holidays, when minimum occupancy has been paid, children under 4 sometimes go free. In other circumstances under-4s receive £15 discount and 4–13s receive £10 discount. On hotel holidays under-4s sometimes go free; 4–13s who share parents' room receive 15% discount for first child and 10% discount for second child in certain circumstances. No child discounts on villa rental only.

Beach Villas

8 Market Passage, Cambridge CB2 3QR
Tel: 0223 311113

Corsica, Cyprus, France, Greece, Italy, Portgual, Sardinia, Spain, Turkey, US

Villas for rent in many mainland and island locations, including Elba. All accommodation is fully furnished and equipped, including linen. At most locations maid service is available. Car hire discounts are offered with local companies if booked in advance. Cot hire is £12 per week.

Prices Prices vary according to property, from £169 to £313.

Discounts There is a well-established discount scheme for families. Details are available on request.

Blakes Villas

Wroxham, Norwich, Norfolk NR12 8DH
Tel: 0603 784141

France

Self-catering holidays with many properties ranging from accommodation with pools to holiday villages and simple country gîtes. All accommodation is fully furnished and equipped, except for linen and towels. Each villa is listed with a photograph, description of the rooms and kitchen (many with dishwasher and washing machine), a list of local amenities and the level of comfort to be expected.

Prices Available on request.

Caribbean Villas & Hideaway Hotels

4a William Street, London SW1 9HL
Tel: 071 245 1235
Fax: 071 245 1250

Caribbean

Traditionally the company has offered luxury staffed villas, most with private pools, in several Caribbean islands. Many are beachfront and the local staff will cook, clean, shop and garden for you.

In addition they offer a range of accommodation including self-catering and a selection of hotels in the moderate to luxury categories. Two/three-centre holidays are a speciality.

Prices Please call for brochure.
Discounts Details available on request.

International Chapters

International Chapters, 102 St John's Wood Terrace, London NW8 6PL
Tel: 071 722 9560

Greece, Italy, Mexico, Morocco, Portugal, Spain, West Indies, France, UK, US

Beautifully restored historic properties in Italy, a wide selection of châteaux, villas and farmhouses in France, and villas in the West Indies, Mexico, Morocco, Portugal, Greece and Spain. Villas and apartments are very well furnished and fully equipped, including linen. Most have swimming pools. Cots can be provided if requested in advance. Maid service is provided in the Caribbean, many properties in Italy and France and Morocco, and elsewhere a cook is available on request. Travel arrangements can be organized if you wish. This company also acts as an agent for Tuscan Enterprises and Cuendet Italia.

Prices (per property per week)
Villas: £200–£2500 (low); £350–£6000 (high)

Chateau Welcome

PO Box 66, 94 Bell Street, Henley-on-Thames, Oxon RG9 1XS
Tel: 0491 578803

France

Over 85 privately owned châteaux all over France offering the opportunity to stay with top professional and noble families. Children are welcome and there are many opportunities to join the family in their favourite sports and pastimes, such as tennis, golf, riding, hunting, shooting and fishing, etc. An excellent opportunity to make friends and practise your French.

Most hosts can provide dinner for their guests. You will be received as a privileged guest and as a friend of the family.
Prices Room with breakfast (double occupancy with private bathroom) from £60. Suites and family rooms are also available.

Corfu a la Carte

8 Deanwood House, Stockcross,
Newbury, Berks RG16 8JP
Tel: 0635 30621
Greece

Villa, apartment and cottage holidays on Corfu, Paxos and Skiathos. Accommodation ranges from the basic to the luxurious, but all is fully furnished and equipped, including linen, hand towels and maid service. Baby equipment, high chairs (£6 per week), cots and playpens may be hired when booking. Cots cost £24 for two weeks. This company is experienced in dealing with family holidays involving babies and young children.
Prices (per adult per week, including flights)
£190–£489
Discounts Under-2s pay £20 per flight (no seat). Children aged 2–5 may receive better deal discounts (please enquire), 5–16 receive a £60 discount off-season and a £50 discount in high season.

CV Travel

43 Cadogan Street, Chelsea, London
SW3 2PR
Tel: 071 581 0851 or 071 584 8803
Greece, Italy, Portugal, Spain

Villas in the unspoilt areas of Umbria, southern Spain, Paxos, southern Italy, Tuscany, Majorca and the Algarve. Properties range from simple Greek village houses to Italian palazzos. All villas are comfortably furnished and fully equipped, including maid service and some have cooks. Gas, electricity and water are included and linen is provided and changed weekly. High chairs and playpens can also be hired in many villas. CV Travel can arrange baby-sitting at most villas. Food hampers can be provided on arrival at the villa and cost between £20–£35. Windsurfers, small boats and cars can also be booked in advance in the UK. The brochure has a useful information section with details of airport hotels and parking and CV Travel also provide all clients with a copy of their holiday booklet with detailed information and advice on the holiday areas, what to take on holiday, national tourist offices and suggested guide books and other reading.
Prices £320–£995 per adult per fortnight, including flights. Details of villa rental only packages on request.
Discounts Under-2s pay £25 for flight (no seat, meals or luggage allowance); children aged 2–12 receive discounts of £10–£40.

Corona Holidays

73 High Road, London E18 2QP
Tel: 081 530 3747
Canary Islands

Winter and summer holidays in luxury villas and apartments. Most properties are in holiday estates, some with half board option. Facilities for children vary; some offer playgrounds, parties, free day-nursery and baby-sitting service. Cots and pushchairs are available for hire. UK flights (charter and scheduled), can be arranged from your nearest local airport on request.
Prices (per person per week, 4 sharing) Apartment in Tenerife: £83.30
Flights extra.
Discounts Under-2s pay £15, cots and meals payable locally. Children aged 2–11, £20 reduction per week on most holidays. 2 days free per person per fortnight on some resorts for all individuals.

Dominique's Villas

13 Park House, 140 Battersea Park
Road, London SW11 4NB
Tel: 071 738 8772
Fax: 071 498 6014
France

A selection of beautiful villas, farmhouses, manor houses, even châteaux in secluded countryside, some in spectacular settings. The villas are situated in the Loire, the Dordogne, Lot et Garonne,

Provence and the Côte d'Azur. All are fully furnished and equipped, most with a swimming pool, some with tennis court, washing machine and dishwasher. Gas, electricity and water are free, but heating is extra. Linen is sometimes included in the cost of the property, and can always be hired if requested in advance. Cots are provided free of charge, but must also be requested in advance. Each villa is listed with a photograph, details of the accommodation and local amenities. Bonded through ATO.
Prices (per villa per fortnight, including short Channel crossing for car and 2 adults) Villas: from £1200
Special rates are available for additional cars and passengers, as well as for Motorail fares. A good selection of hotels en route is available.

Florida Home Owners' Association

Lyndale, Dayseys Hill, Outwood, near Redhill, Surrey RH1 5QY
Tel: 034 284 2623/4155

US (Florida)

Luxury villas and apartments for rent throughout Florida. All accommodation is fully furnished and equipped, including air-conditioning, colour TV, linen and towels. Some properties have their own private swimming pool; others have access to a pool, but there may sometimes be a charge. Cots are generally available for hire. Most properties are within reach of Disneyworld. The company also make fly/drive travel arrangements.
Prices (per property per week) from £255

French Villa Centre

175 Selsdon Park Road, South Croydon, Surrey CR2 8JJ
Tel: 081 651 1231

France

Villa specialists with properties in coastal and rural locations. Some apartments, gîtes and luxury cottages are also available in certain areas. All accommodation is fully furnished and equipped, the gîtes being the cheapest and simplest. Travel arrangements can be made on request, but if you prefer to make your own, you simply pay for the accommodation. Cots can be supplied for a small fee, normally around £5 per week.
Prices Details on application.
Discounts Children receive the standard discounts on ferries and flights.

Greek Islands Club

66 High Street, Walton-on-Thames, Surrey KT12 1BU
Tel: 0932 220477

Greece

Ionian villa and apartment specialists on the 'unspoilt' islands of Paxos, Cephalonia, Ithaca and Kythira. All properties are well furnished and close to beaches and small villages. A daily maid service operates, except on Sundays and a Club creche operates on Cephalonia. Car hire and private boat hire available. Cots, high chairs and baby-sitters can be provided.
Prices Prices range from £400 to £800.
Discounts Children under 2 pay £50, which includes cot and linen. Other child discounts available on request.

Ilios Island Holidays Ltd

18 Market Square, Horsham, West Sussex RH12 1EU
Tel: 0403 59788

Greece, Italy, Turkey

Villa, apartment and hotel holidays in the quieter parts of the Greek islands: Cephalonia, Leskada, Naxos, Skiathos, Skopelos, Meganissi, Tinos, Pelion Peninsula and Zakynthos. All are comfortably furnished (some more so than others), and are fully equipped, including linen. Cots and high chairs may be hired for a small charge. Cot linen is limited, so you are advised to take your own.
Prices (per adult per week in a villa, including return flight and transfers) from £280–£400
2 weeks £350–£550
Discounts Details of discounts for under-12s are available on request.

Interhome

383 Richmond Road, Twickenham, Middlesex
Tel: 081 891 1294

Austria, Belgium, France, Holland, Hungary, Italy, Spain, Switzerland, UK, US

Villas, apartments, houses and hotel holidays. All self-catering accommodation is comfortably furnished and fully equipped. Extra charges may be made for electricity, gas, linen, towels and final cleaning, depending on location. About 10–15% of properties have cots. Please request these on booking.

Prices (per apartment per week)
A good quality apartment for 4 people in high season costs £300.

Discounts None available on accommodation. Under-4s travel free on ferry; 4–13s pay £30 each, which includes insurance.

Laskarina Holidays

St Mary's Gate, Wirksworth, Derbyshire
DE4 4DQ
Tel: 062 982 2203/4

Greece

Villas, apartments and studios in 'unspoilt' resorts on ten Greek islands. All self-catering accommodation is fully furnished and equipped – some rather more luxuriously than others. A maid service for light cleaning only is available at least once a week.

Prices There is a basic charge for a child of £50 (including use of a cot and linen). Other prices on request.

Discounts Children under 2 travel free (no seat). Free places are available in specified accommodation for 2–12s on 2-week holidays from early May to mid-June and during most of October. These free places are limited to one child per 3 full fare paying passengers. Children aged 2–12 receive a £10–£50 discount.

Magic of Italy

227 Shepherds Bush Road, London
W6 7AS
Tel: 081 748 7575 (*reservations*)

Italy

The company specialises in Sardinia, Tuscany and Umbria offering self-catering villas and apartments with 1, 2 and 3 bedrooms. All have a swimming pool. Magic of Italy also offer Motorail holidays – details on request.

Prices and discounts Details available on request.

Meon Villa Holidays

Meon House, Petersfield, Hants
GU32 3JN
Tel: 0730 68411

Caribbean, France (including Corsica), Gozo, Greece, Italy, Malta, Portugal, Spain, US

Very wide selection of villa and apartment holidays, most with car included. According to the brochure, the properties range 'from 3-star comfort to 5-star luxury'. All accommodation is fully furnished and equipped, including linen, hand towels and maid service. Many of the properties have their own private swimming pool.

Meon also offer self-drive holidays to France and Italy, organizing ferry crossings and the accommodation of your choice: villa, cottage, farmhouse or apartment. Cottages and farmhouses do not supply linen and towels. Cot hire can be arranged on request.

Prices (per adult per fortnight, including flight, in a 4-person villa with private pool on Algarve): from £412 to £1602 (1992 prices).

Discounts Children under 2 pay a flat fee of £40 for 2 weeks, £15 for one week, and this includes cost of cot hire. 2–11s receive a reduction of £10–£40, depending on season.

New Century Holidays Ltd

Century House, Unit 15, Kernick Industrial Estate, Penryn, Cornwall
TR10 9EP
Tel: 0326 375959
Fax: 0326 375316

Spain

Self-catering holidays on the unspoilt coast of Almeria in south east Spain, based

in small friendly villages. Accommodation ranges from basic seaside apartments and village houses to luxury villas with private swimming pools. Cots and child seats available (details on application).
Prices From around £189 per person including flights from Gatwick, Bristol, Birmingham and Manchester.
Discounts Details on application.

Palmer & Parker

The Beacon, Penn, Bucks HP10 8ND
Tel: 0494 815411

Caribbean, France, Portugal, Spain

Up-market villas furnished and equipped to a very high standard. All have maid service, private swimming pool and car hire. Cots and high chairs may be hired weekly for £15 each.
Prices A 4-bedroom villa sleeping 8 in Portugal in August costs around £600 per person. Villas can be rented without travel.
Discounts Under-2s pay a flat fee of £30. Children aged 2–11 receive a reduction of £50 in Spain and Portugal. In France there are no accommodation discounts, but children are entitled to a 20% discount on the fly/drive part of the holiday. In the Caribbean under-2s pay 10% of adult price; 2–11s receive a reduction of about £150 each.

The Portuguese Property Bureau

Algarve House, 1A The Colonnade, Maidenhead, Berkshire SL6 1QL
Tel: 0628 770220

Portugal

Spacious self-catering 2-, 3- and 4-bedroom detached and terraced villas in and around Carvoeiro on the Algarve. Most properties are set in a garden and have a private or shared pool. All the villas are fully furnished and well equipped, including linen and towels. Maid service is included in the price, as is water, electricity, gas and water heating. The company looks after the garden and pool. Cots are available for £16 per week, high chairs for £15 per week. A free welcome food pack is provided on arrival. There is a breakage deposit of £50 refunded within 3 weeks after return. Full package holidays are also available.
Prices (per person per fortnight, accommodation only)
2-bedroom terrace villa with shared pool: £100–£300
2-bedroom detached villa with own pool: £140–£410
Discounts Under-2s are charged £50, which includes hire of cot.

Simply Turkey

8 Chiswick Terrace, Acton Lane, London W4 5LY
Tel: 081 747 1011

Turkey

Several privately-owned villas and studios available for rent, all comfortably furnished and equipped, including maid service. Cots can be provided by arrangement. High chairs can be hired for £10 per week. Hotels are small, usually family-run and in resorts with plenty to do throughout the day. Babysitting can be arranged.
Prices (per adult per week, B+B, including flights and transfers): from £269
Discounts Children under 2 travel free; 2–16s receive £20–£50 discount, depending on season and length of stay.

Skiathos Travel

4 Holmesdale Road, Kew, Richmond, Surrey TW9 3JZ
Tel: 081 940 5157

Greece

Small, specialist company offering villas, apartments and hotels on the Greek islands of Skiathos, Skopelos and Alonissos. All accommodation is close to sea and the company particularly recommends Skiathos for children, as it has safe, sandy beaches. All self-catering accommodation is comfortably furnished and equipped, including linen and hand towels. Maid service available in some locations. Cots may be hired.
Prices (per adult per fortnight, including flight)
3-bedroom villa: from £380 to £550

182 Choosing and planning your holiday

Discounts Children under 2 travel free (no seat); 2–11s are eligible for £15–£20 discounts, depending on season and length of stay.
Note: These are 1992 prices given only as a guideline.

Sovereign Italia

Astral Towers, Betts Way, Crawley, West Sussex RH10 2GB
Tel: 0293 599988

Italy, Sardinia

Large selection of quality villas and apartments and family-run hotels in Tuscany, the Italian lakes and the Neapolitan Riviera. Accommodation ranges from rural farmhouses to 14th-century palazzos, many with pool to beach and city hotels. Travel arrangements by car/ferry or flights from Gatwick, Heathrow and Manchester.
Prices (per person for one week in Tuscany, low season, based on five sharing apartment with pool, including flight from Gatwick): from £322
Discounts 20% child reduction throughout season.

Sovereign Just Turkey

Astral Towers, Betts Way, Crawley, West Sussex RH10 2GB
Tel: 0293 599977

Turkey

Specialist operator offering self-catering apartments, villas and classical hotels in select resorts around Bodrum, Marmaris and south west Turkey. Wide range of two-centre holidays including Istanbul or Gulet cruising combinations. Flights from Gatwick, Heathrow, Manchester, Newcastle, Glasgow and East Midlands.
Prices (per person for one week in Kusadasi, low season, based on four sharing a villa with pool, including flight from Gatwick): from £232.
Discounts Free child places in all resorts in early and late season.

Sun Esprit

Oaklands, Reading Road North, Fleet, Hants GU13 8AA
Tel: 0252 816004
Fax: 0252 811243

France

Sun Esprit devote their holidays entirely to the family market. They offer hotel and self-catering options in France, based in Aquitaine (south west coast) and the Alps. Their resort staff include qualified British nannies, who provide babysitting on 6 nights each week. Freedom for parents and excitement for the youngsters are the hallmarks of a holiday with Sun Esprit. No surcharges. Sells direct to the public and through ABTA agents.
Prices and discounts Available on request.

Sunselect Villas

60 Crow Hill North, Middleton, Manchester M24 1FB
Tel: 061 655 3055

France

Self-catering villas and cottages in northern and southern Brittany. Most villas have their own gardens with garden furniture and are a short walk or drive from the nearest beach. All are fully furnished and equipped. Linen and towels are not provided, but can be hired locally. Cots are supplied free in southern Brittany and can be hired in the north. A returnable deposit of £80 is payable with the final invoice. Gas, electricity (except for heating), water, and cleaning before arrival are free. This company also offers canal and river cruising holidays throughout France. If ferry crossings are at night the cost of a cabin is extra. For disabled travellers a suitable villa can always be recommended.
Prices (per villa per 2 weeks, including ferry for car and 2 adults)
Low: from £550
High: from £805
Return fares are payable separately for additional passengers: children aged 4–13 pay £19; adults pay £34 or £44, according to season. Single-week bookings are possible.

Villas and apartments

Discounts Children under 4 are free. A reduction of £25 is offered on all one-week holidays when travelling on Brittany ferries.

Thomson Tour Operations

Greater London House, Hampstead Road, London NW1 7SD
Tel: 081 200 8733 (London)
021 632 6282 (Birmingham)
061 236 3828 (Manchester)

Canary Islands, Cyprus, Greece, Ibiza, Italy, Majorca, Malta, Minorca, Portugal, Spain

Large selection of good quality private villas with pools and small apartment complexes offering excellent value holidays for families. All accommodation is fully furnished and equipped, including linen, hand towels and maid service. Cots are available in most places, some provided free, others are available for about £3 per night. Villa specialist brochures handled by Thomson include *Villas and Apartments* and *OSL*.

Prices (per adult per fortnight, including return flight and transfers)
Low: £90–£400
Mid: £70–£470
High: £500–£600

Discounts Children under 2 pay £15 (no seat on plane). First child receives discount between £20 and £100 or pays price in child column on price panel. Second child receives £10–£35 reduction. Child discounts vary from 2–11 and 2–16 inclusive, depending from which brochure the holiday is booked.

Patricia Wildblood

Calne, Wilts SN11 0LP
Tel: 0249 817023 *or* 081 658 6722

Italy, Spain

A wide selection of comfortable private holiday homes and houses in Menorca. Villas range from simple holiday standard accommodation to large houses set in their own grounds with private pools. Most villas have a pool and barbecue. All are fully furnished and equipped, including linen. Maid service can be arranged. Cots are available and must be paid for before departure. Gas, electricity and water are included. A welcome food pack is provided on arrival. Windsurfing facilities can be arranged. Car hire is available on request.

Prices Available on request.

Discounts From 2–15 years there are various child reductions depending upon age and season.

Walking holidays

Don't disregard this section in the belief that these holidays will only appeal to seasoned walkers. The companies listed grade their walks for difficulty and will advise you which to take if you are in doubt. Some of the people we spoke to said that young children, even babes in arms (or slings) were commonplace on their tours, while others had experienced none younger than eight. Ultimately it's down to you. If *you* are fit and healthy and can tolerate carrying a baby or toddler for several hours, these companies won't turn you away, but they will have a few words of caution for you.

Can you guarantee that your child won't impede the progress of the tour? Unscheduled stops for feeds and nappy changes are not really fair on the other walkers. Also pushchairs are out of the question on all but the most level and simple of walks.

Some larger companies claim that other guests muck in and help look after and entertain children. The same cannot be promised with all groups. The success of the walk boils down to your sensibility and your child's adaptability, so be honest with yourself before making a commitment to this type of holiday. It could be great, and indeed many children love it. On the other hand the fear of your child delaying or upsetting the rest of the party may be too offputting.

Prices
Unless otherwise stated 1992 prices have been quoted. These are intended only as a guide to the type of holiday offered and travellers should check with the company concerned for 1993 pricing details.

HF Holidays Ltd

Imperial House, Edgware Road, London NW9 5AL
Tel: 081 905 9556

Austria, France, Italy, Majorca, Malta, UK

Independent and accompanied walking holidays. Walks are graded from gentle rambles to full-scale mountaineering and include discovery and theme walks for those with special interests in such things as archaeology and birdwatching. The more arduous tours are not suitable for young children. Family walking holidays bear children in mind, so there are organized social events and entertainments in the evening. Entertainments and children's facilities are not necessarily available abroad; ask when booking. Accommodation in the UK is mainly in HF's own 'characterful' houses, with small, family-run hotels in some locations. Abroad, accommodation tends to be in guest houses. Rooms are single, twin or multi-bedded with tea and coffee-making facilities in each. Cots and high chairs can be provided on request, and there are facilities for washing and drying clothes.
Prices A family holiday with one week's full board in UK ranges from £250–£299 per adult.
Discounts Children's discounts available. Call for details.

Ramblers Holidays Ltd

Box 43, Welwyn Garden City, Herts
AL8 6PG
Tel: 0707 331133

Andorra, Austria, Azores, Borneo, Canada, China, Cyprus, Czechoslovakia, Egypt, France, Greece, Himalayas, India, Italy, Japan, Jordan, Madeira, Malta, Mexico, Morocco, New Zealand, Papua New Guinea, Peru, Philippines, Portugal, Spain, Switzerland, Thailand, Turkey, UK, US, West Indies

This company offers escorted walking tours, graded for difficulty, in an astonishing variety of destinations. Some tours are designed for those with a spirit of adventure; they may have only simple accommodation and operate changes of the itinerary – some tours consist of pony trekking with only a sleeping bag for comfort. Only children above 16 years of age are allowed on these holidays.

Prices (per adult per tour, including flights and transfers)
Austria: 1 week half board from £340; 2 weeks from £470
Canada: from £1275

Welsh Wayfaring Holidays

Neuadd Arms Hotel, Llanwrtyd Wells,
Powys LD5 4RD
Tel: 05913 236

UK (Wales)

Guided walks in the Cambrian mountains, Brecon Beacons and Elan Valley. On a 7-day holiday, walking is done on 5 days, and on a 4-day break, walking is done on 3 days. Accommodation is in the family-run Neuadd Arms Hotel. Some en suite rooms are available and are considered particularly suitable for families. Cots can be provided if requested when booking. A launderette and drying-room are available. If you opt out of walking (which you are free to do), arrangements can be made for riding, birdwatching, mountain cycling, fishing and car touring. Details on request.

Prices (per adult per week, including full board with packed lunch and afternoon tea): from £200

Discounts On accommodation only, under-5s are free. Children aged 5–12 receive 75% discount when sharing parents' room; 13–16s receive 40% discount. Small children are entitled to reduced rates on restaurant meals, and there is a special children's menu at the bar's snack counter.

Section 3

Holiday transport

Facilities for families at British airports

This is a survey of the facilities available at airports throughout Britain. We have tried to make each entry as comprehensive as possible, but we would welcome further information, personal experiences or advice from readers.

In this age of long airport delays, it is well worth the travelling family's while to read this section carefully before booking a flight. Although the larger airports may have more lavish facilities, it is worth remembering that they also have to deal with a greater volume of passengers. Smaller airports often have a shorter clearance time for customs and baggage control and this fact alone might mean the difference between a hassled journey and relaxing trip.

It is always advisable to contact the airport information desk before you set off. The staff will be able to give you details of the facilities available and advise you on any special needs or requests.

In general, children love the thought of travelling by plane. You can encourage them to look forward to their time at the airport by helping them to draw pictures and write stories. These might be about what they are expecting to see, or what they think planes and airport buildings of the future will be like. Most airports have viewing terraces where, usually for a small fee, you and your children can have a grandstand view of the planes landing and taking off (contact the airport information desk in advance to check availability). Games involving making plane noises and running around with arms outstretched may be great fun in your garden or front room, but remember that an overexcited child hurtling through a crowded concourse will drive you (and other passengers) to distraction.

There are several colourful children's books on the market about children's visits to an airport and these can be successfully used when calming an anxious child's fears. If adults show fear of flying, children will quickly assume that aeroplanes are frightening things. Try to keep your own misgivings to yourself. Above all, it is important to make children feel from an early age that flying is fun.

By giving this part of your holiday as much advance planning as possible, you may even manage to view your time at the airport as an enjoyable part of your holiday.

Aberdeen

Aberdeen Airport, Dyce, Aberdeen
AB2 0DU
Tel: (0224) 722331

Europe, UK

- Seven miles from city centre, 35 minutes by express bus service (exact fare required on boarding). Aberdeen station is nearest rail connection point. Car parking available adjacent to the

terminal. Disabled parking with telephone to main terminal for assistance.
- Mother and baby room in main terminal; facilities including disposable nappies available on request from information desk.
- Children's meals and snacks available, including baby food.
- Medical facilities in main terminal building, with qualified first aider on duty at the information desk.
- Books and toys for sale in terminal. Allow 20 minutes to clear baggage and customs.
- Child seats available with car-rental.

Belfast

Belfast International Airport, Belfast BT29 4AB
Tel: (08494) 22888

Canada, Europe, UK, US

- 18 miles from Belfast, 30 minutes by shuttle-coach from city centre.
- Mother and baby room in main departure lounge.
- Disposable nappies are available.
- Children's meals, snacks, and baby food available.
- Books and toys for sale in terminal, also audio and video cassettes.
- Allow 40 minutes to clear baggage and customs.
- Child seats available with car-rental.

Birmingham

Birmingham International Airport, Birmingham B26 3QJ
Tel: (021) 767 7145/6

Worldwide

- Excellent bus, coach and British Rail connections; 10 minutes by rail from Birmingham New Street Station.
- Both short-term and long-term parking available. Disabled passenger information available.
- Mother and baby rooms are situated adjacent to the Birmingham Bar landside, and in the international departure lounge.
- Disposable nappies available.
- Children's meals and snacks available, but not baby food.

- Playcare centre, with trained staff, situated on the mezzanine floor of the main terminal, will take without charge 2–8 year olds for 30-minute periods. This facility is situated opposite the American Express office.
- Spectators' viewing gallery, with shop and buffet facilities, is open from early morning to early evening all year round. There is a small charge made for admission.
- Allow 30 minutes to clear baggage and customs.
- A small shopping mall is situated in the main terminal.
- Child seats available with car-rental.

Bournemouth

Bournemouth International Airport PLC, Christchurch, Dorset BH23 6SE
Tel: (0202) 593939

Channel Islands, Cyprus, Greece, Italy, Spain, UK

- 6 miles from Bournemouth. Good road connections.
- Mother and baby room in main terminal.
- Disposable nappies are available.
- Children's meals and snacks, but no baby food.
- Books and toys for sale in terminal.
- 30 minutes check-in time.
- Child seats available with car-rental.

Bristol

Bristol Airport, Bristol BS19 3DY
Tel: (0275) 474444

Canada, China, Channel Islands, Cuba, Europe, India, Morocco, Nicaragua, Tunisia, Turkey, United Arab Emirates, UK

- Eight miles from city centre. A service bus runs to the airport from Bristol Marlborough Street bus and coach station and Bristol Temple Meads railway station (journey time approx 20 minutes).
- Both short-term and long-term parking facilities are available.
- Mother and baby room situated on ground floor of main terminal.

- Disposable nappies available.
- Children's meals available, but not baby food.
- Books and toys for sale in terminal.
- Allow 20 minutes to clear baggage and customs.
- Child seats available with car-rental.

Cardiff

Cardiff-Wales Airport, Nr Cardiff, South Glamorgan CF6 9BD
Tel: (0446) 711111

Canada, Europe, Turkey, UK

- Twelve miles from city centre. Express bus service runs from Cardiff bus station, which is adjacent to Cardiff mainline railway station (journey time approximately 40 minutes).
- Good car parking facilities available.
- The airport shop sells a comprehensive range of baby goods including disposable nappies, travel potties and dummies as well as toiletries such as baby lotion, talc, cream, shampoo, etc. It also sells books and toys.
- Mother and baby room in female toilet area.
- The buffet and self-service area will provide means with which to warm milk and also gives Heinz dinners and desserts free of charge to parents. Highchairs are available.
- Wide range of facilities available for disabled passengers: lifts to all floors of the terminal building, specially adapted telephones and toilets. Ramps and automatic doors are available for easy access. A custom-built wheelchair lift enables travellers to utilize their own wheelchairs when descending two flights of stairs on the International Pier. A limited number of special car parking spaces are available on request.

Coventry

Coventry Airport, Warwickshire CV8 3AZ
Tel: (0203) 301717

Europe, Ireland

- Two miles from Coventry city centre, 20 minutes by bus from city centre.

- No mother and baby room.
- Children's meals and snacks, but no baby food.
- No books and toys for sale in airport.
- Allow 10 minutes to clear baggage and customs.
- Child seats available with car-rental.

East Midlands

East Midlands International Airport, Castle Donnington, Derby DE74 2SA
Tel: (0332) 810621

Belgium, Channel Islands, France, Greece, Holland, Spain, UK

- Eight miles from Loughborough, approximate journey time 30 minutes by coach.
- Parent and baby rooms situated in main terminal and airside.
- Disposable nappies are available.
- Children's meals, snacks, and baby food are available.
- Soft play area situated opposite airport information desk.
- Books and toys for sale in terminal.
- Cartoon booths and TV booths are sited both landside and airside.
- A selection of games is available from the information desk.
- In the event of delays, children's entertainment is organized, and children's play-packs are available from the information desk.
- Allow 30 minutes to clear baggage and customs.
- Child seats available with car-rental.

Edinburgh

Edinburgh Airport, Edinburgh EH12 9DN
Tel: (031) 333 1000
Fax: (031) 335 3181

Canada, Europe, UK

- Express bus from Waverley Bridge rail terminus in city centre (approximate journey time 25 minutes).
- Mother and baby room situated on ground floor of main building near to gate 6, between the bar/buffet and the international arrivals area.
- Disposable nappies available.

Holiday transport

- Children's meals and snacks available, but no baby food.
- Medical facilities available in main terminal building and a trained nurse is usually on duty at the information desk.
- Books and toys for sale in terminal.
- Spectators' gallery available during daylight hours.
- Allow 20 minutes to clear baggage and customs.
- Child seats available with car-rental.

Exeter

Exeter Airport, Exeter, Devon EX5 2BD
Tel: (0392) 367433

Canada, Europe, UK

- Five miles from city centre.
- Short-term and long-term parking available. Disabled travellers are advised to contact the airport prior to arrival. Taxi rank at the airport and at Exeter mainline rail and coach station. Mother and baby rooms situated in the main terminal building and in the international departures lounge.
- Disposable nappies available.
- Children's meals and snacks available, and free baby food.
- Cartoon cabin in terminal and video in international departures lounge.
- Shop selling books, confectionery and papers.
- Allow 20 minutes to clear baggage and customs.
- Child seats available with car-rental.

Gatwick

London Gatwick Airport, West Sussex RH6 0NP
Tel: (0293) 535353

Worldwide

- Excellent public transport connections. InterCity Gatwick Express service runs non-stop to Victoria every 15 minutes and has a journey time of 30 minutes. (A number of airlines have a baggage check-in facility at Victoria.) The Thameslink rail service links Gatwick with The City of London in 35 minutes. Direct train and coach services from many other major British cities. Situated on the M23 and just a few minutes drive from the M25.
- Short-term and long-term car parking always available. Special arrangements for people with disabilities – please contact airport staff in advance. Fully equipped babycare rooms are available for use by mother or father. They are situated before and beyond passport control – look for the baby bottle symbol. Parents with pushchairs may use the disabled toilets found throughout the terminals. Showers are also available in both terminals for a small fee.
- A soft-play area for children aged up to five can be found at Gatwick Village, South Terminal. Youngsters must be supervised by parent or guardian.
- Satellite TV is shown at the TV lounge in Gatwick Village and there is a spectators' viewing gallery overlooking the stands and runways (small charge for admission).
- Medical facilities are available 24 hours a day and the charity Travel-Care have offices at Gatwick Village. Staff offer help to anyone who has a problem whether travel-related or not.
- A wide selection of restaurants offer children's menus, and children's toys and goods can be found at the airport's extensive range of shops – many of them High Street names. A Value Guarantee promises that prices at Gatwick are the same as in the High Street for both shopping and eating.
- Allow 40 minutes to clear baggage and customs.
- Child seats available with car rental.

Glasgow

Glasgow Airport, Paisley PA3 2ST
Tel: (041) 887 1111

UK, Europe, Trans-Atlantic

- Fifteen minutes from city centre on M8. Nearest railway station is Paisley Gilmour Street (2 miles from airport). Regular bus service from Glasgow city centre (Central and Queen Street stations and Anderston Cross and Buchanan bus station) to airport forecourt.

Facilities for families at British airports

- Short-term car-parking facilities are available near to the terminal building.
- Mother and baby room on first floor.
- Disposable nappies available.
- Children's food and snacks, and baby food available from The Granary.
- Medical unit situated in International Arrivals hall.
- Books and toys for sale in terminal.
- Allow 30 minutes to clear baggage and customs.
- Child seats available with car-rental.

Guernsey

Guernsey Airport, La Villiaze, Forest, Guernsey
Tel: (0481) 37766

France, Holland, Germany, Switzerland, UK

- Five miles from St Peter Port town centre.
- Regular bus service to town centre. Taxi rank in airport forecourt.
- Car parking for 280 cars.
- Mother and baby room. Disposable nappies are available from information desk in emergencies only. Nappies, baby food and other supplies may be obtained from shop ten minutes walk away (turn left outside airport entrance).
- Buffet open all day for lunches, snacks and drinks.
- No books and toys on sale in airport. No duty-free shop.
- Allow 15 minutes to clear baggage and customs.
- Child seats available with car-rental.

Heathrow

Heathrow Airport, Hounslow, Middx TW6 1JH
Tel: General enquiries: 081 759 4321
(Terminal 1): 081 745 7702/4
(Terminal 2): 081 745 7115/7
(Terminal 3): 081 745 7412/4
(Terminal 4): 081 745 4540
Fax: 081 745 6261

Worldwide

- Good transport connections. Central London to Heathrow (approximate journey time 50 minutes) by London Underground (Piccadilly Line, every 3–7 minutes). Coach services from all over Britain call at Heathrow (check arrival and departure points with operator in advance). London Transport's Airbus operates a regular service to and from central London (all buses are equipped to carry disabled passengers).
- Short-term and long-term car parks available. Disabled travellers are advised to contact airport information desk for further details.
- Parent and child rooms are available in each terminal. They may be locked very late at night or early in the morning and at these times, passengers are advised to contact the terminal's information desk.
- Disposable nappies available.
- Children's food, snacks and baby food available.
- Medical centre, staffed 24 hours a day, is situated in Queen's Building.
- Playcare Centre, staffed by qualified nursery nurses, is available free of charge for 2–8-year-olds. It is situated in the airside departures concourse of Terminal 4.
- Spectator viewing facilities are available free on the roof of the Queen's Building.
- Travel-Care unit, situated in the Queen's Building, is staffed by professional social workers. A confidential advice service is offered to passengers during office hours.
- Books and toys for sale in terminal.
- Allow 30 minutes to clear baggage and customs.
- Child seats available with car-rental.

Humberside

Humberside Airport, Kirmington, South Humberside DN39 6YH
Tel: (0652) 688456

Channel Islands, Denmark, Holland, Majorca, Norway, UK

- Thirteen miles from Scunthorpe. Two miles from Barnetby Railway Station.
- Mother and baby facilities in conjunction with disabled toilets.
- Facilities for disabled passengers include parking just outside the airport.

194 Holiday transport

- No special children's or baby food.
- Books and toys for sale in terminal.
- Allow 15 minutes to clear baggage and customs.
- Child seats available with car-rental.

Jersey
Jersey Airport, St Peters, Jersey JE1 1BY
Tel: (0534) 46111

France, UK

- Five miles from St Helier. Regular bus service to town centre (journey time approximately 20 minutes). Taxi rank outside arrivals hall.
- Car park adjacent to airport.
- Mother and baby rooms in departure lounge and arrivals hall.
- Nappies not available.
- Children's food, snacks and baby food available.
- Medical facilities are available on site.
- Disabled travellers are advised to contact information desk for details of facilities available.
- No books or toys for sale at airport.
- Spectators may gain access to terminal roof, but best views are from the NW side of the airfield.
- Allow 15 minutes to clear baggage and customs.

Kent
Kent International Airport, P.O. Box 500, Manston, Kent CT12 5BP
Tel: (0843) 823333

Channel Islands, Cyprus

- Two miles from Ramsgate. No public transport connections. Taxis from Ramsgate town centre.
- Parking facilities are good, with some packages providing free parking.
- Children's play area and televised films.
- Disabled travellers are advised to contact airport staff before travelling for information about the facilities available and to make any particular requests.
- Mother and baby room situated in central concourse.
- Children's meals and snacks, but no baby food.

- Books and toys on sale in terminal.
- Allow 15 minutes to clear baggage and customs.
- Child seats available with car-rental.

Leeds—Bradford
Leeds—Bradford Airport, Yeadon, Leeds
Tel: (0532) 509696

Canada, Europe, UK

- Eight miles north-west of Leeds. Six miles north of Bradford. Eleven miles south of Harrogate. No direct rail connection to the airport. Regular bus service from Bradford. Taxis available outside terminal during operational hours.
- Short-term and long-term car parking available for 1,000 cars.
- Mother and baby room situated on ground floor.
- Disposable nappies available.
- Children's food and snacks, but no baby food.
- Books and toys for sale in terminal.
- Disabled travellers are advised to contact the airport duty office before departure for details of facilities available.
- Allow 30 minutes to clear baggage and customs.
- Child seats available with car-rental.

Liverpool
Liverpool Airport, Liverpool L24 1YD
Tel: (051) 486 8877

Europe, UK

- Eight miles from city centre. Regular train service from city centre to Garston Station (10 minutes by bus from airport terminal). Runcorn Station is 10 minutes away by taxi, available from airport terminal entrance.
- Car parking available adjacent to airport terminal.
- Disabled travellers are advised to contact airport information desk in advance for details of facilities which include an Ambu-lift.
- Mother and baby room in main concourse.
- Nappies not available.
- Children's food and snacks, but no baby food.

- Medical aid should be sought at the information desk or from any security officer.
- Books and toys for sale in terminal.
- Allow 30 minutes to clear baggage and customs.
- Child seats available with car-rental.

London

London City Airport, King George V Dock, Silvertown, London E16 2PX
Tel: (071) 474 5555

Belgium, France, Germany, Holland, Sweden, Switzerland

- Easily accessible by public transport from central London. Use RiverBus or Docklands Light Railway to connect with free Airport Bus service from Canary Wharf to the Terminal. London City Airport/Silvertown Station is on the North London Line and is within walking distance of the airport terminal. The airport is five minutes by taxi from Plaistow Underground Station (District & Metropolitan lines).
- Long-term and short-term car parks adjacent to airport entrance.
- Mother and baby room situated on ground floor, near to female toilet.
- Nappies not available.
- Children's meals and baby food not available.
- Toys available in terminal.
- Good facilities for disabled passengers. Check with airport staff before travelling.
- Allow 10 minutes to clear baggage and customs.
- Child seats available with car-rental.

London Luton

London Luton International Airport, Luton, Bedfordshire LU2 9LY
Tel: (0582) 405100 (24 hours)
Tel: (0582) 400505 (office hours)

Channel Islands, Europe, Ireland, UK

- Excellent coach, rail and road connections. Regular service from Central London by British Rail *Luton Flyer Railair Link* (journey time as little as 49 minutes). Regular connections from Luton Station with Intercity trains to the Midlands and beyond. Direct coach services to many destinations in UK, including Stansted, Heathrow and Gatwick airports.
- Both long-term and short-term parking is available. Disabled drivers are strongly advised to write to the Airport Security and Car Park Manager in advance for allocation of suitable spaces.
- Mother and baby rooms are situated in the landside concourse and departure lounge.
- Medical centre in landside concourse (open 24 hours).
- Unsupervised playcare centre for children aged 2–8 years.
- Books and toys for sale in terminal.
- Children's entertainments may be organized, in the event of long delays.
- Good facilities for disabled visitors. Check with airport staff for details.
- Spectators' facilities: Café Expresso and full bar facilities are available in the lounge bar, offering an excellent view of the runway.
- Allow 40 minutes to clear baggage and customs.
- Child seats available with car-rental.

Manchester

Manchester Airport PLC, Manchester M22 5PA
Tel: (061) 489 3000

Worldwide

- Good transport connections by bus, rail and car. A regular shuttle bus service calls at Manchester Victoria railway station and Piccadilly bus and railway stations. A rail link to the airport terminal is currently under construction and is due for completion in Spring 1993.
- Long-term and short-term parking facilities available. Disabled travellers should contact airport staff in advance for further details of the facilities available to them.
- Changing facilities are available in the female toilet areas in both terminal A (domestic arrivals hall) and terminal B (main concourse and international pier C satellite).

Holiday transport

- Children's meals, snacks and baby food available. Ice-cream parlour.
- Playcare centre staffed by qualified nursery nurses where children aged 2–8 years may be left for short periods free of charge. This is situated opposite the duty-free shop in terminal B.
- Books and toys for sale in terminal.
- Chemist's shop in main concourse.
- Spectator terrace available during daylight hours.
- Allow 30 minutes to clear baggage and customs.
- Child seats available with car-rental.

Newcastle

Newcastle International Airport, Woolsington, Newcastle NE13 8BZ
Tel: (091) 286 0966

Europe, UK, US

- Good rail, bus and coach connections. Regular bus service runs from Newcastle Central railway station and Eldon Square bus concourse direct to airport terminal (approximate journey time of 25 minutes). Coach service to and from major cities in England and Scotland. The recently opened Metro rail link provides transport to the airport and to 45 other Metro stations in Tyne and Wear. This rail link enables passengers to travel from any mainline BR station via Newcastle Central, to Newcastle International Airport (journey time is 22 minutes).
- Short-term and long-term parking facilities.
- Mother and baby room in main concourse.
- Disposable nappies available.
- Children's meals and snacks available, but no baby food.
- Books and toys for sale in terminal.
- Roof terrace is open to spectators during daylight hours.
- Disabled travellers are advised to contact the airport prior to arrival to confirm facilities available.
- Allow 30 minutes to clear baggage and customs.
- Child seats available with car-rental.

Norwich

Norwich Airport, Norfolk NN6 6JA
Tel: (0603) 411923

Channel Islands, Holland, Malta, Spain, UK

- Three miles from Norwich city centre. Taxis available to airport forecourt.
- Nursing parents' facilities available in First Aid room.
- Disposable nappies available.
- Snacks, but no baby food.
- Books and toys for sale in terminal.
- Allow 15 minutes maximum to clear baggage and customs.
- Child seats available with car-rental.

Prestwick

Prestwick Airport, Prestwick, Ayrshire KA9 2PL
Tel: (0292) 79822

Canada, US

- Rail, coach and road connections. Prestwick railway station is on the main line from Glasgow to Ayr (approximate journey time from Glasgow 45 minutes). A regular coach link runs from Glasgow city centre (Buchanan Street bus station) and Edinburgh (St Andrews Square bus station).
- Short-term and long-term parking facilities.
- Disabled travellers are advised to contact the airport information desk in advance for details of the facilities available.
- Mother and baby rooms are situated in the main terminal (landside) and the international departure lounge.
- Disposable nappies available.
- Children's food and snacks available, but no baby food.
- For medical services, contact the information desk.
- Books and toys on sale in terminal.
- Spectator terrace situated on the second floor, overlooking the aircraft parking area. Terrace contains an amusement arcade.
- Allow 30 minutes to clear baggage and customs.
- Child seats available with car-rental.

Southampton

Southampton Airport, Southampton
SO9 1RH
Tel: (0703) 629600

Channel Islands, France, Holland, UK

- Situated 2 miles from Eastleigh. Four miles from Southampton town centre by train.
- Parent and child room in main concourse.
- Nappies not available.
- Children's meals and snacks, but no baby food.
- Books and toys for sale in terminal.
- Allow 15 minutes to clear baggage and customs.
- Child seats are available with car-rental.

Southend

Southend Airport, Southend, Essex
SS2 6YF
Tel: (0702) 340201

Channel Islands, Malta

- Three miles from Southend, 18 miles from M25 (junction 29), with bus and British Rail connections. 60 minutes by rail from London (Fenchurch Street and Liverpool Street stations).
- Short-term and long-term parking available.
- Good facilities for disabled travellers but passengers are advised to contact airport staff for further details before travelling.
- Parent and baby room in main concourse.
- Nappies not available.
- Children's meals and snacks available, but no baby food.
- No first-aid room, but airport fire-service staff are trained to give medical attention for minor ailments.
- Books and toys for sale in terminal.
- Allow 15 minutes to clear baggage and customs.
- Child seats available with car-rental.

Stansted

Stansted Airport Ltd, Stansted, Essex
CM24 1QW
Tel: (0279) 680500

Canada, Caribbean, Europe, Mexico, Tunisia, Turkey, US

- Situated 35 miles north east of London. Express trains every 30 minutes between Stansted Airport and Liverpool Street Station (fastest time, 41 minutes). Also, direct trains to Cambridge, Peterborough and the South Midlands.
- Short-term and long-term car parking available.
- Disabled travellers are advised to contact the airport information desk for further details of the facilities available.
- Nursing parents' rooms are available throughout the airport.
- A children's play area is available in the international departure lounge.
- Disposable nappies are available.
- Children's meals and snacks available, and baby food is complimentary.
- Books and toys for sale in terminal.
- Children's entertainment may be organized in the event of long delays.
- Child seats available with car-rental.

Teeside

Teeside Airport, Darlington, Co. Durham DL2 1LU
Tel: (0325) 332811

Channel Islands, Holland, Spain, UK

- Located 6 miles from Darlington. Bus and train connections.
- Mother and baby room in main concourse.
- Nappies not available.
- Children's meals and snacks, but no baby food.
- Books and toys for sale in terminal.
- Allow 20 minutes to clear baggage and customs.
- Child seats available with car-rental.

Airlines

Whether you're planning a short hop across the Channel or a long-haul flight to Australia, air travel can be a bit of an ordeal if you have children in tow. However, airlines do try to help, and many of them have excellent facilities for those needing to feed and change babies, or to amuse toddlers and small children.

Major airlines will obviously have a greater range of services than smaller carriers, but you can expect to find any of the following: bassinets, cots, changing facilities, baby food, milk, bottle-sterilizing and warming facilities, disposable nappies, children's meals, toys, games and books, seatbelt extensions. In addition you will often find you can take the push chair right up to the aircraft rather than having to carry your child, and many airlines allow people with children to board early.

It is essential to request everything you might need when reserving your seats. If you book through a travel agent, do ask for confirmation of your requirements. Airlines are usually willing to supply almost anything, if properly requested, but all too often the tour operator fails to make the request, or does not make it clearly enough. For peace of mind it's worth phoning the airline yourself about twenty-four hours before departure and reiterating your needs.

To find out exactly which childrens' facilities particular airlines offer, it's best to contact them direct. Below you will find the addresses and phone numbers of most major airlines.

Aer Lingus
223 Regent Street, London W1R 8JQ
Tel: 081 899 4747

Aeroflot Airlines
70 Piccadilly, London W1V 9HH
Tel: 071 355 2233 *or* 071 491 1764

Air Canada
7/8 Conduit Street, London W1R 9TG
Tel: 081 759 2636 (*Reservations only*)

Air France
179 Piccadilly, London W1V 9DB
Tel: 081 742 6600

Air India
17–18 New Bond Street, London W1Y 0BD
Tel: 071 491 7979

Air Malta
15 Quadrant Arcade, Regent Street,
London W1
Tel: 071 437 6477

Air Mauritius
49 Conduit Street, London W1R 9FB
Tel: 071 434 4375/9

Air New Zealand
New Zealand House, Haymarket, London SW1Y 4TE
Tel: 081 741 2299

Air Zimbabwe
52 Piccadilly, London W1V 9AA
Tel: 071 499 8947

Alitalia
205 Holland Park Avenue, London W11 4XB
Tel: 071 602 7111

Airlines

American Airlines
15 Berkeley Street, London W1X 6ND
Tel: 081 572 5555

Austrian Air
50–51 Conduit Street, London W1R 0NP
Tel: 071 439 0741

Britannia Airways Ltd
Luton Airport, Luton, Beds LU2 9ND
Tel: 0582 424155

British Air Ferries
Viscount House, Southend Airport, Essex
SS2 6YL
Tel: 0702 354435

British Airways
Speedbird House, PO Box 10, Heathrow
Airport, Hounslow, Middlesex TW6 2JA
Tel: 081 897 4000

British Midland Airways
East Midlands Airport, Donnington Hall,
Castle Donnington, Derby DE74 2SB
Tel: 0332 854854

BWIA (British West Indian Airways)
48 Leicester Square, London WC2H 7LT
Tel: 071 839 9333

Caledonian Airways
Caledonian House, Gatwick Airport, Sussex
RH6 0LF
Tel: 0293 536321

Canadian Airlines International
Rothschild House, First Floor, Whitgift
Centre, Croydon, Surrey CR9 3HL
Tel: 081 667 0666

Cathay Pacific
52 Berkeley Street, London W1X 5FP
Tel: 071 930 7878

Continental Airlines
Beulah Court, Albert Road, Horley, Surrey
RH6 7HZ
Tel: 0293 776464

Cyprus Airways
29 Hampstead Road, London NW1 3JA
Tel: 071 388 5411

Dan Air Services Ltd
Newman House, 45 Victoria Road, Horley,
Surrey RH6 7QG
Tel: 0345 100200

Delta Air Lines Inc
Victoria Place, 115 Buckingham Palace
Road, London SW1
Tel: *Freephone* 0800 414767

Egyptair
29–31 Piccadilly, London W1V 0PT
Tel: 071 734 2395/6 *or* 071 437 6426 *or*
071 437 6309 *(Reservations)*

El Al (Israel Airlines)
185 Regent Street, London W1
Tel: 071 437 9255

Emirates
125 Pall Mall, London SW1Y 5EA
Tel: 071 930 3711

Finnair
14 Clifford Street, London W1X 1RD
Tel: 071 408 1222

Garuda Indonesia
35 Duke Street, London W1M 5DF
Tel: 071 486 3011

Gulf Air
10 Albemarle Street, London W1X 3HE
Tel: 071 408 1717

Iberia
29 Glasshouse Street, London W1R 5RG
Tel: 071 437 5622

Icelandair
Third Floor, 172 Tottenham Court Road,
London W1P 9LG
Tel: 071 388 5599

JAL (Japan Airlines)
5 Hanover Square, London W1R 0DR
Tel: 071 408 1000

Kenya Airways
16 Conduit Street, London W1R 9TD
Tel: 071 409 0277

KLM (Royal Dutch Airlines)
8 Hanover Street, London W1R 9HF
Tel: 081 750 9000

Korean Air Lines
Greener House, 66–68 Haymarket, London
SW1Y 4RF
Tel: 071 930 6513

LOT (Polish Airlines)
313 Regent Street, London W1R 7PE
Tel: 071 580 5037

Lufthansa
23–26 Piccadilly, London W1 OEJ
Tel: 071 408 0442

Holiday transport

MAS (Malaysian Airline System)
61 Piccadilly, London W1V 9HL
Tel: 081 862 0770

MEA (Middle East Airlines)
48 Park Street, London W1Y 4AS
Tel: 071 493 5681

Monarch Airlines
Percival Way, London/Luton Airport,
Luton, Beds LU2 9NU
Tel: 0582 424211

Northwest Airlines
8–9 Berkeley Street, London W1X 5AD
Tel: 0345 747800

Olympic Airways
Commonwealth House, 2 Chalkhill Road,
London W6 8SB
Tel: 081 846 9080

Philippine Airlines
Centre Point, 103 New Oxford Street,
London WC1A 1DG
Tel: 071 836 5508

PIA (Pakistan International Airlines)
45 Piccadilly, London W1V 0DY
Tel: 071 734 5544

Qantas
Qantas House, 395–403 King Street, London
W6 9NJ
Tel: 0345 747767

Royal Air Maroc
205 Regent Street, London W1R 7DE
Tel: 071 439 4361

Royal Jordanian Airlines
177 Regent Street, London W1R 7FB
Tel: 071 734 2557

Royal Nepal Airlines
Butler House, 177–178 Tottenham Court
Road, London W1P 9LF
Tel: 071 388 4314

Ryanair
Barkat House, 116/118 Finchley Road,
London NW3 5HT
Tel: 071 435 7101

Sabena
Geminii House, 10–18 Putney Hill, London
SW15 6AA
Tel: 081 780 1444

SAS (Scandinavian Airlines System)
52 Conduit Street, London W1R 0AY
Tel: 071 734 6777

Saudi Arabian Airlines
171 Regent Street, London W1 5RG
Tel: 081 995 7777

Singapore Airlines
143 Regent Street, London W1R 7LB
Tel: 081 747 0007

Sudan Airlines
12 Grosvenor Street, London W1X 9FB
Tel: 071 499 8101

Swissair
Swiss Centre, 10 Wardour Street, London
W1V 4BJ
Tel: 071 439 4144

TAP (Air Portugal)
38–44 Gillingham House, Gillingham Street,
London SW1V 1JW
Tel: 071 828 0262

Thai Air
41 Albemarle Street, London W1X 3FE
Tel: 071 499 9113

TWA (Trans World Airlines)
200 Piccadilly, London W1V 0DH
Tel: 071 439 0707

US Air
Piccadilly House, 33–37 Regent Street,
London SW1Y 4NB
Tel: *Linkline* 0800 777333

Varig
16/17 Hanover Street, London W1R 0HG
Tel: 071 629 5824

Virgin Atlantic Airways
Ashdown House, High Street, Crawley,
Sussex RH10 1DQ
Tel: 0293 562000

Zambia Airways
163 Piccadilly, London W1V 9DE
Tel: 071 491 0650/8/9

Ferry lines

Travelling by ferry when you have a family *can* be great fun. Children usually find the prospect of a trip on a boat highly exciting. However, what do you do if your journey is delayed? The children are often wound up to the edge of hysteria, while you gnash your teeth in frustration. What sort of welcome can you expect when you step on the quay?

Most ferry ports have only limited facilities for children, relying heavily on those provided on board the ships themselves or on the amenities of the towns they are situated near. Most ports offer baby changing facilities and some kind of food, with many operating children's menus. Some ports have play areas and one or two even organise cartoon screenings.

We have listed the major ports of Britain and ferry operators, together with several continental ports that offer a direct service to the UK. If you have any queries regarding their facilities, ring the enquiries numbers we have included – their staff are only too happy to help.

B+I Line

Reliance House, Water Street, Liverpool
Tel: 051 227 3131 *or*
071 734 4681 *(London)*

Regular sailings between Holyhead and Dublin, and between Pembroke and Rosslare. Children's facilities include a special children's menu.
Prices Available on request.
Discounts Children under 5 travel free. Children aged 5–16 pay 50% of adult fare.

British Channel Island Ferries

PO Box 315, Poole, Dorset BH15 4DB
Tel: 0202 681155

Year-round overnight sailings from Poole, plus day sailings from April to October. Overnight accommodation is offered in comfortable cabins. Ships have many facilities, including children's play areas.
Prices Available on request.
Discounts Under-5s travel free with good discounted fares for children aged 5–16.

Brittany Ferries

Millbay Docks, Plymouth PL1 3EW
Tel: 0752 221321

Regular sailings between several ports: Portsmouth to Caen and St Malo; Plymouth to Roscoff and Santander; Cork and Roscoff (March–October only). 'Truckline Les Routiers' service between Poole and Cherbourg available during summer only. Special facilities on board include a nursing mothers' room, children's play area, TV, and high chairs in the restaurants.
Prices (Portsmouth–Caen standard return, car only)
E: £102
D: £134
C: £170
Each adult pays £28–£56.
Discounts Children under 4 travel free; 4–13s receive a discount on the adult fare.
N.B. These are 1992 prices given as a guideline only.

Color Line

Tyne Commission Quay, Albert Edward Dock, North Shields NE29 6EA
Tel: 091 296 1313

Sailings twice or three times weekly, depending on time of year, between Newcastle, Stavanger and Bergen. Children under 16 pay half adult ferry fare. There are no special on-board facilities for children.

Prices Available on request.

Discounts Children under 4 travel free if sharing berth with parents. Under-16s receive 50% discount. 25% off for students on production of an ISIC, NUS or ANSA card and not in high season.

Hoverspeed

Maybrook House, Queen's Gardens, Dover CT17 9UQ
Tel: 0304 240202/240241 or 081 554 7061 (London)

Regular hovercraft flights between Dover, Calais and Boulogne. Crossings are more frequent during summer months, and fares are highest during July and August. There are no special child facilities on board the hovercraft, but there is a nursing mothers' room in the hoverport.

Prices Available on request.

Discounts Under-4s travel free. Children aged 4–14 pay around 50% of the adult fare.

Isle of Man Steam Packet Company Ltd

Imperial Buildings, Douglas, Isle of Man
Tel: 0624 661661

Daily sailings from Heysham to Douglas, with many additional sailings between Heysham and Liverpool. The crossing takes about 4 hours. Day trips and weekend excursions are also available as well as seasonal crossings to Dublin, Belfast and Fleetwood. Special facilities include a room off the ladies' lavatory where nappies can be changed and a mothers' room where babies can be fed in peace.

Prices Available on request.

Discounts Children under 5 travel free.

North Sea Ferries

King George Dock, Hedon Road, Hull HU9 5QA
Tel: 0482 795141;
Reservations: 0482 77177

Nightly sailings from Hull to Rotterdam and Zeebrugge. Child facilities include cots, if requested when booking, and a children's playroom.

Prices Available on request.

Discounts Under-4s travel free, provided they share a berth with their parents. Children aged 2–13 pay 50% of the adult fare, and must occupy a cabin berth. A special 'Family Cabin Fare' is available, based on 4-berth inside special cabins, 4-berth economy cabins or 4-berth outside special cabins.

Olau-Line

Sheerness, Kent ME12 1SN
Tel: 0795 666666

Two daily sailings between Sheerness and Vlissingen in Holland. On-board facilities for children include a playroom (unsupervised) and a swimming pool. Midweek during summer holidays, there is a children's entertainment programme. (See also Olau-Line in Section 2: Hotels.)

Prices Available on request.

Discounts A special family fare is available on day sailings: up to three children travel free if accompanied by two adults paying £53 (without car).

P & O European Ferries

Channel House, Channel View Road, Dover, Kent CT17 9TJ
Tel: 0304 203388

Daily sailings on the following routes: Dover to Calais/Boulogne and Ostend; Felixstowe to Zeebrugge; Portsmouth to Cherbourg/Le Havre; Cairnryan to Larne. Special facilities on board for the family include: children's play areas; video lounges showing free films; mother and baby changing facilities. Children's menus/portions are available from waiter and self service restaurants. Cabins are available on Portsmouth–Le Havre/Cherbourg and Felixstowe–Zeebrugge and Dover/Ostend.

Prices Please note fares are dependent on route, date and time of travel. Details on request.
Discounts Reduced fares for students, and children (4–14). Children under 4 travel free.

P & O Scottish Ferries

PO Box 5, Jamieson's Quay, Aberdeen
AB9 8DL
Tel: 0224 572615

Daily, and some weekday only, sailings from Aberdeen and Scrabster on the Scottish mainland to the Orkneys and Shetlands. From end May–early September connections from Lerwick to the Faroes, Iceland, Norway and Denmark are offered in conjunction with the Smyril Line. P&O does not specify any particular facilities for children or nursing mothers.
Prices Available on request.
Discounts Under-4s travel free; 4–13s pay half adult fare.

Sally Line

Argyle Centre, York Street, Ramsgate, Kent CT11 9DS
Tel: 0843 595522 *(Ramsgate)*
071 409 2240 *(London)*

Regular sailings between Ramsgate and Dunkirk. Motorail link from Lille (53 miles from Dunkirk) to Bordeaux, St Raphael and Narbonne. All boats have a supervised crèche and special play areas for children, including a 'Sea of balls' – an enclosed room filled with plastic balls which provides hours of fun – video games, TV, children's meals at reduced prices and a room for nursing mothers.
Prices Details of all Sally Line prices (including 60-hour returns and day-trip rates) are available on request.
Discounts Children under 4 travel free. If motoring, up to 3 children under 14 always travel free.

Sealink Stena Line

Charter House, Park Street, Ashford, Kent TN24 8EX
Tel: 0233 647047
(reservations/enquiries)

Sealink Stena Line offers a network of 8 routes: 4 to France; 1 to Holland; 2 to Ireland; and 1 to Northern Ireland.

In addition, the company offers links to Scandinavia through the services of its sister company, Stena Line, which operates 7 routes on the Baltic.

Sealink Stena Line has invested almost £200 million on new and refurbished ships in the past year and has simplified its fare structure giving the option of new 'all-in' fares for a car, minibus or motorised caravan.
Prices Details available on request.

Scandinavian Seaways

Scandinavia House, Parkeston Quay, Harwich, Essex CO12 4QG
Tel: 0255 241234

Regular sailings between UK, Denmark, Sweden and northern Germany, with connections to Norway. Special on-board facilities for children include cots, playroom (unsupervised), organized games and cinema.
Prices Available on request.
Discounts Children under 4 travel free if sharing parents' accommodation; 4–16s pay 50% of adult fare and are given separate berths.

Swansea Cork Ferries

Ferry Port, Kings Dock, Swansea
SA1 8RU
Tel: 0792 456116

Crossings between Swansea and Cork between March and October. Four sailings in each direction in low season and six between mid-June and mid-September. In high season the outward journeys are at night, returns during the day. In low season all are at night at 9 p.m. from Swansea. The crossing takes 10 hours. Cabins, berths and pullman seats can be booked, but cost extra. There is a snack bar, restaurant and a swimming pool. Cots are bookable in advance.
Prices Details on request.
Discounts Children under 5 travel free at all times, 5–15s are half price as foot passengers. In a car, two children = one adult.

Ferry ports

Boulogne (Sealink and P&O)

Gare Maritime, Quay Chanzy BP309
62204 Boulogne Sur Mer
Enquiries 010 33 21 30 25 11 (Sealink)
           010 33 21 31 78 00 (P&O)
           010 33 21 31 68 38 (Tourist Office)
Sailings to Folkestone and Dover

- There is a baby changing room, however nappies are not available. Other amenities include a shop that sells books and small games and a cafeteria. The port is close to shops and restaurants in the town of Boulogne.
- A railway station is nearby and taxis can be telephoned from the terminal.
- Bureau de change facilities are located within the terminal.
- Wheelchairs are available on request.

Caen (Brittany)

Enquiries 010 333 196 8600
Sailings to Portsmouth.

- Caen port has a large car park. Amenities include a baby changing room, gift shop/newsagent, and a cafeteria which serves snacks.
- The railway station is a 20-minute bus ride away, and there is a taxi rank within the terminal. Ask at the toutist information desk for bus and rail timetables etc. A car rental telephone number can be obtained at the information desk.
- Bureau de change and a post office are both located on site.
- Wheelchairs and staff assistance are available on request for disabled passengers.

Calais (P&O and Sealink Stena Line)

Car Ferry Terminal,
62226 Calais Cedex
Enquiries 010 33 21 46 10 10 (P&O)
           010 33 21 96 70 70 (Sealink)
Sailings to Dover

- Parking is free of charge.
- Amenities are limited to a cafeteria which serves snacks, and a newsagent.
- The railway station is nearby and a local bus service is available.
- Bureau de change facilities are located within the terminal.
- Assistance is available on request for disabled passengers.

Cherbourg (P&O, Brittany, Sealink Stena Line)

Enquiries 010 33 3344 2013
Sailings to Portsmouth, Southampton and Poole.

- Cherbourg ferry port has a large car park with space for caravans.
- Amenities include a newsagent and a cafeteria which sells snacks.
- A rail service runs from Cherbourg to Paris and Caen. A taxi rank is adjacent to the port. Cherbourg town is a 10-minute walk away.
- A bureau de change is located within the terminal.
- Wheelchairs are available for disabled passengers on request and there is also a lift.

Cork (Brittany)

Car Ferry Terminal, Ringaskiddy,
County Cork
Enquiries 010 35 321277705
Sailings to Roscoff

- There is no overnight or long-term parking, but daily parking is free of charge and at owner's risk.
- Amenities include a baby changing room, souvenir shop and a cafeteria which will warm baby food on request. Other shops are located approximately five minutes' walk away in the local village.
- Brittany ferries provide transport by bus to the bus terminal in Cork City (10 minutes from the railway station) approximately mid-May to mid-September – but it is not free. The city centre is approximately 10 miles from the port. Car rental is not available on site and must be pre-booked. Taxis are situated adjacent to the port.
- Bureau de change facilities are situated on site.
- Assistance and wheelchairs are available for disabled passengers on request.

Dieppe (Sealink)

Enquiries 010 3306 3320
Sailings to Newhaven

- Parking is available on the seafront only.
- There is no baby changing room or shop, and catering is limited to a snack bar which serves sandwiches etc.
- Rail, bus and taxi facilities are nearby, and an Avis car rental office is situated at the port.
- There is no bureau de change or bank. The port is near the centre of town, however, so access to banks and restaurants is fairly easy.

Dover (Sealink Stena Line and P&O)

Dover Harbour Board, Harbour House,
Dover, Kent CT17 3PU
Enquiries 0304 240400
Sailings to Boulogne, Calais, Ostend

- Parking facilities in the port are not actually restricted to periods of 72 hours despite the signs.
- Amenities include a baby changing room with disposable nappies, a shopping area that can provide for 'most travel needs', and two Welcome Break restaurants that cater for children's meals.
- A courtesy bus service runs nearby and a taxi rank is situated adjacent to the port. Car rental is available from within the terminal and passengers are advised to check with the company in advance about the availability of child seats.
- Banking and bureau de change facilities are both located within the passenger terminal.
- Toilets and ramps are in situ for the benefit of disabled passengers, and wheelchairs can be provided on request.
- All staff are trained in first aid.

Dun Laoghaire (Sealink Stena Line)

Dun Laoghaire Harbour, County Dublin
Enquiries 010 35 312808844
Sailings to Holyhead

- There is no in-port parking.
- A baby changing room is available, but disposable nappies are not. The shop sells newspapers and sweets, but other things can be bought in the town centre which is only a few minutes' walk away.
- A caravan snack bar sells tea and snacks only at the car ferry check-in/St Michael's pier.
- The local railway station is only a few minutes' walk away and trains run every 10–15 minutes. A special bus runs from the port to Heuston Station to facilitate early morning arrivals and a taxi rank and Hertz car-hire office are nearby.
- Banking is available in the nearby town, and a bureau de change is situated at the tourist office at the harbour roundabout.
- Wheelchair assistance available on request.

Felixstowe (P&O)

Car Ferry Terminal, The Docks,
Felixstowe IP11 8TB
Enquiries 0394 604040
Sailings to Zeebrugge.

- Parking facilities are available at a charge of £3.00 per 24 hours.
- Amenities are limited to a self-service restaurant that caters for children's meals. There is also a licensed bar in the passenger terminal, which is not open to children.
- A local bus service runs close to the port, and there is a service from the port to the railway station. A taxi rank is also available. There is no car-hire office within the terminal, but cars can be delivered to the port if the company is contacted by telephone.
- There is both a Natwest and a TSB on the dock. These offer bureau de change facilities as well as those on ship.
- Toilets and ramps are in situ for the benefit of disabled passengers, and wheelchairs are available on request.
- To occupy the children, the port provides a soft adventure play area and a lego table.
- There is a medical centre situated within the port which has an accident and emergency department with paramedic team on call.

Folkestone

Folkestone Harbour, Folkestone, Kent
CT20 1QH
Enquiries 0303 220544
Sailings to Boulogne.

- In-port parking facilities are available (check details with port office).
- Amenities are limited to a restaurant which will provide children's meals on request.
- Transport connections include British rail and local bus service, with a taxi rank adjacent to the port. Car rental facilities are on site (check with company about child seats).
- Bureau de change and banking facilities are also available within the passenger terminal.
- The port is equipped with toilets for the disabled, ramps and wheelchairs which will be provided on request.
- All staff are trained in first aid.

Harwich (Sealink Stena Line)

Parkeston Quay, Harwich, Essex
CO12 4SH
Enquiries 0255 243333
Sailings to Hook of Holland

- Short-term parking from 50p per hour. Long-term parking from £3 per day.
- Harwich benefits from a new passenger terminal which was built in 1988.
- Amenities include a baby changing room with disposable nappies and washing facilities, a well-stocked shop which sells toiletries, soft toys, books and magazines and also a restaurant that caters for children's meals.
- British rail, a local bus service and a taxi rank are adjacent to the terminal building. There is a freefone car rental facility within the terminal, and customers should check with the firm in advance about the availability of child seats.
- Bureau de change facilities are located within the terminal building.
  Toilets and ramps are in situ for the benefit of disabled passengers, wheelchairs can be provided on request and there is a lift available.
- The terminal operates a 'hostess' service which provides assistance or information on request. The 'hostesses' and stewards are easily identifiable by their uniform, and will welcome passengers on arrival at the terminal.
- A children's play area is situated in the car waiting hall.
- All staff are trained in first aid.

Holyhead (Sealink Stena Line)

Enquiries 0407 762304
         0407 766762
Sailings to Dun Laoghaire

- Holyhead has facilities for both long- and short-term parking. Up to two

hours in the short-term car park are free of charge. Long-term parking is charged at £1 per day/£5 per week (pay-and-display machines). Car parks are signposted from the main approach road to the port and are patrolled regularly by the port security staff.
- There is a nursing mother's room in the main passenger terminal with facilities for the purchase and disposal of nappies.
- Holyhead has a snack bar within the terminal.
- Inter-city and provincial trains operate from the main passenger terminal, with direct services to Crewe, Birmingham, London, Manchester and West Yorkshire. Local buses pick up and drop passengers adjacent to the main terminal building. Car hire is available from Hertz and Eurodollar – a freefone line is available in the main passenger terminal.
- Holyhead town centre is approximately a quarter of a mile from the port and banking facilities etc. are available there. A bureau de change can be found at the Sealink ticket office, which is situated in the main terminal building.
- Disabled toilets are located in the passenger terminal. On request, arrangements can be made for a member of the port staff to meet disabled passengers and escort them to and from ferries. The port is equipped with a number of wheelchairs.
- All staff are trained in first aid.

Hook of Holland (Sealink)

Enquiries 010 31174 783944
Sailings to Harwich

- Parking is free of charge.
- There is a cafeteria which sells snacks in the railway station.
- A railway station is situated within the terminal, and, 10 minutes away, a bus service runs back and forth to the Hague. A taxi tank is nearby.
- A bank and bureau de change are both on site in the station.
- Wheelchairs are available on request and all areas have ramps.

Larne (P&O and Sealink Stena Line)

Larne Harbour, Larne, County Antrim
BT40 1AQ
Enquiries 0574 274321
Sailings to Cairnryan, Stranraer

- Parking within the port is free.
- Amenities include a baby changing room with disposable-nappy dispenser, and a restaurant that caters for children's meals and will be happy to warm baby food etc.
- The railway station is joined to the terminal and there is a bus service to the local town. A taxi rank and car rental office are located within the terminal.
- Bureau de change and banking facilities are available on site.
- The terminal is built on one level and wheelchairs can be provided on request.
- A television area is also provided for passenger use.
- All staff are trained in first aid.

Newhaven (Sealink Stena Line)

Newhaven Harbour, Newhaven, East Sussex BN9 0BG
Enquiries 0273 514131
Sailings to Dieppe

- Parking facilities include two pay-and-display car parks within the port security area and close to the passenger terminal.
- Amenities are limited to a buffet snack bar and vending machines for the provision of hot or cold drinks and snacks.
- Public transport includes Network Southeast connections to London, Brighton and all south-coast resorts. Local buses stop near the port. Car rental is available from B.V.H. which is situated adjacent to the port.
- There are no banking facilities within the port, however, all major banks are located in Newhaven town centre which is less than a quarter of a mile from the harbour. Bureaux de change are situated in both passenger and car ferry terminals.
- All the terminals are at ground level and the foot-passenger gangway is especially wide to accommodate

wheelchairs. There are special toilet facilities for disabled passengers and these are also available on all ships. Special assistance for car and foot passengers is available on request.

Plymouth (Brittany)

Millbay Docks, Plymouth, Devon
PL1 3EW
Enquiries 0752 221321 (reservations)
Sailings to Roscoff and Santander

- Parking facilities are, at present, available free of charge, but cars are left at owner's risk.
- Amenities are limited to a cafeteria which will provide children's meals on request.
- Although there is no baby changing room the toilets are quite large.
- British Rail is a quarter of an hour away and a taxi rank is situated within the port. Car rental can be arranged for passengers coming off the ferry. The rental office is adjacent to the port.
- Bureau de change facilities are located within the passenger terminal.
- Help can be arranged for disabled passengers on request and there are lifts and ramps in some areas.
- All staff are trained in first aid.

Portsmouth (P&O and Brittany)

The Continental Ferry Port, Mile End, Portsmouth PO2 8QW
Enquiries 0705 827701 (Brittany)
         0705 772244 (P&O)
Sailings to Caen, St. Malo, Le Harve and Cherbourg

- Parking facilities within the port are limited, however a short-term, privately owned car park is situated adjacent to the port.
- Amenities include a baby changing room, newsagent, cafeteria and a licensed bar which admits children when accompanied by adults. Children's meals can be provided on request.
- A shuttle service transports passengers to and from the local station and a local bus service goes into the town centre. A taxi rank is situated at the port.

- Bureau de change facilities are located within the terminal building.
- Disabled passengers have access to most areas of the terminal which is built on one level. Wheelchairs can be provided on request.
- A first-aid room is located within the terminal.
- General enquiries should be made at the portacabin. Tourist information is available at limited times in high season in the departure area.

Rosslare (Sealink Stena Line), B+I and Irish Continental

Enquiries 010 35 35333115
Sailings to Fishguard

- The large car park is free, but cars are left at owner's risk.
- Amenities include a newsagent and a restaurant and bar which will provide children's meals.
- A railway station is close by, and local buses run to and from the port. Taxis can be telephoned from the terminal and there is a choice of three car rental companies on site.
- Bureau de change facilities and Tourist Information are located within the terminal.
- Disabled facilities include ramps and toilets. Wheelchairs are available on request.
- The passenger terminal is relatively new and is furnished with ample comfortable seating.

St. Malo (Brittany)

Enquiries 010 3399 817385
Tourist Office 010 3399 566448
Sailings to Portsmouth

- The port has a large car park.
- Amenities include a baby changing room, a shop that sells small toys and books and a snack bar.
- The railway station is a ten-minute walk away, and there is a local bus service nearby. Taxis are available from the terminal, and there is a choice of four car-rental firms on site.
- Bureau de change and banking facilities are located within the terminal.

- Wheelchairs are available on request.
- The tourist information desk is open from June to September only.

Southampton (Sealink Stena Line)

Travel Centre, European Way, Eastern Docks, Southampton, Hants SO9 1XH
Enquiries 0703 233 973
Central Reservations 0233 647047
Sailings to Cherbourg

- Short- and long-term (small area) parking facilities are available at a charge of £2 per day, payable in travel centre.
- Amenities include a baby changing room with feeding area, a snack bar which sells sandwiches, hot drinks etc. and a newsagent. For those passengers who check in early, there is easy access, on foot, to the 'Ocean Village' which offers a wide selection of shops and restaurants.
- British Rail is 20 minutes' walk from the passenger terminal and a taxi rank is nearby.
- Bureau de change facilities are located within the terminal building.
- Wheelchairs are available on request and there is a separate area in the car park for disabled passengers waiting to embark. Lifts operate to all levels of the terminal.
- All staff are trained in first aid.

Stranraer

Enquiries 0776 2262
Sailings to Larne

- Stranraer has a large car park which is free of charge to passengers.
- Amenities include a baby changing room with disposable nappy dispenser and a tea bar which serves snacks and will supply baby food on request. High chairs are also available.
- A British Rail station is situated within the port and a city-link bus service operates to major cities. Taxis can be hired from within the terminal. The local town is only a short distance away.
- Car rental facilities are provided by Hertz from their office in the passenger terminal. (Ask about child seats when booking.)
- A bank is situated only five minutes' walk from the port, and a bureau de change is located within the terminal.
- Toilets and ramps are in situ for the benefit of disabled passengers. Wheelchairs are available on request and there is a bus service to and from the ship.
- All staff are trained in first aid.

Zeebrugge (P&O)

Enquiries 010 32 50 54 22 22
Sailings to Felixstowe and Dover

- There is a short-stay car park.
- Amenities include a cafeteria which provides snacks, hot meals, breakfast and children's portions at a reduced price.
- A tram service runs along the coast and is situated approximately 300m from the port. The railway station is approximately 2km away. Car hire can be arranged by telephone and vehicles can by delivered to the port.
- A bank and bureau de change are located on site and van/bus rental is available from this office.
- The passenger terminal is built on one level and a mini-bus is available to transport disabled passengers to and from the ship.
- A first-aid room is situated within the terminal building.

# Coach stations

National Express

National Express coach stations are normally located centrally and, as a result, tend to rely on the facilities afforded by the town in which they are located.

Coach stations normally house a newsagent, but no other shopping facilities.

Catering services are usually limited to a small snack bar or café. There is a baby changing room in Digbeth coach station (central Birmingham).

Contact the site in question for details of facilities available for disabled travellers.

See also **Coach transport and coach tours** in Section 2.

Service stations

When travelling by car, a long journey with children can quickly turn into a nightmare. Extremely urgent requests for a toilet, persistent hunger pangs and unexpected medical calamities can turn patient parents into neurotic wrecks.

During the last few years the service stations located on the sides of Britain's motorways and major trunk roads have improved beyond all recognition, with many of the necessities of family travel more than adequately catered for. Not only are baby changing facilities now regarded as the norm rather than the exception, we are now expecting sophisticated amenities such as book shops, automatic cash dispensers and record shops. Some sites even house branches of up-market confectioners, business bureaux, and flower shops.

We have listed below the facilities offered by some of the major service station groups. It is well worth planning a substantial journey and plotting your pit stops along the way. Letting toddlers tear around a play area for half an hour may save your nerves for the next three.

Forte Travelodge

Tel: 0800 850950 (freefone)
Tel: 0345 500400 (North America)

There are nearly 100 Forte Travelodges in England, Wales and Scotland and the number is still rising. Forte Travelodge is part of the Forte Hotels group. There are also 450 travelodges in North America – contact the number listed above for further details. The lodges are designed with motorists in mind and you pay one standard charge per room per night. You pay for your first night's accommodation on arrival and can check in anytime after 3 p.m. – you will be welcomed even into the small hours. By using a major credit card to book, you will guarantee your room no matter how late your arrival.

This price covers accommodation in a room that will sleep a family of up to three adults, a child in a cot (please book in advance) and a child under 12 – a good deal for those travelling with children. As Forte Travelodges are usually located within close proximity of a Little Chef, Happy Eater, Harvester or Welcome Break, meals can be purchased easily. All rooms are comfortable and well equipped: the double bed has a feather duvet, there's a single sofa bed and a child's 5-foot bed if required. Cots should be requested when making your reservation. Other room features include ensuite bathroom facilities, personal central-heating controls, a colour TV with radio/alarm clock, and tea- and coffee-making facilities. The lodges have been designed with security in mind and the reception is constantly manned during the hours of darkness.

Every Forte Travelodge has at least one room with special facilities for disabled people. Containing a single bed to ease wheelchair access, Forte Travelodge recommend two adults and one child as the maximum booking for these rooms (please check availability and reserve a disabled room in advance).

See table in this section for locations and phone numbers.

Granada Service Stations, Hotels and Lodges

Service Stations

Granada service stations are located throughout Britain on motorways and major roads. Facilities vary from site to site, but some amenities are standard.

- Baby changing facilities are available at most service stations
- Service-station shops sell small toys, books, magazines, etc.
- Restaurants will provide baby food and heat milk on request
- High chairs are provided
- One child under twelve receives a free main course when an adult spends £3.99 or more in the restaurant
- Special parking spaces are reserved for disabled customers
- Disabled toilets and ramps are provided
- A 'service call' device is available for disabled customers. This allows them to alert the petrol station attendant to fill up the car as necessary
- Outdoor play areas at most service stations

Granada Hotels and Lodges

Granada Motorway Services Ltd,
Head Office, M1 Service Area, Toddington, Bedfordshire, LU5 6HR
Tel: 0800 555290 (Central Reservations)
05258 73881 (Head Office)
Fax: 052 555 602 (Central Reservations)
05258 75358 (Head Office)

Granada Hotels and Lodges are situated on the motorways and major trunk roads of Britain, enjoying easy access to both transport systems and some of the most scenic and interesting parts of the country. All establishments have either the AA three star and RAC hotel rating or the AA Lodge Stamp of Approval. Granada operates a Weekend Budget Break package in both their hotels and lodges, in addition to offering single night accommodation.

Weekend Budget Breaks

In Granada Hotels:
Prices include accommodation in a single, twin, double or family room (sleeping two adults and two children up to the age of 16), full English breakfast and, for half board, dinner up to a standard limit (currently £13) in the hotel's restaurant. All rooms enjoy private bathroom, colour TV, radio, wake-up alarm, tea- and coffee-making facilities, direct-dial telephone, hairdryer and trouser press. All hotels have their own fully licensed restaurant and lounge bar. Rooms specially adapted for disabled travellers are also available – check for availability before booking.

In Granada Lodges:
Prices include accommodation in a single, twin, double or family room (sleeping two adults and two children up to the age of 16), full English self-service breakfast in the Country Restaurant of the Granada Service Area. All bedrooms have private bathrooms, colour TV, radio, wake-up alarm, and tea- and coffee-making facilities. Rooms specially adapted for disabled travellers are also available – check for availability before booking.

Pavilion Service Areas and Lodges

Pavilion Services Ltd, 38 Market Square, Uxbridge, Middlesex UB8 1NG
Tel: 0895 233333
Fax: 0895 239635

Pavilion Service Areas

Pavilion Service Areas are located throughout England and Wales. They are open twenty-four hours a day for food, fuel and public facilities. Many also have twenty-four hour shops.

Each service area has disabled toilet facilities, baby change and feeding rooms and, during office hours, fax facilities.

Two of Pavilion's eleven service stations feature a restaurant with full table service, while eight have self-service restaurants (with separate children's menus) and a fast-food outlet. Many provide such amenities as cash dispensers and play areas. At Hilton Park and Swansea there are Motorway Superloos, which offer you the luxury of a self-contained private bathroom with toilet, vanity basin, hairdryer and even a trouser press.

Pavilion Lodges

Pavilion Lodges are situated in the grounds of service areas and provide convenient accommodation facilities when you wish to break a long journey. The reception desk is staffed twenty-four hours a day and, as a result of their service area locations, you are able to obtain food at any hour of the day or night.

All rooms are double glazed for insulation and contain double or twin beds, private bathrooms with shower and toilet, colour TV, radio and alarm clock, tea and coffee-making facilities, electric trouser press, hairdryer and individual heating controls. Every Lodge has specially equipped rooms for disabled guests.

Accommodation is available throughout the Pavilion network at a standard rate, with prices based on a charge for single-room occupancy and carrying a surcharge for each extra adult accommodated. Both prices include continental breakfast and VAT.

Granada Service Stations and Lodges

Site Name	Road	Junction	Telephone	Accommodation
Birch	M62	J 18/19	Tel: 061 655 3403	(Lodge)
Blyth	A1		Tel: 090 959 1836	(Lodge)
Exeter	M5	J 30	Tel: 0392 74044	(Lodge)
Ferrybridge	M62/A1	J 33	Tel: 0977 672767	(Lodge)
Frankley	M5	J 3/4	Tel: 021 550 3261	(Lodge)
Grantham			Tel: 0476 860686	(Lodge)
Heston	M4	J 2/3	Tel: 081 574 5875	(Lodge)
Kinross	M90	J 6	Tel: 0577 64646	(Lodge)
Leicester	M1	J 22	Tel: 0530 244 237	(Lodge)
Leigh Delamere	M4	J 17/18	Tel: 0666 837097	(Lodge)
Magor	M4	J 23	Tel: 0633 880111	(Lodge)
Musselburgh			Tel: 031 653 2427	(Lodge)
Saltash	A38 Bypass		Tel: 0752 848048	(Lodge)
Sheffield			Tel: 0742 530935	(Hotel)
Southwaite	M6	J 41/42	Tel: 06974 73131	(Lodge)
Stirling	M9/M80	J 9	Tel: 0786 815033	(Lodge)
Stoke			Tel: 0782 777000	(Hotel)
Swanwick	A38/A61	J 28 of M1	Tel: 0773 520040	(Lodge)
Tamworth	M42/A5	J 10	Tel: 0782 260123	(Lodge)

Pavilion Service Areas and Lodges

Site Name	Road	Junction	Telephone	Accommodation
Aust Pavilion	M4	J 21	Tel: 04545 632851	(Lodge)
Bangor Pavilion	A5	J A55/A5	Tel: 0248 361234	(Lodge)
Cardiff West Pavilion	M4	J 33/A 483	Tel: 0222 891141	(Lodge)
Farthing Corner Pavilion	M2	J 4/5	Tel: 0634 23343	(Lodge)
Forton Pavilion	M6	J 32/33	Tel: 0524 791775	(Lodge)
Hilton Park Pavilion	M6	J 10A/11	Tel: 0922 412237	(Lodge)
Knutsford Pavilion	M6	J 18/19	Tel: 0565 634149	
Newark	A1	J A1/A46/A17	Tel: 0636 610010	
Rivington Pavilion	M61	J 6/8	Tel: 0204 68641	
Scotch Corner Pavilion	A1	J A1/A66	Tel: 0325 377719	(Lodge)
Swansea Pavilion	M4	J 47/A483	Tel: 0792 896222	(Lodge)

Welcome Break Service Stations and Lodges

Service Stations
Tel: 0909 617766

Welcome Break service stations are operated by the Forte group and are located throughout the country along motorways and major roads. Their facilities vary from site to site, but some amenities are common to all:

- Shops carry a full range of toys, books, etc. Some sites also house a bookshop
- Restaurants (The Granary and Little Chef) cater for children, with special meals, baby food, etc.
- Restaurants will warm baby food on request
- High chairs are available on request
- Most sites house a Julie's Pantry outlet, which offers hamburgers, french fries, milk shakes, cola, etc.
- Most sites have baby changing rooms with disposable-nappy dispensers
- most service stations have a well-equipped children's play area
- Cashpoint facilities are available at some service stations
- Ramps and special toilet facilities are provided for disabled travellers

Welcome Lodges

Charnock Richard
Mill Lane, Charnock Richard, Chorley, Lancashire, PR7 5LR
Tel: 0257 791746 (reservations)
Fax: 0257 793596

Oxford
Peartree Roundabout, Woodstock Road, Oxford, OX2 8JZ
Tel: 0865 54301 (reservations)
Fax: 0865 513474

Newport Pagnell
M1 Motorway, Newport Pagnell, Buckinghamshire, MK16 8DS
Tel: 0908 610878 (reservations)
Fax: 0908 617226

London – Scratchwood
M1 Motorway, Hendon, London NW7 3HB
Tel: 081 906 0611 (reservations)
Fax: 081 906 3654

The four Welcome Lodges are run by the Forte Hotels group and are situated in the grounds of Welcome Break service areas. They offer comfortable facilities, including lounges and bars and, as you would expect, have ample parking facilities. All sites (except Charnock Richard) feature family rooms, which will be allocated on request when available. However, all Welcome Lodges offer free accommodation for children under 16 when they share their parents' room – meals will be charged as taken. Each room has its own bathroom and TV. Welcome Lodge often operate special weekend packages, which have a standard price per room per night and include English or Continental Breakfast – contact central reservations number for further details: 0800 40 40 40.

Welcome Break Service Stations & Forte Travelodges

Site Name	Road	Telephone	Accommodation
Abington	A74/M74	Tel: 0800 850 950	
Burtonwood	M62	Tel: 0925 710376	(Travelodge)
Charnock Richard	M6	Tel: 0257 791494	(Welcome Lodge)
Fleet	M3	Tel: 0252 815587	(Travelodge)
Gordano	M5	Tel: 0275 373709	(Travelodge)
Grantham	A1	Tel: 0476 77500	(Travelodge)
Gretna Green	A74	Tel: 0461 37566	(Travelodge)
Hartshead Moor	M62	Tel: 0274 851706	
Newport Pagnell	M1	Tel: 0908 217722	(Welcome Lodge)
Oxford	A34	Tel: 0865 54301	(Welcome Lodge)
Sarn Park	M4	Tel: 0656 659218	(Travelodge)
Scratchwood	M1	Tel: 081 906 0611	(Welcome Lodge)
South Mimms	M25/A1(M)	Tel: 0707 665440	(Travelodge)
Sutton Scotney	A34	Tel: 0962 761016	(Travelodge)

To reserve a room in a Forte Travelodge phone: 0800 850 850.

Eating on the move

Hungry children can make long car journeys a misery. Luckily there are a large number of family restaurants located throughout the UK which cater specifically for children and provide reasonably priced food, fast.

As people tend to associate fast-food restaurants with fatty and unhealthy food, many of the chains are now promoting a healthier image, offering a choice of low-fat and vegetarian dishes as well as traditional burgers and chips, pizzas and so on.

Many chains offer special deals. It is often cheaper to plump for an all-in meal price (eg burger, chips and a soft drink) than to pay for items separately. Be canny and check prices before ordering. Offers for children's meals may involve small toys or specially designed packaging.

If you are sticking primarily to the motorways then service stations usually provide snack and restaurant facilities (*see* our section on service stations).

In this section we have listed some of the major restaurant chains which cater for families and are situated along 'A' roads and high streets in cities and large towns. They aim to be as accessible as possible to family parties and most of them are open all day, seven days a week, providing rooms for nappy changing, special children's menus and high chairs. It is even possible to arrange for special children's birthday meals in advance at many of the restaurants. The restaurants usually provide hats, ballons and other goodies – an ideal way to celebrate a birthday on the move!

Happy Eater Restaurants

Head Office
Tel: 0256 812828

Happy Eater Restaurants are located along major 'A' roads around the country. They are open 7 days a week from 7 a.m. to 10 p.m. (including Bank Holidays). There are information packs available at every restaurant containing details of locations and telephone numbers of Happy Eaters in the UK, along with vouchers and special offers.

Other facilities include:

- Varied menu including 'healthy' options and all-day breakfast
- Special children's menu and an assortment of baby food
- High chairs, baby changing room and facilities for the disabled
- Children's play area with swings, slides and seesaws outside and a Lego table inside
- Shop selling travel games, sweets, gifts etc
- A Forte Travelodge situated nearby at many sites

Little Chef

Head Office
Tel: 0256 812828

With 358 restaurants, Little Chef is Britain's biggest roadside restaurant group and largest table-service operation. Open from 7 a.m. to 10 p.m., seven days a week, it offers a fast and friendly waitress service with freshly cooked food to order.

The Little Chef 'Big Choice' menu features drinks, snacks and meals to suit every taste. The menu includes burgers, fish, chicken and vegetarian options as well as a selection of hot and cold desserts.

Little Chef provides special facilities for families with young children and babies. There is a subsidised children's menu with a space travel theme, plus a selection of free baby and junior dishes. Courtesy newspapers are provided and every child is given a lollipop on leaving the restaurant.

Other facilities include:

- High chairs
- Baby changing facilities in most restaurants
- Little Chef roadmap available free at all restaurants – gives details of all the Little Chefs around the country including the provision of baby change and disabled facilities.
- Spacious car parks
- Little Chef shop selling travel items, confectionery and gifts

McDonald's Restaurants Ltd

Head Office
Tel: 081 883 6400 (all enquiries to the PR Department)

There are 449 McDonald's outlets situated throught the UK (including Northern Ireland). About 70 offer drive-through facilities and these tend to be situated in city suburbs and on the outskirts of large towns. (A map showing the location of McDonald's Restaurants can be obtained from the Information Board in some restaurants or on request from the Head Office.)

All branches are open for breakfast from 7.30 until 10.30/11 a.m. After this the standard menu applies. The restaurants stay open until about 11 p.m.

McDonald's restaurants are family-orientated and offer a range of facilities for parents and children. Depending on the size of the restaurant there may be a separate play area and many have a trained lobby hostess who will provide children with balloons and hats.

Children's parties can be arranged by contacting the restaurant manager. The charge is presently 70p per person plus the cost of the food eaten. For this the children will receive bags, hats and balloons and the lobby hostess will supervise and entertain the children.

Menus and promotions may vary from restaurant to restaurant and as well as the standard burgers, fries, fish, chicken, soft drinks and desserts, many restaurants sell salads and sundaes. Happy Meals are frequently on offer and usually consist of a soft drink, burger, fries and a toy.

Other facilities include:

- Information Board in each restaurant offering McFood Fact leaflets, with maps available in some branches
- Baby changing rooms in many restaurants
- Smoking and non-smoking areas in most branches

Pizza Hut (UK) Ltd

Head Office
Tel: 081 959 3677

Pizza Hut has over 300 outlets situated primarily in high streets throughout the UK (details of locations can be obtained by phoning the Head Office). Pizza Hut provide dine-in, take-away and delivery services. They aim to appeal to families, with a full range of facilities available. They boast a huge choice of pan and 'Thin 'N' Crispy' pizzas with thousands of topping combinations, as well as a comprehensive range of starters, pastas, desserts and a fresh salad bar.

Other facilities include:

- Children's chairs and menu suggestions
- Colouring mats and crayons
- Children's parties – can be arranged by contacting the restaurant manager in advance
- Baby changing facilities
- No-smoking sections in each restaurant

Pizzaland International Ltd

Head Office
Tel: 0895 811 911

There are Pizzaland restaurants located in towns and cities throughtout the UK, with 73 in England, 2 in the Channel Islands, 2 in Wales and 14 in Scotland. Most of them are open in the mornings for coffee etc. The menus are centred around freshly baked pizzas and related items and there are salad bars in each restaurant.

Other facilities include:

- High chairs
- Children's menu of soup, garlic bread, baked potatoes, pasta, pizza, soft drinks and desserts. There are puzzles, and pictures to colour in on the back of each children's menu.
- Children's parties can be catered for at most restaurants by contacting the restaurant manager. There is no fixed menu but each child will receive free party items.
- For the benefit of those who prefer not to eat meat or fish, these items are marked on the menu.

Wimpy International Ltd

Head Office
Tel: 0628 891655

There are 231 Wimpy Table-Service Restaurants situated throughout the UK. All the restaurants are located in the high streets of major towns and cities. Most are open from 10 a.m. to 6 p.m. and many open for breakfast at 7 a.m.

Other facilities include:

- Menus offering vegetarian and fish dishes, salads, desserts and tea cakes as well as burgers and grills. Breakfast is served all day.
- Children's menu – 'Mega Bites Kids' Specials' for children aged 12 and under
- High chairs – available at most restaurants

Section 4

Getting there

Long journey? No problem...

A parent's solution to the problems of travelling with children

Sheila Sang

You've packed your bucket and spade, holiday fever is upon you and you're all eager to get there. Problem is, between you and your holiday home, there's the journey. Anyone who's ever travelled with children will know: kids and travel don't mix. If they're not squabbling or wanting something to eat, they're being sick or thinking of new ways of asking 'how much longer?' If you don't have the skills of a children's entertainer and contortionist, it helps to travel well-prepared.

Tool kit for travel

There are some basic things which no travelling parent should be without – tailor them to suit your children and your journey.

★ **A favourite toy** You may find that your child will play with only one or two things, however long the journey. Take a firm favourite that you know has long-playing power.

★ **A new toy** If there's something your child has been longing for, consider making a present of it before or during a tedious journey.

★ **Food** fills their mouths and minds at the same time – fewer complaints! Avoid anything messy, like chocolate or bananas, and steer clear of pastry which is at best messy and may bring on travel sickness. Try them with fresh and healthy nibbles like carrot sticks, cubes of cheese or apple slices rubbed in lemon and kept in a polythene bag. Crisps and individual packets of raisins are an easy option and more certain of a warm reception. If you're providing a picnic you'll obviously need to expand on this – but keep it simple and stick to finger foods.

★ **Drink** Throwaway cartons of fruit juice and squash are great when travelling. Bring plenty – it could save you a fortune. Baby 'carton cups' that fit over a drinks carton and have a spout are great for kids who can't yet cope with a straw. Cans, on the other hand, are a downright nuisance – few young children can drink the lot in one go, and they spill easily. You can now buy small resealable bottles of fizzy drinks if your child insists on these. It is also possible to buy mini 'trial cans' of certain brands – these may also be useful.

★ **Clothes** Bring a change of the basics for everyone who might need them (this might mean a spare tee-shirt for your eight-year-old, but two complete changes for your three-month-old baby). Spare pants are a must for toddlers.

★ **Baby wipes** for all ages. They're an easy and gentle way of keeping hands and faces clean. Many are

available in travel packs; save money by refilling from the larger size.
- ★ **Tissues** They're invaluable for everything from a runny nose to an inadequate public loo, and take up less space than toilet roll. And of course they'll mop up accidental spills too.
- ★ **Books** Invaluable. For a baby or toddler three or four slim paperbacks should get you through most journeys (on a long journey they'll bear repeating), for older children a new book in a favourite series or by a favourite author (Roald Dahl, Enid Blyton?) should be a winner. *See* **Holiday reading for kids**, in section 4.
- ★ **Personal stereo** Well worth having for any child from the age of about three, particularly if you're travelling a long distance. While they listen to absorbing story or song tapes, you can get a few minutes' hard-earned peace and quiet – particularly important if you're travelling alone, or you're outnumbered by your kids. For children under the age of about four you can get sturdy children's personal stereos with easy-to-use controls and, more importantly, a device that allows you to set a maximum volume to protect your child's ears. Otherwise, go for the cheapest you can find. You can also buy an adaptor that allows two children to listen at the same time. With any luck, this will save you from noisy arguments!
- ★ **Children's backpack** How you carry the rest of your essential en-route supplies will vary depending on how you're travelling, but it's a good idea to let each child take care of at least some of their own things in an easy-to-carry backpack. It'll keep their belongings where they can find them (in principle, at least!) and lighten your own load a bit. But beware of leaving children in charge of anything that's breakable, expensive or irreplaceable.

Baby on board

If you've got a baby, you already know that they need more than their own weight in paraphernalia. You'll need all the usuals – nappies, wipes and so on – and more. If you've got a good changing bag that will help, if not, consider buying one. Go for one with a detachable changing mat, lots of separate pockets for small items and a zipped and lined pocket for dirties. If you want to heat bottles or baby food in jars, a restaurant or buffet with a microwave may oblige, or think about buying a travel bottle warmer (the type that doesn't need plugging in). It's also worth using disposable bibs – no yukky dirties to store all journey.

Airlines may provide babyfood (ask in advance) and will always warm bottles and jars.

Toys and games

However you're travelling, some toys and games are no luxury. But avoid games with lots of easily-lost small parts, and don't bring anything that makes a lot of noise – a talking doll or bleeping computer game will drive you all mad.

For babies bring along one or two teethers, soft toys or cloth books to look at and suck, and some building beakers if you'll have access to a table.

Board jigsaws are great for toddlers (you can often find puzzles that relate to travelling); pencil and paper are essentials for all ages. A wipe-clean magic slate will be fun to use, and handy for scoring games. Once children are older special travel versions of popular board games, such as Connect 4 or draughts, will be popular.

But you're likely to find that it's your company and powers of invention that are most in demand. Try some of these 'I spy'-type games to cheer flagging spirits.

Granny went to market and she bought
. . . A memory game where each of you

adds a new (and preferably irrelevant) item after repeating her shopping list so far.

Yes/no game Try to have a conversation without saying yes or no.

What is it? Describe an everyday object to someone who must guess what it is.

Animal, mineral or vegetable One of you picks an object and tells the others whether it's animal, mineral or vegetable. The rest of you try to guess what it is by asking questions that can be answered by yes or no.

Watch this space Before you go, make up a list of things you're likely to see on your journey and cross them off as you spot them. For non-readers, draw pictures (of a cow, horse, bus, lorry, etc).

The never-ending sentence Each person adds a word in turn. The idea is never to finish the sentence. The loser is the one who does!

Travel sickness

Travel sickness can make long journeys a problem for the whole family. Train travel is the least likely to induce it, while coach journeys and choppy ferry crossings are most likely to bring on an attack. Plenty of fresh air can help, and travelling later in the day is generally better than first thing in the morning. Try to make sure the child has eaten, but avoid greasy foods and fizzy or milky drinks. Apples, barley sugar, and dry biscuits make good snacks.

Be prepared for the worst. Keep a stock of disposable nappy sacks handy for your child to be sick in. They're scented and easy to dispose of. A damp cloth sprinkled with bicarbonate of soda will remove the worst of the smell when the worst inevitably happens – and bring a damp flannel for faces and hands.

Try your child with 'Sea Bands'. They're stretchy cloth bracelets with a firm plastic bobble that you place over acupuncture pressure points in the wrist to prevent nausea. Trials indicate that they do help adults, although it's hard to get exactly the right point in a child's wrist.

By car

Chances are your children will be quite familiar with travelling by car, so you don't have any of the benefits that the novelty of a voyage can bring. They're going to need entertaining from hour one.

On the plus side, you're not so restricted in what you can take with you, you can stop whenever you like (or whenever you need to), and your children will be comfortably seated, albeit irritable and restrained. With a bit of luck they may even doze off.

★ Do pack as much as possible in the boot, to keep the car clear and uncluttered. If you restrict the children's leg room they'll get uncomfortable sooner, and then you'll all suffer.

★ Do keep snacks and drinks to hand – but not within reach of the children.

★ Do stop regularly. Plan for an unhurried journey and encourage the children to go to the loo when you stop, to cut down on irritating unscheduled breaks. Many motorway service stations and roadside restaurants (e.g. in the Happy Eater chain) have children's playgrounds or indoor play areas.

★ Do encourage each child to use a bag for rubbish – otherwise their playthings could soon disappear under a mound of crisp wrappers, drinks cartons etc, and they'll become uncomfortably cluttered.

★ Do use motorways where you can. They're faster, less likely to cause travel sickness, and the kids are more likely to doze off. Most motorway service areas have baby changing and feeding rooms.

★ Don't encourage children to read in the back of the car. It can cause headaches or make children feel sick.

By train

This is probably the easiest and most comfortable way of travelling with children. You'll usually have a table for games or drawing on, which can double as a changing mat at a pinch (the loos in most trains are too small, though some trains do now have baby changing facilities). Your children will also be free to wriggle in their seats to their hearts' content, and wander up and down the train (with a supervising adult if need be). They may even make friends with other children in the carriage.

★ Do use a good shoulder bag with a wide strap to carry all the clutter you're likely to want at hand.
★ Do reserve your seats, even when you're travelling at off-peak times. You could consider asking to be near the buffet, which has the advantage that food and drinks are at hand, but buffet prices are often high and these carriages are usually crowded and have a lot of through traffic. If you're waiting for a train to stop at your station ask a guard whereabouts the carriage you'll be travelling in will stop – no panic as you try to get your luggage and your children on.
★ Do travel off-peak whenever possible. Tickets are often cheaper, and trains less crowded. You're also more likely to find sympathetic fellow-travellers.
★ Do try to break up a long journey for your child – even if it's just walking up and down the carriage.
★ Do keep children away from the carriage doors. People have died falling from these.
★ Do take a rubbish bag. British Rail have removed many bins for security reasons.
★ Don't rely on the buffet. They're sometimes closed for parts of the journey, and sometimes not open at all.

By air

Air travel's fast, but the journey to get to the airport and long waits once you get there often mean that's not the way it feels. Many charter flights require that you turn up two hours before take off – and that's assuming the flight's not delayed. And children quickly become fractious once they've got over the excitement of being on a plane, because they're confined in a small space and have little to look at.

★ Do make sure the airline knows if you're travelling with a baby. You can often book a small cot in advance and order baby food. But most under-twos will be expected to spend the journey on your lap – trying for both of you.
★ Do arrive in time to get seats together, preferably at the front of the plane or behind the safety exits, where there will be more leg room and space to store your belongings. This also has the advantage that your child won't annoy the person in front by raising and lowering their table and kicking and jogging the seat back. A window seat will give you more privacy if you want to breast-feed, but children will have more room to wriggle next to the aisle.
★ Do bear in mind that you're allowed only one piece of handluggage each, and that most of your things will go in an overhead locker that will be hard to get at unless you sit at the aisle. Try to keep the things you *know* you'll need in one small bag that you can keep by your feet.
★ Do pick a good spot in the departure lounge, near a window that looks out onto the planes, if possible. Some airports have supervised play areas for children – ask when you check in (*see* Airports in section 3).
★ Do use mother and baby facilities at the airport – they're usually comfortable and it'll be easier than trying to feed or change your baby on the flight.

* Do make sure the children use the toilet at the airport before boarding. They won't be allowed to use the one on the plane until after take-off.
* Do be prepared for the pressure changes at take-off and landing to hurt your children's ears. Sucking and swallowing can help, so offer babies the breast or bottle and have some sweets ready for older children. (Fruit polos are good because they minimize the risk of choking. Chewy sweets work well because there's extra jaw action!) Yawning also helps – you could encourage your child to do so by yawning yourself.
* Don't rely on a personal stereo or Game Boy as in-flight entertainment for your child. Most airlines don't allow their use in case they interfere with radio frequencies. In the light of recent terrorist attacks, all personal stereos, radios etc. must be declared before boarding.
* Don't check-in your buggy or carry-cot before you need to. Most airlines will take them from you just as you get on the plane. Make sure they're clearly labelled.
* Don't count on children eating aeroplane food – if you're travelling at a time they need to eat, bring something you know they like. And don't count on getting fed in the plane if you're travelling single-handed with an under-two. It's a physical impossibility with a toddler on your lap in a crowded plane.

By ferry

Travelling by ferry can be great fun for the kids – what other type of transport lets them explore level by level, then run around and let off steam on the deck? Many ferries now boast children's playrooms and video rooms, where you can sit and relax with your kids, or even leave reliable older children while you do a bit of exploring yourself, and many restaurants offer free baby food. But again, long waits and long journeys can dull the novelty, and at peak times overcrowding can lead to far from pleasant journeys.

* Do check out how long you're likely to wait to get your car on the ferry. If you've got a long wait you'll do better to go to the ferry terminal, even if it's very crowded, particularly after a long journey.
* Do take some toys and games to the ferry terminal or be prepared to play with your children. They're often crowded and there's little specially laid on. (*See* Ferry ports and ferries in Section 3.)
* Do make sure you have a small bag packed with everything you're likely to want on the trip, because you won't be allowed to go back to your car during the crossing.
* Do use the baby changing facilities on the ferry. They're often a cut above the usual ladies' loos, and fairly spacious.
* Do travel overnight and book a cabin if you're taking a long sea crossing. Put your children in pyjamas before you leave home so they can sleep in the car as well.
* Do bring some drinks and snacks with you. Ferry facilities can be expensive, there may be queues, and on night crossings bars and restaurants may shut.
* Don't let young children explore unattended. They could easily get lost or get excited and climb up on the rails on deck.
* Don't let older children discover the many tempting slot machines on board and in the ferry terminals unless you can face the arguments about disappearing holiday spending money.

By coach

Travelling by coach is the least attractive option if you have children, even in a modern and luxurious coach. Indeed, some companies refuse to take reservations for children under twelve. The problem is that they will be expected to sit still for what is often quite a long journey – there simply isn't anywhere for them to go, and any undue restlessness or wandering is going to disturb other passengers. On top of that, rest stops are likely to be fairly few and far between, and certainly not at the intervals that a child would like.

★ Do go by luxury coach if you can afford it. These provide refreshments and videos that will help distract your child.
★ Do check in advance that there's a toilet on the coach to avoid disasters.
★ Do arrive early, to get good seats. If your child is prone to travel sickness the middle of the coach is far better than the back.
★ Do try to keep your essential travel kit in one big bag, which you can then store in the luggage rack overhead, or under your seat. You won't want to be cluttered up with loads of bags.
★ Do travel by night if possible. Your child should then sleep for most of the journey.
★ Don't leave anything you'll want on the journey in your checked-in luggage. It will be packed away in the hold.

Holiday reading for kids

 Wayne Jackman

If you are about to embark on a family holiday then it is essential to consider entertainment for the children, particularly on the journey. Remember the old Latin saying, 'bored and restless children cause *parentus nervosa breakdownum*' (I'll leave you to work that out). So, before you set off clutching your road map and insect repellent, arm yourself with pens, pencils, drawing paper, activity books, puzzles and story books for the children. The books recommended on the following pages are organized into four age groups and are available as lightweight paperbacks.

Another useful distraction is a selection of story tapes which can be played on the car cassette or on a personal stereo. There are hundreds of stories, rhymes and songs available on tape and Roald Dahl story tapes never fail to amuse. With a little careful planning the journey will seem like part of the holiday! Bon vacances.

Books for the under fives

★ *Alfie Gives a Hand*, by Shirley Hughes (Fontana Picture Lions, 1985).

Shirley Hughes is one of the most popular authors for younger children and the reason is obvious. Each of her books relates directly to a child's experiences and is beautifully illustrated. In this one, Alfie has been invited to a birthday party, but should he take his comfort blanket?

★ *Burglar Bill*, by Janet and Allan Ahlberg (Little Mammoth, 1990).

Another sure-fire winner from this prodigious husband and wife team. Colourful colloquial language and colourful colloquial pictures. Burglar Bill finds a baby 'on a job' and then finds true love, settling down to an honest life as a baker. Just like in real life really.

★ *Each Peach Pear Plum*, by Janet and Allan Ahlberg (Picture Puffins, 1989).

Written by the well-known husband and wife team, this is a simple but ingenious book. The charming, detailed illustrations combine perfectly with the rhyme on each page. One of my family's favourites, which is read again and again.

★ *Hand Rhymes*, collected and illustrated by Marc Brown (Picture Lions, 1987).

A wonderful collection of poems with simple hand-actions to accompany them. Illustrations that are simple to follow; a delightful introduction to poetry which will provide hours of entertainment.

★ *Happy Birthday Moon*, by Frank Asch (Picture Corgi, 1990).

Quite brilliant tale of a bear who asks the moon what he wants for his birthday. Is it the moon answering or just an echo of his own voice? The text is simple but very engaging and is a great 'read aloud' book.

★ *Look Out, He's Behind You*, by Tony Bradman and Margaret Chamberlain (Little Mammoth, 1989).

A brilliant re-telling of Little Red Riding Hood without any of the gruesome bits. What's more, the bad wolf is constantly hiding behind lift-up flaps so that our heroine can't see him stalking her. Silly girl! All my kids spotted his hiding places straight away and loved shouting out 'Look out, he's behind you'.

★ *Mrs Armitage On Wheels*, by Quentin Blake (Picture Lions, 1990).

Splendid solo effort by Quentin Blake more usually known for his illustrations of other people's stories. Mrs Armitage invents things on wheels aided by her dog Breakspear. Unfortunately they fall to bits! Lively and full of humour.

★ *My Naughty Little Sister At The Fair*, by Dorothy Edwards and Shirley Hughes (Little Mammoth, 1991).

Cautionary tale of a little girl who wanders off and gets lost. Marvellously well observed illustrations and considerable suspense before a happy ending. This might prevent any budding Scott of the Antarctics to at least tell you before they hike off to Malaya.

★ *Pre-Reading Activity Book*, developed by Oxford University Press for The Early Learning Centre (1987).

Terrific value for money, this activity book contains stacks of line drawings to colour in and is thick enough to last any holiday bar an expedition up the Zambezi river. As one young critic said, *'It's got great "colouring-in" things and matching-up stuff and it's brilliant.'* Enough said!

★ *The Snowman*, by Raymond Briggs (Picture Puffins, 1980).

This is a timeless classic already familiar to many children through the beautiful film based on Raymond Briggs' drawings. No words, just beautiful illustrations that any young child can follow on their own.

★ *The Very Hungry Caterpillar*, by Eric Carle (Picture Puffins, 1974).

This book is popular with children between the ages two and five year after year. A clever folding-page arrangement, with a text that is easy to remember and shout out, creates another book that bears constant re-reading. It also helps a child to count and learn the days of the week.

★ *What's That?* (Campbell Books, 1991).

This is an excellent simple and straightforward book for the babies. Forty clear colour photos of everyday objects which the young child can easily recognize and point at and name. It certainly encouraged my youngest to start talking.

★ *What's The Time Mr. Wolf?*, by Colin Hawkins (Fontana Picture Lions, 1986).

The bright and bold illustrations in this book are both humourous and occasionally deliciously scary. With a clock face on each page and a cleverly repetitive text this is a brilliant introduction to learning to tell the time.

Books for children aged 6–8

★ *Amazing Mazes* by R Heimann (Hippo Books, 1990)
★ *The Great Puffin Joke Dictionary* by Brough Girling (Puffin, 1990).

Two activity books stuffed full with the sort of quizzes and jokes that children of this age love. These will keep them occupied for hours but be prepared to lend a hand occasionally and suffer end-

Holiday reading for kids

less repetition of their favourite jokes, most of which are actually funny – the first time round!

★ *Big and Little*, by Sue Limb and Siobhan Dodds (Picture Lions, 1990).

Delightful stories about two pals – one a gentle giant and the other a Tom Thumb impersonator. Illustrated with gentle comic pictures which match the irreverent, quirky text and funny rhymes. Ideal for those with some reading fluency.

★ *The Boy who Turned into a Goat (and other stories of magical changes)*, by James Riordan (Piper Books, 1988).

This is a wonderfully imaginative collection of magical folk tales from around the world. It's ideal for reading aloud or for the reader who is progressing from picture books to longer stories.

★ *Dirty Beasts*, by Roald Dahl (Picture Puffins, 1986).

Brilliant irreverent and absurd comic verse about a menagerie of dirty beasts doing extraordinary things. Lively cartoon illustrations by Quentin Blake.

★ *E.S.P.*, by Dick King-Smith (Young Corgi, 1989).

Old Smelly the tramp can hardly believe his luck when he meets Eric the pigeon who can forecast winners at the horse races. This is vintage stuff from this humorous and well-loved author, being easy to read and full of whimsy.

★ *How Tom Beat Captain Najork and his Hired Sportsmen/A Near Thing for Captain Najork*, by Russell Hoban and Quentin Blake (Young Piper, 1988).

Two books for the price of one with the title nearly as long as the stories. Only kidding! My seven-year-old read (most of) it alone and giggled away to himself especially when the flowers drooped as Mrs Fidget Wonkham-Strong passed by.

★ *Jimmy Tag-Along*, by Brian Patten (Puffin Books, 1989).

Here are plenty of fun-filled stories with marvellous characters such as Mrs Battyhats. Ideal for those who have developed reading fluency.

★ *The Magic of the Mummy*, by Terry Deary (Simon & Schuster, 1990).

Here's a funny, jokey story about a very spooky subject. Children will love it and it's perfect for them to read alone.

★ *Now we are Six*, by A.A. Milne (Magnet Paperback Editions, 1979).

Go on – you read this when you were young, now pass on the delights of Christopher Robin and Pooh Bear. Learn them by heart and have a family recitation whilst waiting at Gatwick for Flight 216 to Majorca!

★ *The Old Joke Book*, by Janet and Allan Ahlberg (Picture Puffins, 1987).

You might think these are 'old chestnuts' but try telling that to the kids. They'll think these beautifully illustrated jokes are the funniest thing since Gazza cried for his mummy. Try to avoid pre-empting the punchlines which you'll almost certainly know. After all, we were all young once, and how do you think Bruce Forsyth got started!

★ *Orlando's Home Life*, by Kathleen Hale (Puffin Books, 1991).

Facsimile reproduction of wartime book which retains the vivid colours of the original detailed illustrations of Orlando cat. There is more text than is usual nowadays in a picture book which assumes an interest in language often missing in modern books. Good stuff for those learning to read.

★ *You Can't Catch Me*, by Michael Rosen and illustrated by Quentin Blake (Picture Puffins, 1982).

Very funny poems and quite brilliant cartoon drawings make this a sure-fire winner.

Books for children aged 9–11

★ *Alix and the Tigers,* by Alexander McCall Smith (Young Corgi, 1988).

This is a hilarious story of two elderly tigers who make a run for it from the local zoo. Alix helps them in their escape from a cruel hunter sent to kill them. Luckily she is a pretty canny lass.

★ *The Castle of Adventure,* by Enid Blyton (Piper Books, 1988).

What are those flashing lights coming from the castle on the hill? A gang of children set out exploring and uncover a sinister plot. Bags of adventure here in this classic yarn.

★ *Charlie and the Chocolate Factory,* by Roald Dahl (Puffin Books, 1985).

Here is the ultimate Dahl book about the adventures of a group of children let loose in Willy Wonka's chocolate factory. It is fast-moving and humorous, exciting and unputdownable.

★ *The Lion, The Witch and the Wardrobe,* by C.S. Lewis (Lions, 1988).

This has to be one of the most exciting and enchanting adventure stories ever written. Follow Peter, Susan, Edmund and Lucy as they go through the wardrobe into the magical land of Narnia. Compulsory reading for every child.

★ *Little Sir Nicholas,* by David Benedictus (BBC Books, 1990).

This is a classic story of mistaken identity, jealousy, romance and intrigue in an historical setting. Don't be surprised to find kids reading this one under the sheets with a torch.

★ *Matilda,* by Roald Dahl (Puffin Books, 1989).

This is the tale of Matilda, a genius with gormless parents, who discovers that she can make trouble for the uncaring grown-ups in her life. Illustrated by Quentin Blake, this is Roald Dahl at his best and it is a funny and delightful book to read.

★ *The Spy Before Yesterday,* by Catherine Storr (Young Puffin, 1991).

'My dad was a spy . . .' and so begins the story of Ben's dad's past. Or is it just in Ben's vivid imagination? A real 'Billy Liar' for children, full of comedy and fantastic whoppers.

★ *Stories for Tens and Over,* edited by Sara and Stephen Corrin (Young Puffin, 1986).

What is it like to lead a dog's life? Why did Lady Godiva ride naked through the streets of Coventry? (Doesn't everyone!) Was Roman Britain a bundle of laughs or like one extended Latin lesson? Find out in this collection of stories to suit all tastes and which excites the imagination.

★ *Why the Whales Came,* by Michael Morpurgo (Magnet-Mandarin, 1987).

Here's a mysterious and captivating tale of an old eccentric called the Birdman and his relationship with two children on a small island community. The Birdman has a secret which is revealed when a whale is washed ashore.

★ *Writing Jokes and Riddles,* by Bill Howard (Armada Original, 1988).

This is a must for any child who enjoys making people laugh – or trying to! With plenty of examples, this book actually tries to explain the principles behind making jokes and will provide hours of distraction on long journeys.

Books for children aged 12–14

★ *Byker Grove,* by Adele Rose (BBC Books, 1989).

Based on the BBC TV series of the same name, this is the story of a teenagers' dream hideout where they can do any-

Holiday reading for kids 231

thing they want (within reason!). Good, 'street cred' language and a story full of pertinent teenage issues and characters such as 'good-looking but moody Martin Gillespie'!

★ *The Chestnut Soldier*, by Jenny Nimmo (Mammoth, 1989).

This is the third in her Smarties Prize-winning trilogy of mystery and magic in the Welsh mountains and maintains the very high standard. (The others being *The Snow Spider* and *Emlyn's Moon*.) 'Perhaps soon, when I am 13, the wizard in me will fade away and I will be an average boy,' ponders Emlyn our hero. No such luck, mate!

★ *Courage Mountain – The Further Adventures of Heidi*, by Fred and Mark Brogger (Puffin Books, 1990).

This is a thrilling and nail-biting wartime adventure full of suspense. A group of children must flee for their lives across the mountains.

★ *Finders, Losers*, by Jan Mark (Orchard Books, 1990).

Here's a very clever book containing six separate stories that are connected to each other in some way. Each story poses a puzzle that will only be understood once all six stories have been read. Well, that's one way to make sure you finish the book, I suppose!

★ *Going Solo*, by Roald Dahl (Puffin, 1986).

This is the second part of the great man's autobiography in which he describes daring deeds, fantastic adventures, wartime exploits, African safaris and deadly man-eating snakes. Every bit as entertaining and fantastic as his stories. Read it yourself afterwards – I did.

★ *Just William*, by Richmal Crompton (Macmillan's Childrens Books, 1990).

This is the very first 'William' book ever written and this new paperback edition reproduces the original cover and illustrations beautifully. William's pranks are the scrapes of every healthy youngster and very, very funny. Buy it for yourself, never mind the children!

★ *Odysseus II – The Journey Through Hell*, by Tony Robinson and Richard Curtis (BBC/KNIGHT, 1987).

Don't be put off by the title, this is a fabulous book by two members of the Blackadder TV team. It's a very clever and funny adaptation of the old Greek myth – a sort of Junior Hitchhiker's Guide To The Galaxy.

★ *Perhaps You Should Talk to Someone and other stories*, edited by Julia Eccleshare (Viking, 1990).

Here's a great collection of short stories to appeal to all tastes. It's entertaining and perceptive, containing work by a sparkling list of top contemporary writers such as Graham Greene and Alice Walker.

★ *Stalky & Co.*, by Rudyard Kipling (Puffin Classics, 1987).

Classic schoolboy pranks involving three little scamps called Stalky, Beetle and M'Turk. Witness their duels with their unfortunate housemasters whose lives are made a misery by a whole string of boyish larks. This was based on Kipling's own schooldays and has been a favourite since 1899!

★ *The Young Person's Guide to Saving the Planet*, by Debbie Silver and Bernadette Vallely (Virago Press, 1990).

Nearly all young people are concerned about the environment and this book answers all their questions from acid rain to CFCs. It is full of positive actions young people can take, including certain holiday souvenirs which should be avoided!

Wayne Jackman has been a children's TV presenter for ten years on such programmes as Playschool, Allsorts *and* Saturday Starts Here. *He has written numerous children's books and for many of the children's TV programmes. He has three children himself and knows full well the relief that a good book can bring to a child on a long journey.*

Renting somewhere to stay

Kathy Rooney

If taking the family to a hotel is too expensive then renting a cottage, villa or apartment is a practical, flexible and more economical option. You don't have to get organized for set mealtimes, and often the cost of a holiday is much reduced. The disadvantage of renting is that meals still have to be cooked, so parents don't escape from the daily chores of life at home.

When you are booking, either through a travel company or privately, make sure you check out *exactly* what you are getting for your money.

Many brochures are well laid out and informative, but it is always worth trying to find out what may have been omitted from the details – is there a noisy, busy road nearby, for example? The recommendations of reliable friends and relatives are also very useful.

Here are some suggestions for points to check out:

Accommodation

★ How big is the accommodation? How many bedrooms? Will children have to share rooms?
★ Are cots supplied? Is there an additional charge? Does the children's room contain bunk beds and if so, are they fitted with safety rails?
★ Is the kitchen big enough to cook comfortably in? In many villas and apartments kitchens are minimal and it is not easy to cook a family meal. Does the cooker have an oven? Is there a toaster?
★ Is linen supplied? Does 'linen' include towels? What about tea-towels? It is well worth taking your own efficient drying-up cloth rather than relying on what you might find. How often are the sheets changed?

Cleaning

★ Does the rent include cleaning? If so, how often is the property cleaned? If cleaning is not included, can this be arranged for an additional fee?

Fuel and heating

★ Are there gas and electricity? Are they included in the rent? If not, how are they paid for? If electricity and gas are metered how much is charged per unit? What sort of coin should you use to feed the meter? Surprising amounts of holiday spending money can disappear into gas and electricity meters.

Facilities

★ What equipment is there? Do you need to take a tin opener, corkscrew or sharp knife? It's probably a good idea to take those essentials anyway.
★ Is there are washing machine and/or tumble drier? Are they coin operated?
★ What drying facilities are there? This is especially important if you are holidaying in the UK or northern Europe where the weather is unreliable. There is nothing worse than taking the smell of slowly drying sweaters home with you as your main holiday memory.

Swimming pool

★ If there is a pool, check how big it is. Also ask whether there is a shallow end or special children's pool for young non-swimmers. Check if the pool is filtered and how often the filter is examined. (Be wary if the pool is not filtered.) If the pool is shared, check how many people share it (approximately). Peak summer-time use can strain even the most efficient filtration unit.

Safety

★ Is the property near a busy road? Check exactly how far away the road is – 20 metres or 200 metres can make all the difference to your holiday relaxation.
★ Is there a garden? Does it have a fence or hedge? Is the back garden a safe place for children to play in unsupervised?
★ Is there a river or stream near the property?
★ If the property is near the sea, how far away is the beach? Do you have to cross the road to get there? How far away are any cliffs?

★ Is there a doctor nearby? When abroad, is English-speaking medical assistance available?

Transport

★ Is a car essential? Question carefully to make sure your idea of 'essential' equates with that of the person trying to sell you the holiday. A couple of miles' walk may be fine for an adult but is less practical with a baby in a buggy and a toddler trailing along behind.
★ Is public transport available? How often do buses/trams/trains run, and to where? What is the time of the first and last bus etc. each day? How expensive and easily available are local taxis?

Shopping

★ How far away are the nearest shops? How near is a big supermarket? Do you need a car to get there?
★ Are the local shops open on a Sunday? This is important if you are travelling on the Saturday and get delayed.
★ Does the travel company supply a 'welcome pack' of tea/coffee, bread, eggs etc. to tide you over the first evening/breakfast?

Eating out

★ The chance to eat out *en famille* is a great plus of renting your own accommodation. Abroad, restaurants are usually very tolerant of children, and even in the British Isles restaurants and pubs are beginning to cater more for families.
★ Before you book the accommodation check how far away the nearest eating places are. Are they within walking distance – even for young legs?

Special requirements

★ Do you and your family have any special requirements? Always check what facilities are available before you confirm your booking.

Insurance

★ Insurance is usually available when you book through a travel company. However, it is worth considering even if you book privately, as cancellations can still occur and it is best to be on the safe side. *See* **Holiday insurance** in this section for more information on insurance.

Baby-sitting

★ Can baby-sitting be arranged locally? How much does it cost?

Change-overs

★ What is the change-over date and time? There is no point in driving hell for leather all night to arrive in the morning only to find that you can't get access to the property before 5.30 p.m.

Getting there

★ Check directions very carefully, especially if you are driving. Some of the most interesting holiday properties gain their charm from being off the beaten track. That charm wears rather thin if it is late at night and your weary and fractious family have been travelling all day in the back of a hot and stuffy car.

★ If you are flying and hiring a car from the airport, check what will happen if your flight is delayed. Where will the hire company leave your key and documents?

Don't be put off by this list – when travelling with a family it is worth checking out as much as possible in advance, so you know what to expect (and what to pack) and can enjoy the holiday even more once you get there.

Whether a caravan in Scotland or a villa in the West Indies, renting your own accommodation can give you a terrific family holiday. Make sure you enjoy it!

A half-term holiday break at Butlins

Sheila Sang

My parents never took me to Butlins. Determined that my children shouldn't suffer the same deprivation, half-term break sees me heading for Butlins Southcoast World, Bognor, with Robert (ten) and Dorothy (six), three women friends and sundry offspring and friends – fourteen children in all, aged five to sixteen. I feel rather outnumbered.

Sunday

The tide of excitement from the children is rising around the house. Long lists are written (phonetically by some), numerous phone calls check who's bringing what. We're going self-catering, none of us confident about holiday camp menus. Two enormous suitcases are packed, one with our holiday gear and loads of towels, the other full of other holiday requisites such as baked beans, chocky biscuits and cornflakes. We're taking no chances.

Monday

The big day. The children are running around in circles, so we plan to leave early and make use of the fun fair until 4 o'clock when we can check in. But like all good plans this one isn't hitch-proof. The phone rings to bring the bad news that Glynis's five-year-old with recently-fitted grommets has indescribable gunge pouring from his ears, while Sue's ten-year-old is found to have a lump in her groin (which fortunately turns out to be a pulled muscle). Two of us start our hols in the doctor's surgery.

Undaunted, Martha comes over to my house and we chat rather apprehensively over a cup of coffee as our combined car loads of children wreak havoc around the house and the noise rises above the danger level. However, by lunchtime Sue and her family are off on the train, and the three cars containing me, Martha and Glynis have set off in a 40mph convoy certain to cause apoplexy to any would-be speeders heading south from London.

It's our turn for harassment next as we face booking in at Butlins. The children are roaming the funfair under supervision from Martha, but even without them it's bedlam. Reception is a seething mass of children and noise. There are huge queues to get chalet keys. But the worst is to come. The Southcoast complex is fairly large and the car parks, sensibly, are at one end, accommodation at the other. But the provision of transport to get people and luggage from one side to the other consists of one or two desultory vans and a small and erratically-arriving 'train'. It ends up as a free-for-all, with those with young children and lots of bags (yes, me) permanently stationed at the pick-up point with no relief in sight.

Once we get there, the accommodation is good (we've chosen County Suites, Butlins's best). It looks absolutely what you'd expect from a holiday camp – row upon row of terraced rooms

– but the flats are comfortable, the kitchens well-equipped with all the basics, there's a TV and plenty of cupboard space. But, just when I think it's safe to relax, I realize I can't find my handbag. Panic searching produces nothing: during the scrum to get transport my bag and I have parted company. That takes care of most of my evening entertainment, with trips to security/reception/security/car park in the vain hope of being reunited with all my holiday spending money, cheque book, cards, etc. Look on the bright side, at least it makes it worthwhile having gone for the holiday insurance.

Fortunately, the kids aren't affected by my plight as they can play with their friends during this interlude, but the stresses of the day have taken it out on me, and I'm beginning to wonder if my parents didn't have the right idea after all . . .

Defeated, we retire to the fish-and-chip restaurant for a pleasant meal and a much-needed bottle of wine, then brave an unexpected fog (produced by the weather, not the wine) to go to the ballroom, for entertainment and a disco where those children so inclined can work off whatever energy they have left. By now our party is fairly scattered, the older children are making the most of the freedom Butlins offers – we've agreed they can come and go, so long as they show up at set times. They set off to explore the Arcades (strict cash limits have to be set early), children's club, teens disco and generally find their way about while the younger ones put their minds to mastering the Birdie Dance (or some slightly more sophisticated variation well beyond my abilities). The ballroom is huge, with a large dance area surrounded by a sea of tables, and shops around the edges blocking off what might have been a sea view. Fortunately the kids don't spot the shops this evening (they sell everything any fairly optimistic child might be prepared to ruin the evening by asking for). But the day is taking its toll, and even the older ones are too tired for Karaoke at the teens' club, so it's back to the chalets for all, ready to face the next day's strenuous activities. Will it be the funfair or the swimming pool? We're not qualified to take part in the 'Dad and his Lad' competition . . .

Tuesday

Woken by Scott knocking on the door, wanting Robert to go swimming. Think, 'go away'. Say, 'in a minute'. A minute later, Scott knocks on the door . . .

Once we're awake, there's constant traffic between our four adjacent first-floor chalets, connected by a long covered terrace. From downstairs the sound is volcanic.

Five-year-old Aled now has flu as well as gungy ears, so is chalet-bound. The rest of us are raring to go, and those who aren't already at the swimming pool or funfair are studying Butlins's on-screen information about today's activities. We've missed Keep Fit and decide against the Junior Princess competition so I head for the Aquasplash, Butlins's 'Sub Tropical Waterworld' with an assorted handful of left-over kids. The pool is amazing, with three water flumes, a large aqua-glide slide, children's slides, jacuzzi, rapids and of course a wave pool. The children are thrilled, and it's very well-supervised. Just as well because it's impossible to keep your eye on more than one of them at a time – the place is awash with bodies.

Back for lunch, and the startling revelation that between the four of us, no-one has remembered the tomato ketchup. The afternoon poses another problem – the children are all agreed on going to the cinema, which in this environment we reckon they're old enough to do without our help. That leaves us adults with time on our hands – will it be the whist-drive or Bingo? A quick bit of research reveals certain gaps in our whist skills ('is that the game where you make tricks?') so we go for a refreshing seaside walk outside the gates to remind ourselves that the real world still exists, then try our hand at Bingo. Sue won 50p. We're hooked!

Tonight we try the 'family venue' for entertainment – the larger ballroom has a band playing big band music, and with children into Vanilla Ice and Michael Jackson, this is not likely to be the biggest hit. But the place is packed and our potential disco stars are more likely to do other children an injury than strut their stuff. Nor does anyone fancy the organized evening children's activities. The older girls didn't even want to go to the 'Blind Date' show or judge the Mr Smooth competition. So it's back to the chalets by 9 p.m., where Glynis has been busy preparing a feast for the grown-ups.

The children choose not to go to bed. (Surprise! We let them stay up, they're on holiday.) We split up into two adjacent chalets – one for girls, one for boys. So much for years of attempting a non-sexist upbringing.

Wednesday

Wake to the continuous thunder of tiny feet past the window, and gratefully drink a cup of tea brought in by Glynis. After some dithering, we head down to the funfair. Most of the children thoroughly explored all the rides here yesterday, and they are amazing; as good or better than most funfairs, with the added bonus of inspiring confidence in the standards of maintenance and supervision (both necessary, since children are allowed to go to the funfair on their own). The Butlins staff are obeyed unquestioningly, they somehow combine a friendly attitude to children while tolerating no nonsense. Wish I could get the knack.

Much of the funfair is undercover, a wise precaution, but the resulting gloom leaves the onlooker rather less exhilarated than the children.

A return trip to the swimming pool is cut short at lunchtime for a session of water aerobics. I would have joined in, but Dorothy's natural modesty prevents her waving arms and legs before the audience collected around the gallery of the pool . . .

In the afternoon we persuade some of the children to follow us for a seaside walk and to explore the seaweed and breakwater pools of Bognor's own natural themepark, the seaside. Not surprisingly, this second visit is rather less restful than our first adult exploration, and the children take some persuading not to return with several pounds of shells and interesting stones (or boulders) in their pockets.

Back on site, we decide to make the most of what Butlins has to offer, the children go off to play table-tennis, check out the trampolining, and (with a fourteen-year-old to accompany them) to play snooker. Martha wins the jackpot (£5.30) at Bingo. I wonder if we'd have won at whist?

After supper we take a quick trip to the family venue to check out if we've finally got the hang of dancing to 'Aga Du, Du, Du'. We should have gone to the dance class offered on the first day.

On return to the chalets we discover much excitement among our teens. Two of them are now 'going out' – they've been for a walk. I wonder if they noticed the giggling shadows hiding behind lamp posts in their wake? Chaperoning from the adults is clearly unnecessary . . .

Thursday

A day of brilliant southcoast sunshine draws a group of us back to the beach, though some opt out in favour of the Butlins go-karts (£1) and Arcade.

Dorothy seems a bit listless and doesn't want to do anything much, and despite the wonderful time she has had at the pool is unwilling to brave the changing room again, which is both chilly and crowded. No such qualms for the others, who head off to spend some time at the splash.

For one of them this turns out to be a very long time indeed. She loses the others, and then can't find her locker because she gets confused by the identical changing rooms on either side of the pool. Our forays don't find her immediately either, and it's a very cold and upset ten-year-old that emerges three hours later.

Young love has floundered early, and out teenagers' romance is off. Couldn't have been the intense scrutiny could it?

I put in some time on packing – wary of the return struggle to the car with the cases, I'm taking as much as I can back now. For the evening we decide to opt for the ballroom, rather than the elbow to elbow dancing with Redcoats in the family venue. Now they have the room and freedom, the children's main entertainment seems to be chasing Aled (now recovered from flu) and requesting expensive extras – hats, crowns, facepaints, trolls. But everyone is having a great time, and our plans for an early night go the way of many of our other schemes this holiday, where organization is quickly replaced by entering into the non-stop on-the-go spirit of things.

Friday

Chaos again as we pack our bags and empty the chalets to join other families in the struggle back to the car parks. The vans have given up playing hard to get, and don't appear at all, the able-bodied carry what they can and we head back to the cars and a last go on the funfair rides.

Has it really been just four days? Would we do it again? Ask the kids: it's been wonderful.

The *best* thing about Butlins: the funfair, the swimming pool, and the freedom the children have to wander around without constant supervision. And the cost. Out of season, self-catering, this is a very cheap holiday.

The *worst* thing: the struggles booking in and out.

Home now to put my feet up!

Holiday insurance

Ernest R. Jones

Have you ever been bitten by a camel? Or had a head-on collision with a llama whilst riding a bike? No? Well some people have, and all in the name of a 'good holiday'. There is an old idiom that says 'it will never happen to me'. However, such calamities happen frequently to all manner of travellers and unexpected incidents can range across the whole spectrum covered by a Comprehensive Travel Insurance Policy. They may include personal accidents, all manner of minor and major medical problems, losses and thefts of all baggage, personal effects and money, personal liability (where you cause injury or damage to a third person or their property), delayed baggage (where the airline in their wisdom deliver it to the airport at Corfu when you happen to be staying in Tenerife), delay of your plane or ship, cancellation of your holiday due to events beyond your control such as family illness, bereavement or redundancy, curtailment where you need to return home early from your holiday due to illness of one of the travellers or illness at home of a close relation, and hi-jack compensation.

It is a fact that one in every fifteen persons travelling abroad will make a claim on their policy from the very minor and mundane type of incident such as Spanish Tummy (perhaps requiring a Doctor's visit and a prescription costing perhaps £45 in all) to multiple accidents, heart attacks, hepatitis, which may cost scores of thousands of pounds in some cases for treatment, surgery and repatriation.

Believe it or not, the UK Travel Insurance market is probably the most sophisticated in the world, possibly due in some part to the earnest desire of the British public to escape from the English weather. As a result, the market has diversified to include tailor-made travel policies. People travelling on business can have their own policy which, while covering most holiday eventualities, may also cover the cost of replacing that person at an important conference or meeting if he or she became ill, or covering the value of goods or samples which are not normally covered by holiday Travel Insurance. Perhaps the baggage limit should be increased as he or she would need to carry several suits and extra luggage which would need to be adequately insured.

Single parent families or indeed any family with young children should look around for policies which may give free cover to children under a certain age (such as Extrasure Travel Policy) or at least reduced premiums for children. Groups of physically or mentally disabled children or adults may require different types of cover, although very often the supervision of these groups is so efficient that claims are kept to a minimum. Nevertheless, a policy should be sought which will cover unforeseen illnesses, or events that may well stem from their disability.

Different types of holiday require different cover. Holidays can vary from teenage lager-drinking contests in Benidorm to an exotic Kenyan safari, scuba diving off the Florida Keys, or indeed a coach tour of Italy for Senior Citizens. Many policies are designed for particular types of holiday such as special winter-sports insurance. Long-haul holidays, such as those to Africa or Australia, may last for several months at a time and will therefore require greater baggage and personal-effects cover.

Specific holiday activities, such as scuba diving, caving, canoeing etc. attract many young people, but most of these activities are excluded from normal insurance cover. As a result Activitysure has brought out a policy designed specifically to cover such otherwise excluded sports. Motorists' needs while travelling are different, not least as there is a greater exposure of the risk of theft from their vehicles, but also a greater risk of personal accident. When hiring a car abroad, you will also need collision-damage insurance for the rented vehicle. Americasure has combined all these aspects into one policy at a considerable saving to the traveller.

Claims on all types of policy and for all types of traveller vary enormously. Contrary to some opinions, insurers do actually *want* to settle valid claims – efficiently handled claims are their best form of advertising. Many claims are handled by third party Administrators. However, it is necessary to supply to the Claims Handlers all the information, documentary or otherwise, which will enable them to settle the claim quickly. It is not in the Claims Handlers' interest to prolong a claim by protracted correspondence – it costs money, and holds up the settlement of a claim. Therefore, for example, if a traveller is robbed or loses items of value, they should report the incident to the local Police Station and obtain a report if possible. Some Police will not give a written report, and in such cases it would be useful to obtain a report from the hotel manager or courier to confirm that the Police were contacted. (Some companies will not meet a claim for theft or loss of valuables if there has been no attempt to contact the local Police.)

The 'It will never happen to me' philosophy is a myth. Something *will* happen to you sometime, somewhere, but travellers can increase their chances of a trouble-free holiday by following a few pieces of common-sense advice.

★ Be careful with your goods. Use hotel safes and never leave items unattended in vehicles, locked or otherwise. Some policies will not reimburse you at all for goods stolen from cars, while others insist that they should always be hidden under blankets or in a locked boot. However, it is safer *never* to leave your possessions unattended than to gamble with your policy.

★ Drink only bottled water and soft drinks when there is any doubt as to the safety of the local water supply. This is essential when travelling with young children and many travellers advise giving babies and toddlers only boiled water.

★ Do not over indulge in food or drink that your stomach may not be used to. The effects of rich or spicy food when combined with too much alcohol and sun can ruin your holiday.

★ Carry a basic first aid kit to deal with minor problems. (A useful addition to such a kit is a flat pack of soft toilet paper!)

Medical claims generally require immediate treatment and a traveller will normally obtain that treatment locally and be reimbursed on their return home. Such compensation depends on the production of a medical invoice and a fully completed claim form.

However, almost all Travel Policies also include the services of a specialist

Emergency Assistance Company. Their job is to help the traveller when faced with an emergency they cannot handle on their own. Serious accidents and illnesses account for most calls to such services. In cases which require hospitalization and evacuation to the home country, the Emergency Assistance Company can help. They must be contacted immediately a problem arises in order that the company's Doctor can discuss with local medical staff the details and diagnosis of an illness and decide when a patient will be fit to travel back to their home. In most cases, bills from Medical Centres can be settled direct by the Assistance and Claims Company without the patient having the added worry of having to find the cost of treatment in advance of compensation.

Many travellers who receive medical treatment are able to continue their holiday. However, some unfortunate holiday makers have to return home due to illness. Some are well enough to travel in a wheelchair, with perhaps another member of the family escorting them, but others need to travel on a stretcher, and will usually require a medical escort. Air ambulances are available but are only used when no other means of transport can be used, perhaps in a life-saving situation or where an immediate operation is required that is not available in the country of travel and the patient cannot wait for airline availability. Air ambulances are generally small aircraft, and therefore subject to slower flying times and air turbulence. They will have very little room, with space only for a stretcher, medical team and medical equipment. They are used only when absolutely necessary but are nevertheless essential in certain circumstances.

The Assistance Company is there to help in any situation and their emergency number should be used to seek positive help and guidance. In certain circumstances the British Consulate can help a traveller with advice or emergency passports but they are not usually able to help with financial assistance to pay bills or replace lost cash, tickets or personal effects.

Medical Insurance claims frequently exceed £25,000 and sometimes can exceed £100,000. Even claims under other sections of the Travel Policy can be expensive. Recently a man lit and threw a firework at a celebration which exploded in the face of another man, partially but permanently impairing his sight. The liability claim for the third party injury was £14,000.

It is essential therefore that the potential traveller does his homework when it comes to choosing travel insurance. He should look for a policy:

1. Underwritten by a well known insurance company or Lloyd's of London.
2. With medical benefits of not less than £100,000 for Europe and £250,000 for the rest of the world.
3. With adequate benefits for baggage and personal effects.
4. With cancellation limits that will cover the cost of most holiday purchases.
5. That includes an Emergency Assistance Company. The services of this company should be available 24 hours a day and it should employ bilingual staff to cope efficiently with any emergency.

Without doubt, the travel insurance premium is the best investment a traveller can make in deciding his holiday arrangements. Astonishingly, about one-third of the travelling public are not insured at all. These people take the enormous and unacceptable risk of jeopardizing their and their families' future by being responsible for enormous medical bills and repatriation expenses. Holiday insurance is not expensive, especially when compared with the whole cost of the holiday.

So go away and enjoy yourself and forget about claims and losses, secure in the knowledge that your travel insurance premiums provide you with the peace of mind that if you do get bitten by a camel, dive into an empty swimming pool, or simply drink too much Sangria, your treatment costs will be taken care of. An old advertising slogan advised viewers to 'Get the strength of Insurance around you' – nowhere is that more relevant today than in holiday insurance.

November 1991   Ernest R. Jones,
*Chairman and Chief Executive,
Mercury Insurance Services Ltd.*

Section 5

Family travel know-how

Family travel know-how

Air travel

We're all familiar with the speed and efficiency of air travel, but having children in tow presents a number of different considerations. In this entry we have compiled a quick checklist of airline services you can request when booking, plus various tips gleaned from parents who have hard-won experience of travelling by air with children.

Checking in: Allow ample time for this procedure and remember that babies do not have a separate baggage allowance, although a pushchair, carrycot and flight bag are permitted. Ask at the check-in desk about any special request that you have made.

Pushchairs: Some airlines allow pushchairs to be used until you board the plane, with some even stowing it in the passenger cabin if space permits. This might not seem a priority, but airports often have miles of corridors which can be an arm-aching distance to carry a squirming toddler.

Boarding: Most airlines allow families with young children to board before other passengers. They may also automatically give you seats near a bulkhead (the partition dividing the cabin) so that you have more room. Be sure to request this position if it's not offered; it also has the added advantage of giving adults more legroom.

Carrycots/skycots/bassinets: If you take your own carrycot it will be stowed with the luggage in the hold. Note that airlines prefer the collapsible variety. Most airlines will supply a cradle, skycot or bassinet, but you must request this when booking. They are normally suitable for babies up to nine months old with a maximum weight of 24 lb/11 kg.

Nappies: The thought of changing nappies in a confined space at 25,000 feet is not an appealing one, but some airlines now have changing tables in toilets and may even supply disposable nappies. Most stress that these are for emergency use, i.e. when your own supply runs out, and they are usually the smallest size, suitable only for very young babies.

Breastfeeding: Altitude has no effect on the flow of breast milk. Some airlines (usually Oriental ones) will supply a screen or curtain if you want privacy during feeding.

Bottle-feeding: If your child is bottle-fed it's a good idea to take several prepared (plastic) bottles with you which the crew will store and heat up as required. If necessary, the crew will also sterilize bottles and teats, usually in boiling water from the coffee-making facility rather than a special sterilizing unit. Many airlines keep a stock of bottles, milk and powdered feeds, but check in advance that the food they carry is suitable for your child.

Food: A selection of prepared baby foods is available on many airlines so you don't need to carry heavy bags full of tins and bottles. Ask in advance for any specific dietary needs and most airlines will try to meet them or offer suitable alternatives. Older children may be offered children's meals, usually of the burgers, beans and chips variety. Again, these must be requested when booking. Vegetarian food must always be booked in advance, both for children and adults.

Help: On every flight there will be at least one attendant designated to offer particular help with children. With any luck, he or she will seek you out and hold the fort with toddlers if you have to change the baby.

In-flight entertainment: This can take many forms: toys, colouring books, games, stories, puzzles, comics, films, recorded music, computer games and nursery rhymes. Most are offered free of charge and, as a rule, the longer the flight, the greater variety of things to do. However, it's a good idea to bring your child's favourite toy or game.

Take-off and landing: All modern aircraft are pressurized for passengers' comfort, which means that we can breathe and function normally at high altitudes. During take-off and landing, however, it's not unusual to experience discomfort from pressure on the ears – a popping sensation sometimes leaving a temporary deafness. Children, particularly young babies, can find this very distressing. Their normal reaction is to cry and this, in fact, is the best thing they can do, as it relieves the pressure on the ears. Alternatively you can try to pre-empt their distress by bottle or breastfeeding them during take-off and landing, as the swallowing action can also provide relief.

For toddlers, sucking a dummy or a boiled sweet can help relieve the pressure, while for other children coping with popping ears can be turned into a game by yawning and blowing their noses.

If your child has a cold it would be wise to seek your doctor's advice before flying as catarrh can increase the discomfort of pressure on the ears and even lead to ear infections.

In-flight: If you are worried about keeping your baby or young child quiet during the flight, mild sedatives such as Phenergan can be purchased from the chemist, but do speak to your doctor first. Phenergan reduces the chance of travel sickness and will help your child to sleep. If you have not had time to request a special seat earlier you can always ask at the airport when you check in if there is a spare seat you could have next to you; on the plane you can ask the flight attendant to move you if you spot any spare seats. Another tip is to ask for flight meals to be served separately if there are two adults. That way you can take it in turns to enjoy your meal. If the flight is relatively short it is probably worth giving your baby a double nappy to avoid the need to change – space is short on airplanes. But by and large the range of facilities provided for mothers and babies can be quite impressive, even on smaller airlines. (Full details are given under **Airlines**, Section 3.) Although games and toys are supplied, it's wise to take a few favourite things for children under three as the toys provided by most airlines are only suitable for children over four. *See also* **Nappies**.

Disembarking: Usually airlines that allow early boarding allow you to get off first – a boon in getting through customs and immigration quickly!

The Air Traveller's Code: In April 1990 the Civil Aviation Authority published a leaflet giving advice on safety and security for air passengers. *The Air Traveller's Code* will tell you how much baggage you can take into the cabin, what you can and cannot carry by air, what security precautions you should take with your luggage, and how you can help to make security checks as quick and painless as possible.

Copies of *The Air Traveller's Code* can be obtained free of charge from: Air Traveller's Code, Freepost (GL1776), Cheltenham, Glos. GL50 2BR. Tel: (0293) 573 924 (24-hour answerphone service).

Airports

Thousands of families pass through airports every year, and the growth of family travel has provoked an excellent response from airport authorities and architects. Facilities for families can now be quite impressive. We have carried out a survey of British airports and have listed all those features a travelling family might find useful (*see* Section 3: Holiday transport).

Mother and Baby rooms are becoming more and more common. My only gripe is that very few airports provide *Parent and Child* rooms, with the result that lone males travelling with children are forced either to struggle in a cubicle in the men's toilet or boldly to march into the women's toilet to use the Mother and Baby room. It seems a shame that so many airport designers have ignored the fact that children will not *always* be looked after by their mothers. However, things are changing and several airports have changed their policy. Another solution is to try the disabled toilet as these are mostly unisex.

Airport information desks are generally staffed by very helpful people. It is always worthwhile phoning the airport in advance to find out about the facilities available and to make any special requests. During the summer, when there are extensive delays, some airports organize entertainment for children, but you must never rely on this to avoid infant boredom. Always take a small selection of books and toys with you and be prepared to launch into your best comedy act when things seem at their most dire (*see* Section 4: Holiday reading for kids; and Keeping the kids amused . . . and keeping your sanity).

A pair of reins can be useful when you are trying to keep hold of a mischievous youngster and watch your cases while monitoring the departure board. Try to arrange in advance a point in the airport where you can meet up with any of your children (and other adults!) should you get separated – with any luck, this might minimize your headache should disaster strike.

Flight delays mean delays with everything else, so if you suspect that your departure might be delayed it is well worth taking emergency supplies of food and drink that you can ration out. Extra nappies and a roll of soft toilet paper should also be packed. It is also important to warn the children not to be frightened of the numerous armed security guards they will see at the airport – they might be necessary, but they don't always add to the holiday atmosphere!

See also Section 3: Holiday transport.

Baby-sitting

Some parents (especially if they both have jobs outside the home) may look forward to a holiday as the time they can spend devoted to being with their children. But for most people a holiday is the time to relax and enjoy yourself – and this will often mean without the children's company! For this reason baby-sitting facilities or children's entertainment are features which many parents look for in making their holiday choice.

The most extensive facilities are offered at large hotels in popular resorts. Although the hotel may not be your ideal choice, if there is plenty for the children to do it may just mean you really can have a holiday yourself. Some of the smaller hotels may also say that a baby-sitting service is available on request. 'Baby-listening' services are often available in hotels of all sizes. Although this means you have to eat either in the hotel or very close by, it does offer the option of eating something other than that supplied by room service. In practice we have found that hotels can often provide you with a babysitter whom you pay directly and by the hour. Although you can ask about qualifications, this is largely irrelevant as it is likely that the sitter is the doorman's cousin or sister anyway. In Paris and Wiesbaden we found three very suitable women who turned up to look after our son. One was an exquisitely dressed postgraduate student who came with her own toys, another was a grandmother in jeans who had planned a ride in her brother's taxi as the entertainment for the day, and another, the trainee-receptionist in the hotel who spoke English and taught our child some German. All you can realistically do is make up your mind when you meet the sitters whether you will trust them or not. We did in each case and had no complaints. The rule of thumb is that even if the hotel does not mention baby-sitting, ask whether it is available.

Some of the tourist boards were helpful when we asked if they had information about particularly suitable hotels for families. The Swiss Tourist Office produces a very useful brochure which is free on request called 'Special Hotels for Families'.

Boredom

It's impossible to say whether travelling before the advent of the jumbo jet and package holidays really was more leisurely and pleasurable, but it is certainly almost always the opposite of that now. Boredom is a problem to be dealt with when travelling with children and we offer a few suggestions below.

Airports/airlines: If you are delayed in boarding your plane (all too likely these days) at least you still have freedom of movement and can go for walks around the airport. A lot of the airport shops have inexpensive toys on sale and this may be the moment to splash out. If you're not already seated near a window looking out at the planes, find one. Most children are fascinated by what they can see and this should pass the time fairly pleasantly. Once you're on the plane and it's delayed in taking off, you will have to be a bit more imaginative. Several parents suggested taking along small presents (new books or toys) to give to the children at intervals, possibly wrapped up so that they have the excitement of tearing off the paper. A lot of the airlines have special children's packs which include drawing books and crayons. On long-haul flights they often have a children's station on the in-flight audio entertainment which could save the day.

Car: If your children do not sleep in the car you will have to entertain them if the journey is long. To save your voice bring along their favourite tapes and books. You can always play 'the first one to spot ten blue cars' kind of games as well. Be prepared to stop to allow everyone a breath of fresh air, so build extra time into your journey.

Rail: Bring games, but not the sort with too many small pieces which could roll away and get lost. Card games work well, though they can be noisy, and there is the continuing interest of watching the scenery as you whizz through the countryside. In some of the more exotic places trains go slower and there is so much new to see that boredom is lessened.

Coach: Although coach travel can never be ideal for children, the advent of the modern coach journey is a great improvement on what came before. Fares are kept low, but in order to compete the companies have to offer at least what is available from the other services. Often video is on offer and that should keep most children happy. For smaller ones you may have to sit and read quietly to them while the film is on. Looking out of the window should take up at least some time. If you have a Walkman or personal cassette player, bring it. It will be worth the extra weight. The same thing applies to coach journeys as for cars and planes – make sure you bring some books, old favourites and new ones.

If you've tried everything and the children still complain of being bored, remind them you are too and you're all just going to have to live through it. After all, the aim is to reach the place where you are going when all boredom will be banished and the fun begins!

Breastfeeding

If you are breastfeeding already you will be aware of the benefits of this transportable method of nourishment. It is positively tailor-made for travel with babies. If you are travelling a long way, try and rest if you can to ensure there is no interruption to the milk flow. Most major airports, railway stations and ferries have mother and baby rooms for changing and feeding (ask staff for their location). Often they have bottle-warming facilities too. People do seem to respect these facilities so you will mostly find you are alone in the room and can feed in peace. The changing facilities are a godsend and useful for all children in nappies.

In hot climates and underdeveloped countries breastfeeding has the advantages of avoiding dehydration and conferring immunity. In the countries of southern Europe, and indeed in most countries with warm climates, it is normal to breastfeed, but this does not mean you are at liberty to be indiscreet. In Greece, for example, while no one would be surprised at your breastfeeding, they would not expect to see it in a restaurant. In Italy women have moved away from breastfeeding to such an extent that you are almost congratulated if you breastfeed. In France, of course, you can be very open about it. Nobody minds anything. Breastfeeding is very common in Spain, but is done in private. Spaniards would not be surprised to see tourists breastfeeding in public, but they would not do it themselves. In Saudi Arabia nobody minds you breastfeeding as long as your face is covered!

If you are travelling through hot countries and will be feeding in the car it is a good idea to bring a white sheet to put up over the window – not for modesty's sake but to keep out the glaring sun. A few muslin nappies will provide you with instant cover-up or some privacy if required.

If you want more information about breastfeeding or have very specific questions, you can contact the Association of Breastfeeding Mothers or the La Leche League (*see* Section 7).

Clothing

Always adopt the layer approach to clothes when travelling with children regardless of the time of year or where you're going. The temperature inside aircraft varies from cool to very warm. Airports are often air-conditioned, but remember that the night temperature in some countries at certain times of year can be very different from the daytime temperature. On top of all that it is always possible that the weather will change at your destination. It is best to be prepared, although we don't suggest taking skiwear to Jamaica just in case!

Natural fibres are best, but some synthetics do have the distinct advantage of drying quickly. If you're going to a place where the washing facilities are likely to be non-existent, you should consider taking a travelling clothes line, pegs and washing powder or liquid.

Don't overpack: remember if you're going to a sun and beach holiday the children will spend a lot of time either in their swimming things or in teeshirts and shorts. Most European resorts will be happy to sell you teeshirts, plastic sandals, hats, toys and so on if you want to buy them there. We do suggest hats for children if you're going to spend a lot of time in the sun.

The layer approach to clothes has been the standard recommendation to skiers for decades. For a winter holiday waterproof or water-resistant clothes are a must. If you are going skiing and don't want to fork out for the very expensive children's clothing, you can hire them from ski shops in the UK or possibly at the resort.

You don't want to be prevented from going for a walk with the children just because there happens to be an unseasonal downpour; having experienced this in New York in July we were glad to have brought along waterproof overtrousers and a hat which kept out the worst of the weather.

Coach travel

This is not one of the most popular modes of transport for travel with young children. In fact, a few operators do not allow children under 12 on their coaches. The problems of boredom and frustration tend to be greater in coaches because there's little room to move around and work off pent-up energy.

Coping in cold weather

In **Clothing** we advise dressing the children in layers and this particularly applies in cold weather and for winter sports. It is the air trapped between the layers of clothes that keeps you warm. Your child should wear a hat, gloves and water-repellent clothing whether you are actually going to ski or not. Glare on the ski slopes can be great so children (and adults) should wear goggles. Do not expose babies' eyes to this harsh light.

You will have to invest in proper footwear, unless you can hire warm, lined boots at the resort.

A barrier cream and moisturizer are essentials for all children, especially babies, to protect them from the wind.

Coping in hot weather

Although most children love the sun and heat, it is necessary to protect them against its potential dangers. Unless your children are very used to being out in the hot sun, you should limit the time they spend in direct sunlight in the first week of your holiday to the early morning and late afternoon. They should always be protected by covering them in sun screen lotion, and wearing a hat and sandals in case the sand or pavements are baking. You may also want them to keep a teeshirt on for a while. It's worth taking a clip-on sun cover for the buggy as hats can be hot and don't always stay on.

If you are not near a café or shop, take along bottled mineral water or juice to protect against dehydration. One mother wrote to us suggesting that you take a lot of the individual juice cartons to your destination as they are a) portable, b) you can mix them with mineral water to last longer, c) they are expensive (if available at all) at most resort areas. You should see that your children get salt in their diet to replace that which is lost very rapidly through sweating.

Prevent heat rash by making sure that your child's clothes are loose and made of cotton. Rubbing him or her with baby powder can help. If a heat rash does appear, don't panic – wash the child frequently with cool water, drying him thoroughly and put on some soothing calamine lotion.

A tip for breastfeeding mothers: if it is very hot and the last thing you feel like doing is clasping a sweaty body to yours (even the baby's!) insert a nappy between you and the baby. It will end up soaked in sweat but at least the two of you will be less clammy.

Mosquitoes can be real pests in hot climates, particularly at night when there are ample opportunities to bite unconscious bodies. Mosquitoes have a preference for female flesh but it's wise to protect both sexes and all ages. You can, of course, take your own mosquito nets but these can be tricky to set up. A sheet of muslin draped over a cot affords some protection, but could be stuffy on a hot night. There are some excellent repellents on the market.

'Autan' is available in sticks or sprays and we have found both very effective. Mosquito coils can be bought in camping stores. Once lit they burn very slowly and give off a vapour which repels mosquitoes. A more modern version of this is an electric device which resembles a two-pin plug. It has an indentation on one side in which you place a tablet of repellent. When the device is plugged in, the tablet warms up and gives off an odourless vapour – very effective we're told in keeping insects at bay. You can buy these devices in most chemist's shops abroad. Beware of complacency about mosquitoes nearer to home. We know of a child who was very badly bitten in Paris.

Ferries

People have conflicting opinions about ferries: some have unfailingly enjoyed their crossings, while others have loathed them. It does seem to depend on whom you travel with. We have heard good reports of Sally Lines, but it's obviously a matter of personal choice.

When you plan your crossing, choose a route which will make your car journey shorter when you have reached land. If you are already wincing at the amount of driving your journey involves, consider taking a longer ferry trip, e.g. to Spain direct. Children are easier to occupy on a boat and there is space to walk them around. The less time spent in the close confines of a car, the better.

British ferry companies are trying hard to improve their services, and it is now standard to find video lounges, bars, self-service restaurants and shops on board. Several companies offer special children's clubs, however, facilities for young children are pretty thin on the ground.

Once on the ferry, it can be very stuffy inside and on a rough crossing it's the worst place to be. If you think you or your child may be seasick, sit quietly on deck; most ferries have partially covered seating outside. A pair of reins can be useful to keep an active toddler under control.

There are always ample toilet facilities on board, but it can be hard to find peaceful and congenial surroundings to feed or change your baby. On a journey of more than two hours where no mother's room is available, it's worth booking a cabin as a retreat from the hordes.

See also Section 3: Holiday transport.

Getting lost

Being involuntarily separated from your child is the ultimate nightmare, and it is an even more uncomfortable experience in an unfamiliar place. Since it is impossible to legislate against it, and since no amount of instruction or physical restraint will necessarily prevent it, here are the sorts of things to keep in mind so that when it happens, it is as painless as possible.

When you go on holiday always make sure that each child carries an identity badge with your local address and telephone number in the local language. An identity bracelet such as those used at swimming pools would be the most sensible device. Always try and teach your children your name and address, with as much information as they can sensibly absorb. Do not be afraid to talk to your children about getting lost and above all teach them not to panic. If children get lost in public places there are only two rules to remember, and to impart. Tell children, if lost, to stay in the same place and wait to be found. Do not walk around. Getting lost is no fun, but it comes to an end.

Homoeopathic medicine

There are many proprietary medicines for dealing with common holiday ailments. However, if you prefer to use 'natural' remedies, there are many homoeopathic preparations that can provide effective relief without fear of side-effects or overdoses. Many high street chemists have free leaflets listing remedies for common afflictions, from colds to travel sickness. If you want specific advice, go to a homoeopathic chemist where the pharmacist will suggest some holiday standbys.

Insurance

See Section 4: Holiday insurance.

Long-haul travel

If you're determined to get off the beaten track and go somewhere outside Europe you will more than likely go on a holiday which is known in the travel trade as 'long-haul'. Unfortunately when travelling with children this description is all too apt! The keys to successful or at least sane, long-haul travel are:

1 Allow yourself plenty of time to get to the airport, to check in and make connecting flights.
2 Adopt an easy-going, or what one experienced traveller called 'laid-back', attitude. There are bound to be delays, frustrations and worse, but if you're relaxed everything will be easier to cope with – an obvious point perhaps, but worth remembering.
3 Cut down drastically on hand luggage.

A few golden tips were passed on to us by a very experienced travelling mother: ask other people to help you if you need it. Most will be glad to do so, but probably wouldn't 'bother' you to offer first. Carry a bag of drinks, nibbles and hard fruit (apples are best) to tide the children over in case of delay, or if the food is long in coming on the plane. Don't take soft fruit – the pressure inside the cabin once you are in the air will cause it to bruise at three or four times the rate it would on the ground. We found this out when the pears we brought turned into a soggy mess an hour or so into the flight. Also bring a change of clothes, either to allow for accidents or just to have something clean to change the children into when you arrive. You might convince yourself that way that you and they feel fresh.

We explained in **Air travel** that it is very important a few days before travelling to book the bulkhead seats on the aircraft. They are the only seats where you can have a bassinet for the baby. Because they have more legroom and are just the other side of the service station you have the added benefits of more space and quick service for meals and drinks. The drawback is that you are smack in front of the screen and cannot watch the movie but this is not important for young children. If you are a family of four you'll have the whole row to yourself which allows you to feel you have your own little area.

In **Clothing** we advise you to dress the children in layers for travelling and this certainly applies to long-haul flights. The temperature inside the aircraft alternates between cold and very hot and you will want to be prepared. The airlines provide little blankets and pillows which are useful.

You will need one nappy per hour for the young ones and it is best to bring your own despite the bulk. Even if the right size is available from the flight attendants they may not be free to give it to you when you need it. Changing a baby of up to about one year works well enough in the bassinet. If your child can stand up, then changing him or her in the loo is not too bad given that the sink is near at hand. Some planes have a nappy-changing table in one of the toilets. The flight attendants will tell you which.

If you have more than one child it is a good idea to put labels on them with their name and address. The older ones are more than likely to explore the airport and getting lost is a possibility. (*See* **Getting lost**.) Also using reins for toddlers and attaching them to the pushchair or trolley can save a lot of chasing around.

Generally, it is easier travelling west than east in coping with time changes and consequent jet lag. We talked to many parents and there is no overall recommendation to make about when children should sleep and when they should not. It is best to let them drop off as and when they feel like it. It would be counterproductive to impose a schedule. A useful tip if you are returning to the UK from the east coast of the US: take the daytime flight if possible. It means you arrive at 7 or 8 p.m. here having 'missed' the day but as you'll be tired from travelling anyway you'll probably find everyone goes to sleep normally and wakes up feeling fine.

Medical emergencies

An emergency is by its very nature something for which you cannot prepare. You can however, with a few precautions, ensure that you are in the best possible position should your child fall ill or have an accident.

If you are visiting an EC country you will need a certificate called an E111 to benefit from the reciprocal medical arrangements available. The EC countries are: Belgium, Denmark, France, Germany, Gibraltar, Greece, Irish Republic, Italy, Luxembourg, Netherlands, Portugal and Spain. You can obtain the E111 from your local Post Office. It is valid in most cases for two years and only applies for urgently needed treatment. What is free and what you pay for in each country can be discovered by reading the leaflet SA30 available from the DoH.

There are countries outside the EC where some kind of reciprocal medical treatment arrangements are also in force, but often the cover given is not as comprehensive as in the UK.

The E111 is not a substitute for good travel insurance. Several things, such as repatriation, are not covered in any of the reciprocal arrangements, and in some countries injury from motor accidents may not be. Also, of course, in countries such as USA and Canada where all medical treatment has to be paid for you will need comprehensive insurance in any case. If you are in any doubt about what you need ask your travel agent, insurance company or insurance broker for advice.

Some parents we spoke to suggested that you take along your child's medical records just in case. You may feel happier doing this, but it is probably only really necessary if your child has a congenital condition or is taking a course of treatment.

Motorail

Driving across the Continent with baby and toddler in the back can be an exhausting experience, yet many families are understandably loth to dispense with the freedom and advantages of having their own car on the other side of the Channel. One compromise is to put the car on a train. From the Channel ports you can get as far south as Biarritz, Nice and Milan. From Paris you can reach Madrid or Lisbon; services through Belgium will take you southeast through western Germany to Salzburg, Villach or Ljubljana while from Gothenburg you can go north through Sweden to the Gulf of Bothnia. Norwegian state railways also run limited services.

Motorail isn't cheap. If you take only immediate and tangible costs into account it will certainly be cheaper to drive yourself. However, Motorail undoubtedly gets you there quicker and in a more relaxed frame of mind.

See also Section 3: Holiday transport.

Motoring

There is no doubt that self-drive holidays are among the most popular. You have independence to travel as little or as much as you like, and you can stop whenever it takes your fancy. As with other holidays, the secret of success is in the planning and you must be certain to comply with international regulations.

Essential documents are a valid driving licence and/or International Driving Permit (available from the AA), vehicle registration and proof of insurance. The latter can be in the form of an insurance certificate or a Green Card which is obtainable from your insurers. It is also essential to display a sticker on your car denoting nationality, to carry a warning triangle in case of breakdown, and to adjust your headlights to dip to the right (except for Ireland). Requirements about such things as seatbelts, wing mirrors, fire extinguishers and first-aid kits vary from country to country, but it is your responsibility to ensure that your vehicle complies with local laws. Note too that children under certain ages are not allowed to travel in front seats. The AA can provide details about this and all other requirements.

Before you set off, make sure your car is safe with good brakes and tyres (including the spare), and efficient windscreen wipers. It's also wise to carry a spare fan-belt and some bulbs. The AA offer excellent breakdown insurance throughout Europe – worth the expense for peace of mind. In many Eastern European countries there are cars which patrol the roads to assist motorists if they break down. The AA and RAC have all sorts of useful information about motoring in different countries, and some of the national tourist offices have special leaflets available.

There is much debate about the best way to organize a Continental motoring holiday from the UK, particularly if you have a long drive to make after your ferry crossing. Is it best to travel in daylight hours or overnight? Should you drive direct to your destination or make several stops en route? Each method has its advocates and we outline here those recommended to us by seasoned travellers with children.

Travel during daylight hours means that your children are awake and in need of distraction or entertainment for a big part of the day. This is fine if you have lots of energy and your children don't suffer from travel sickness. You'll also see more of the countryside and perhaps your children will find this adequate diversion. Frankly, we doubt it and can only recommend daylight travelling if your journey is fairly short.

Overnight travel has many champions. Those we know keep their children up late, allow them to 'help' with packing and then set off to the port with the children tucked up in sleeping bags. With any luck, they fall asleep in the car and sleep through the crossing. Normally the parents can have a snooze too. When the boat docks, the children can be carried to the car in their sleeping bags and get the rest of their ten hours. By the time they wake, it's possible to have driven several hundred miles. In order for this to work it's really essential to have two drivers.

To stop or not to stop – this depends on the length of your journey. If it's only a few hours to your destination then it's probably all right to drive direct. If it's more, then it's advisable to stop a few times en route. You and your co-driver need the rest, your children appreciate the break from the car and if you choose your stops carefully, they can be highlights of the holiday rather than hardships.

See also Section 3: Holiday transport.

Nappies

The invention of the disposable nappy has made every parent's life easier, and this is especially so when travelling. Despite the inconvenience of bulk, it is worth carrying enough to see you through your journey as nothing can be worse than running out at a crucial moment. Although the airlines may have nappies, you can't count on it. Railway stations, ferry ports and coach stations are unlikely to stock them. A rule of thumb is one nappy per hour of your journey plus a few to allow for delays. Take a roll of nappy sacks (plastic bags) to dispose of the soiled nappies. Nothing is more unpleasant in toilet cubicles than used nappies just thrown in with the paper towels. Also the bags are useful for uneaten food.

Disposable nappies can be bought almost anywhere now, judging by what our writers tell us in Section 1 (Choosing your destination) but they may be pricey.

One parent wrote to us to suggest that if you take a large pack of nappies with you, treat it as a piece of luggage. Put a label on the nappy pack (sixty–seventy should last a fortnight) and throw into the aircraft hold. Remember that although on beach holidays the children will probably go without nappies by day, because of the amount they drink they will consequently be changed more in the evenings.

Necessities

Parents will have their own views on what is essential to take on a journey and what we offer below are only suggestions gleaned from our experience and research.

For babies: A large bag with outside pockets is the best sort. Put the baby's toys in the pockets. Those made of canvas or nylon are strong and can stand up to the rugged treatment they will receive. The cotton bags lined with plastic for nappy-changing are very useful, but small, so you should bring this packed into the larger bag. It is not always possible, but try to repack the bag in the same order so you don't have to rummage through the whole thing every time you want a tissue. Plastic bags of the small binliner sort are a must. You can put soiled nappies, food or bottles in them. If you are bottle-feeding the baby and travelling by air, it is best to bring the bottles already made up as the flight attendants can warm them for you. If travelling by rail or car it may be better to bring the separate ingredients and count on getting boiled water in the restaurant car or when you stop on the road. Always bring several teats, two pairs of plastic pants and a couple of muslin or towelling nappies. You should also take washing things, a wet flannel, a small towel, kitchen roll, tissues or toilet roll and perhaps also moist towelettes in individual packets. Sterilizing tablets, baby wipes, lotion, cotton wool, several bibs and a supply of nappies should see you through.

Some companies charge an exorbitant price for hiring cots, in which case you may like to consider taking a travel cot with you.

For toddlers: If you bought a strong bag when the baby was small it should still be with you for this stage and can hold the following: jars of food (preferably something the child can eat cold if need be, such as fruit), several plastic spoons, bibs, cartons of juice with the straw attached (they all seem to love these), plastic cup with spout, plastic bags, a few favourite (small) toys and books. If the child is in nappies you will still have your nappy-changing bag and this is a must. We found the kind of collapsible seat which you screw on to any table top invaluable for children up to the age of 2. It lies flat in your case so takes up little space and its versatility means you can go out and eat with your child almost anywhere. A potty could be useful, depending on what stage of training your child has reached. At least one change of clothes should be brought for each child in nappies, and don't forget the socks!

Open spaces, parks and shopping

Wherever you travel to you will be interested in what facilities the country provides for children. These vary enormously. Most cities and towns have parks of some sort or other and a lot of them have play areas reserved for children. We were surprised to find in Paris that there were lots of little parks with swings, slides and sandpits, often in a small area of the churchyard.

Parents become adept at sniffing out areas where they will be welcome with their children and where they can sit for a relaxing half-hour. In that ultimate consumer society, Japan, it is made easy for you to shop in their department stores because you can often leave your child under surpervised care while you go and spend money!

If you are in a city and the weather is inclement, it's worth knowing that many museums have 'hands on' exhibitions where the whole idea is for children to touch and handle the displays. This is a welcome development and could pass many a happy hour.

Passports

It's probably a good idea to get your children their own passports rather than adding them to yours or your partner's; if one of you has to fly home with an injured child this could cause chaos. Acquire a passport for your child as soon as you can to avoid last-minute panics. But if you have left things to the last minute you can always get a British Visitor's Passport over the counter at the post office on production of two photographs. The snag with this is that under-8s can't have their own – if travelling solo they have to have a proper passport.

Preparation

There are two schools of thought on this subject. One has it that you should all go to bed early, with the children relaxed and calmly ready for their journey. The other has it that you should keep them up later than usual, packing and getting excited. Set off early and don't let them sleep until you are on your way – in the plane, car or train. This may work with the over-3s if you are driving from say, the UK to the south of France, but if you are going by air it is probably best to be as rested as possible when you start your day. There is normally delay, lots of walking with hand luggage and standing in queues.

The more you can explain and involve your children in the holiday the more they will enjoy it and get from it – and you likewise. For the child who can read a bit lots of *Junior Guides* and so on are available for foreign countries. You can also buy delightful picture-book foreign dictionaries. Even if your child isn't at the reading stage, you can still teach a few basic words in the appropriate foreign language – please, thank you, hello, good-bye. The locals will love it and your children will feel wonderfully clever.

In general, we cannot overstress the importance of research. It may have been fun going to new places trusting to serendipity and your initiative when you were childless; with a baby it really isn't on. Write to the embassy and the tourist office of the places you're going to; read the books listed at the end of the country descriptions; browse through the travel section of your local bookshop or library; ask the AA or RAC for motoring advice.

Rail travel

By enjoying the special rates available using a Family Railcard, a Young Person's Railcard or a Network Southeast Card, British Rail becomes a viable means of family transport. Most children love trains and will positively relish a shortish rail journey. Make sure you take something to keep them amused (*see* Section 4), a small supply of non-messy food, and some cartoned drinks.

Any kind of substantial journey (either in length or number of changes required) should, wherever possible, be split by an adequate break to avoid boredom – with any luck, each part of the journey will then seem like an adventure. When travelling overnight, it is advisable to book a sleeper – contact British Rail for details of current offers (eg 'buy one get one free'). Children usually find sleepers very exciting, and you never know, you might even get some sleep!

Reservations around bank holidays are essential (and often free of charge). The boundaries of such holiday periods are often wider than you think and free reservations may begin several days earlier than you expect – check before booking for accurate details.

The introduction to Rail Travel and listings given in Section 2 will help you to plan a train journey while abroad. We have listed below the main stations of England, Scotland and Wales. On-station toilet facilities are improving and many sites have baby changing rooms and most have disabled toilets. We would advise you to contact the relevant station before departure.

Sightseeing

The most important point to bear in mind about sightseeing is not to overdo it and put your children off for life. Make the most of the baby stage; it gets harder later. It ought to go without saying that you are asking for trouble if your haul your 3-year-old round St Peter's or the Uffizi for hours on end – or get resentful if you find you can't. It can be frustrating for adults in a foreign country where they've never been before to find they can't go off and see the sights as they used to; if this applies to you then one solution is to organize a holiday with another family so that you can take it in turns child-minding. You and your partner can always take it in turns too.

This is not to say that you can't get away with a limited amount of sightseeing – with appropriate promises of ice creams or treats and some common sense. Don't drag the children round a city on a blazing hot day when they would rather be on the beach or in the pool. On the other hand, in reasonable weather conditions, most children will enjoy ruins – though it may be difficult if they're not allowed to climb on them.

If your child is old enough you can get him or her involved with suitably modified explanations of what you're going to look at and why – though don't be too surprised if your child's enthusiasm wanes when you get to your destination. Above all, keep it short; it's much better to finish the experience on a high when you're all enjoying yourselves.

But even if you decide sightseeing is not for your family you will want your children to get something more out of a holiday abroad than a suntan/sunburn. With a little help and guidance from you older children will get a lot of interest and fun out of noticing the differences between abroad and back home – differences ranging from houses, streets, dress, shops, vegetation to language, food, ways of behaving etc. Learning to notice such things will be of a great deal more educational value to your child than a formal museum, as well as being part of what 'going abroad' is all about.

Skiing

Ski holidays are becoming increasingly popular with families. As a result, many resorts throughout Europe are adding services especially for young children.

Most resorts now have plenty of self-catering accommodation close to the ski slopes and many companies offer baby-sitting and child-minding services so that parents are able to ski alone for at least a few hours. It is very common to find child-minding in the form of kindergartens. They offer organized activities, supervised lunches and children's ski schools, usually for those aged four and above. The under-4s are normally looked after in nurseries, but where this service is not available, the local tourist office will help you find a baby-sitter.

For those worried about the expense of kitting out the children, it's possible to hire ski wear from shops like Moss Bros. The smallest sizes fit children around four years old. Organizing a skiing holiday with children under five can be very difficult. It may be possible to arrange for supervision with a local family via your holiday company and some travel companies are beginning to organize full-time supervision for infants while adults ski. Ask your travel agent's advice before booking.

Travel-cots

If you plan to travel widely with your small children to places where cots may not be provided, or are very expensive, it may be worth buying a travel-cot. A wide variety is available, the lightest weighing about 5kg and ranging upwards in price from about £50. They all fold into reasonably compact bundles and will fit into a car boot (or behind the back seat). Some are sturdy enough to take a child up to four and all will accommodate a 20-month-old. For some families a travel-cot might even be a more sensible buy than an ordinary one – particularly as some have mesh sides and can double as a playpen (another plus on holiday).

Before you splash out, though, you need to decide how often you're going to use the cot and what your priorities are. The sturdier ones that will last your child until he or she goes into a normal bed are, of course, heavier and more expensive than the lighter, flimsier ones which are basically a piece of material suspended on a frame. Some mattresses are more substantial than others and you might want to choose a travel-cot that would take a normal cot mattress for everyday use.

Walking

Walking is fun, healthy and a perfect way to spend a holiday for all the family. There are several companies offering walking holidays, but some do not permit very young children. However, there's nothing to stop you organizing your own rambles. Just make sure they're within your children's ability or you'll end up carrying them. About two miles is usually manageable. Make sure you have a good map, preferably one prepared by Ordnance Survey, as paths are not always clearly signposted. Try to make frequent stops on the route – not only for food and drink breaks – but to look at the flowers, animals and trees. Take some books along to help identify flora and fauna. Children love to keep records, so get them to note down what they see and perhaps draw some pictures. If you can, pick a route that has a hill, or better yet, a stile or two to climb. Some farms in Britain have 'farm trails' – walks around the farm where you and your children can see the animals.

Remember that stout shoes and sensible clothing are necessary for a serious ramble, but don't make it hard work or you'll put your children off walking for life. A good backpack with well-padded hip belt is essential for carrying toddlers; it's worth paying more to get a good one as the cheap ones are very uncomfortable to wear.

Section 6

Medical know-how

Medical know-how

The illnesses and health problems that occur most often on holiday tend mainly to be minor ones. Every effort to prevent and anticipate them is well worthwhile however; a minor illness affecting just one family member – especially if that member is a child – can spoil enjoyment quite effectively for the whole family. More serious illnesses also occur, and good medical treatment is sometimes difficult to find.

Almost all of the likely problems are preventable, and here is a quick guide to some of them.

Accidents

Accidents are by far the most common cause of death or serious injury in travellers of all ages. Much of the problem is that most people lower their guard on holiday, taking risks they would never consider at home. Most accidents occur on the roads, and there are almost always important preventable factors. Common sense precautions that are usually observed at home are often ignored abroad. These include wearing seatbelts, observing speed limits (and not driving too fast when there aren't any), not drinking and driving, avoiding driving at night or when you are tired or suffering from jet lag, and looking both ways before crossing the road! Child seats and restraints should also be used, and children kept on the rear seat, even in countries where there is no legal compulsion to do so.

In an accident that results in injury to anyone else, prison custody for the driver is the rule in many countries, until the case is eventually heard – no way to begin a family holiday. Always make sure that you are adequately insured for driving abroad. Car insurance, and insurance for rented cars, should cover bail bonds, fines and legal costs. It can be difficult to raise the money for such expenses at short notice from the inside of a prison cell. The risk of illness in prison may also be high.

Not the least reason for doing everything possible to prevent accidents abroad is that emergency services and medical care of a good standard may be difficult to find, especially if you are in an island resort, or in a remote place.

Important potential hazards to children abroad include insecure balconies and balustrades; hotel lifts of the continental type, in which the lift cage has only three sides – enabling hands and clothing to become trapped against the lift shaft; and unsafe electrical and gas appliances in rented accommodation.

Air travel

Travel sickness should be prevented in advance, especially if you know that your children are prone to it (see Travel Sickness). Pressure changes make the ears 'pop' during ascent, and may cause earache during descent. Babies who cry on descent should be fed, or given a dummy to suck, and older children in discomfort should be told to pinch their noses and blow, or to suck sweets. Small babies usually tolerate air travel well, and are easiest to look after when breastfed; toddlers do not like being restrained, and easily become restless and irritable; on the longest journeys it may be worth travelling with a mild sedative like Phenergan in reserve (see Antihistamines). Older children may get bored – so travel prepared.

Allergy

Anyone who has had a serious allergic reaction in the past should travel with everything they might need in an emergency, as well as appropriate identification (such as a Medic-Alert necklace or bracelet, available from the Medic-Alert Foundation, 17 Bridge Wharf, 156 Caledonian Road, London N1 9UU. Tel: (071) 833 3034).

Animal bites

Worldwide, dogs are the worst culprits. In the USA, over a million people are bitten each year, badly enough to need hospital treatment; in the UK there are about 250,000 cases, and abroad, on holiday, the risks are much higher. Rabies is a hazard in many countries, and prompt medical treatment may be more difficult to obtain.

All animal bites – and human bites – carry a high risk of infection. Bites should therefore be scrubbed with soap or detergent in running water for at least five minutes, and any dirt or debris carefully removed. An antiseptic capable of killing viruses – such as iodine or alcohol (whisky and gin are also suitable for this purpose!) – should then be applied. Mercurochrome and other brightly coloured antiseptic dyes, popular in some countries, and hydrogen peroxide, are not suitable. At hospital, a medical attendant should further cleanse the wound; stitches are best avoided if possible, and antibiotics are often a valuable precaution. Children who have received their routine immunizations will be immune to tetanus; adults need a booster dose every ten years to maintain immunity. A booster dose of tetanus toxoid should be given after injury; severe wounds in a previously unimmunized person require additional treatment with tetanus immunoglobulin, which may not be readily available in smaller hospitals. Precautions against rabies may also need to be considered (see Rabies).

Antibiotics

If you are prescribed antibiotics abroad, make sure that the full course is completed. If your child is allergic to penicillin (or any other drug) make sure that any doctor consulted abroad understands this. Tetracycline antibiotics should never be given to children, and should not be taken by pregnant women.

Antihistamines

Antihistamines are used to treat mild allergy and itching from widespread insect bites or stings. Antihistamine creams and ointments should be avoided – sensitivity to them can occur following exposure to strong sunlight. Tablets are preferable; Piriton is suitable for children and adults, but tends to cause drowsiness; newer antihistamines like Triludan do not, and can be used by adults and children over six. Another antihistamine, Phenergan, is often used solely for its soporific effect, as a mild sedative for small children, and is occasionally useful for fractious children on long and tedious journeys.

Bee stings

Stings should always be scraped out; grasping them with tweezers will inject more venom. Ice provides valuable relief for pain and swelling, with paracetamol for pain, if necessary.

Blisters

Blisters are a common problem on holiday, especially on an activity holiday. It is best to break in any new footwear gradually, well before going away.

Protect the skin of the feet with talcum powder, especially if long walks are planned. Clean and tape over smaller blisters; larger blisters should be snipped open to drain, cleaned with antiseptic, and covered with a non-adherent dressing such as Melolin.

Blood transfusion

The AIDS risks have received much publicity, but are not the only hazard of blood transfusion abroad. Other risks include hepatitis A, hepatitis B, syphilis and malaria. Furthermore, transfusion with badly matched blood causes severe reactions and can sometimes be fatal; it also has important long-term consequences in girls and women, leading to serious antibody reactions between mother and foetus in a future pregnancy; allergic reactions and febrile reactions may also occur, and the risk of these is greater when storage conditions are poor.

In most of western Europe, North America, Japan and Australasia, all donated blood is now screened for HIV antibodies and prospective blood donors are questioned carefully about their lifestyle and risk factors for AIDS. Elsewhere, however, adequate facilities for screening donated blood and selecting donors are the exception rather than the rule, and the risks from blood transfusion may be high.

Blood transfusion should be given only when medically essential. The risks from an unscreened blood transfusion must, however, be put in perspective, and balanced against the more immediate risks of not having a transfusion when there has been serious blood loss and there is a clear medical need. It is foolhardy to risk death from blood loss by refusing transfusion; most people who need transfusion in an emergency may find they in fact have little choice in the matter.

Accidents are the commonest reason for travellers to need a blood transfusion, and avoiding accidents is the most effective measure that any traveller can take to avoid a blood transfusion.

Breastfeeding

Breastfeeding on the move is in many ways more convenient than bottle feeding. In a hot climate, nursing mothers should make sure that they have plenty of extra fluid to drink. Additional fluid is not generally necessary for the baby, unless there is fluid loss from diarrhoea, vomiting, or a fever; but it is safest to travel with a bottle for use if needed, and to have already introduced a bottle for occasional feeds at home before travel.

Cold climates

Children lose heat relatively much faster than adults, and need to be wrapped up well in cold climates. Clothing should be well-fitting, and in several layers. Gloves and footwear in particular should be of good quality. It is not possible to acclimatize to cold in the same way as it is to heat.

Colds

This is one of the commonest ailments abroad. The usual symptomatic remedies are helpful, and a decongestant nasal spray should be used to prevent discomfort from pressure changes during air travel, though anyone with a really severe cold should not fly.

Contraception

This is a subject that is almost always neglected both by travellers themselves and by their doctors. Diarrhoea reduces the absorption of the Pill. Long flights across time zones often result in doses that are missed or much delayed. The margin of safety, especially with low-dose Pills, is small, and protection is easily lost. An alternative method should be taken in reserve.

Constipation

Dehydration, readjustment of bowel habits after crossing time zones, a change of diet and initial reluctance to use dirty toilets, can each contribute to this problem, which is surprisingly common. Plenty to drink, and a high fibre diet, are preferable to medication; it may be worth travelling with a small supply of natural bran, or a child's favourite variety of breakfast cereal.

Creeping eruption

Creeping eruption is caused by hookworm larvae of dogs, cats, and other animals, which burrow into human skin by mistake. The larvae migrate aimlessly under the skin, producing a painful, raised, itchy, linear reaction. The condition is acquired by contact with soil or sand contaminated with animal excrement. Beaches in Africa, Asia and the Caribbean are often contaminated; also North America, along the Atlantic and Gulf coasts, as I recently discovered to my own cost . . . Sand below the high water mark is safe; everywhere else wear shoes or sandals and make sure children do the same. If the rash occurs, freezing the skin with ethyl chloride is a simple and effective remedy, but medical advice should be sought.

Dehydration

Dehydration can occur rapidly in small children, especially in extreme heat, or if there is diarrhoea, vomiting, or a fever. Almost all cases respond promptly to treatment with oral rehydration solutions (see below); dehydration can be prevented by ensuring an adequate fluid intake in the heat.

Diarrhoea

Although diarrhoea is the commonest of all holiday ailments and is the medical problem most clearly associated with travel in most people's minds, it is not inevitable and can be prevented by careful selection and preparation of food and drink (see Food Safety and Water).

If diarrhoea occurs, the most important aspect of treatment is replacement of lost fluid and salt – especially in children, and especially in warm climates. The best way of doing this is by drinking as much as possible of a solution containing salt and sugar (sugar promotes rapid absorption of the salt). The ingredients are available in correct proportion in sachets (such as Dioralyte, from any pharmacy) that can be made up with water. A child who is vomiting should still be given sips of this solution – and will usually absorb most of it.

Most cases of diarrhoea clear up without specific treatment within three days, though that is sometimes of little consolation. One suitable symptomatic remedy for adults and older children is Arret, available without a prescription in both capsule and liquid form. Diarrhoea with fever or blood (*see* Dysentery) requires skilled medical attention.

Antibiotics are often prescribed inappropriately for diarrhoea; antibiotics can themselves cause diarrhoea, and many cases persisting on return home are due to just this. Antibiotics should generally be reserved for cases in which fever is present, or a serious infection is suspected.

Dogs

Strange dogs should *never* be handled abroad. In many countries, they are in any case not used to being treated as pets, and dogs should never be left alone with small children. Apart from the risk of rabies that is present in many countries, dogs are able to transmit around eighty different diseases to human beings, all of which are more common in hot countries.

Drugs and medicines abroad

Take with you an ample supply of any medication anyone in your family is already using, or is likely to need. Keep all medicines out of reach of children, and preferably in childproof containers. To avoid possible customs problems, it is best if all medicines are kept in their original containers and are clearly labelled. Carry a prescription for anything unusual. Prescribing habits vary widely between different countries, especially within Europe. These variations apply to the form in which a drug is used – injections and suppositories are used in many countries in preference to oral treatment; and also to the particular choice of drug or preparation. In many countries, for example, steroid medications (cortisone-like drugs) are liberally over-used – in tablets, eye-drops and creams. These are powerful drugs that should not be abused, and it is not always easy for a lay person to identify them easily from their names alone. It is best to be cautious of any medication you are given, especially if it contains numerous different active ingredients; always ask for a detailed explanation of anything you are prescribed, if you are worried. Drugs that have been incorrectly stored, or have simply expired, may lose their effectiveness; some drugs, like tetracyclines, actually become toxic when out-of-date. All medications produced by reputable manufacturers carry an expiry date, and this should always be checked before use.

Dysentery

Dysentery is diarrhoea with blood and/or a fever, and always requires skilled medical attention. Until skilled attention is found, take careful steps to prevent dehydration.

Earache

Ear infections are relatively more frequent on holiday – perhaps from more frequent swimming. Ears should be towel-dried after swimming – not probed with cotton buds. Antibiotic treatment is usually necessary if it is prolonged or associated with fever, and swimming should then be avoided until after recovery.

Fever

Most of the causes of fever abroad are the usual causes of fever at home, such as colds, respiratory infections and tonsilitis. When travelling in any country with malaria, the possibility of malaria should be considered in any case of fever. Fever with diarrhoea suggests dysentery, and fever with neck stiffness, headache and inability to tolerate bright light, suggests the possibility of meningitis.

First-aid kits

Consider including the following items:

- diarrhoea remedy, such as Arret
- oral rehydration sachets, such as Dioralyte
- travel-sickness remedy, such as Kwells or Stugeron
- painkiller, such as paracetamol
- mild sedative, such as Phenergan syrup
- antimalarial medication as appropriate
- calamine lotion and antihistamines for insect bites
- insect repellents
- sun screen
- lip cream
- cold remedy
- antiseptic for cuts and grazes
- sticky plasters, steri-strips and wound dressings, bandages
- thermometer
- sterile needles/syringes, if appropriate
- water purification supplies

Food safety

Diarrhoea may cause anything from embarrassment and inconvenience to misery that may wreak havoc on your travel and business plans; other diseases spread by poor hygiene – and prevented by similar precautions – include dysentery, giardiasis, hepatitis A, typhoid, polio and parasitic infestations.

Careful choice and preparation of food offer the best protection; unfortunately, contaminated food can seem most appetizing, and the urge to eat what's available and what you have paid for when you are hungry can be irresistible. High risk foods include:

- raw or inadequately cooked shellfish or seafood
- raw salads and fruit that have not been thoroughly washed in clean water, or that you cannot peel yourself
- food that has required intricate preparation with much handling
- food that has been stored and reheated after cooking
- food left out in warm temperatures – such as hotel buffet lunches in most Mediterranean resorts – bacteria multiply fast under such conditions
- food on which flies may have settled

Eating safely means that you won't always be able to eat when, where and what you want. Food that has been freshly and thoroughly cooked is safe.

Heat stroke

Children generally adapt well to the heat. Adaptation is improved by drinking plenty, well beyond the point of thirst-quenching; it is important to drink enough to keep the urine a consistently pale colour. Heat stroke results in failure of the body's heat control mechanisms; sweating diminishes, and body temperature rises; headache and delirium also occur; in this situation, prompt treatment is essential. The priority is to lower body temperature; remove clothing and cover the victim with a wet bed-sheet, while arranging transfer to hospital.

Hepatitis

Hepatitis is a viral infection of the liver. The illness itself consists of fever, chills, headache, followed by nausea, vomiting and jaundice; the liver becomes acutely inflamed with tenderness or pain in the upper-right portion of the abdomen. The illness can be relatively minor in some cases; in others, it may lead to liver failure, coma and death.

Hepatitis A is a disease of poor hygiene. It is commonest outside northern Europe, North America, Australasia and Japan, in warm climates. It is spread mainly by contaminated food and water; some foods, such as prawns and shellfish, are particularly likely to harbour the virus, so careful attention to food hygiene reduces the risk of catching the disease. Gammaglobulin injection provides protection for four to six months. Anyone who has spent long periods abroad may have acquired natural immunity to hepatitis A, and therefore may not need repeated injections of gammaglobulin for travel; a simple blood test for this can now spare those already immune from further unnecessary gammaglobulin injections.

Hepatitis B is spread in almost exactly the same ways as AIDS – sexually, by blood, and contaminated needles and medical or dental instruments. A safe vaccine is available, though it is an expensive one. It is currently used mainly to protect those at some special occupational risk, and people going abroad to live in high-risk areas.

Ice

Ice is only as safe as the water it is made from, and freezing itself does not kill germs. In places where the water supply is unsafe, don't use ice.

Injections

Hepatitis and AIDS are important hazards that are referred to elsewhere. Hepatitis B has occurred in numerous travellers who have received injections with contaminated needles and syringes, and the AIDS risk from this route of infection is also high. Disposable pre-sterilized needles and syringes are not widely available in many poor countries. If you are going abroad to live in one, or expect that you will need any other medication by injection (including dental anaesthesia) while you are away, you should either satisfy yourself that any needle or syringe used has been adequately sterilized (i.e. boiled for at least ten minutes) or you should take your own supply. Remember that if you have an accident when driving abroad, a blood test for alcohol may be compulsory – in Turkey, for example.

In the UK, needles and syringes are available without prescription at the discretion of a pharmacist. They are also available in kits that include other medical items, from MASTA, London School of Hygiene & Tropical Medicine, Keppel St, London WC1E 7HT Tel: (071) 631 4408; and from SAFA, 59 Hill St, Liverpool L8 5SA Tel: 051-708 0397. In the USA and most other countries, a prescription is necessary and a prescription should always be carried when travelling.

Insects

The sheer nuisance value of mosquito bites should alone be incentive enough to take careful precautions against them, but mosquitoes also spread disease: growing drug-resistance of the malarial parasite has lead to increased awareness of the importance of precautions against insect bites in preventing malaria, and careful precautions reduce the likelihood of insect-borne disease by a factor of ten.

Speculation about mosquitoes bearing AIDS is ill-founded, but malaria is by no means the only risk; there are many unpleasant insect-borne viral diseases for which there is neither a vaccine nor drug treatment, such as dengue fever, which occurs in Southeast Asia, Africa, the Caribbean and Latin America.

Mosquitoes bite especially around dusk, when it is important to wear long sleeves, trousers and socks. Apply a chemical repellent to clothes and exposed skin; diethyl-toluamide (DEET), hexane-diol and citronella are active ingredients to look for, available in sticks, sprays and gels. DEET can be used to impregnate clothing. In your hotel room use a spray insecticide early in the evening; if the room is not screened or air-conditioned, a smouldering mosquito coil, or a 'vaporizing mat' – its electronic equivalent – will release an otherwise harmless insecticidal vapour through the night.

Insurance

British travellers off to EC countries can obtain form E111 from their local Post Office, giving proof of entitlement to free or reduced-cost medical care; apply at least one month before you intend to travel. Additional medical insurance is always advisable, however, and should cover the cost of emergency repatriation by air ambulance if necessary. *See* Holiday Insurance, Section 4.

Jellyfish

Jellyfish are common in the Mediterranean, and it is always worth asking local people if there are jellyfish about when you arrive at a resort. The sting produces a prickly sensation at first, going on to a raised, itchy, red rash that lasts four or five days. Fragments of tentacles should be removed promptly, and vinegar or lemon juice applied to inactivate any of the stinging capsules that may be left. Ice, calamine lotion, antihistamine tablets and paracetamol can be used to provide relief. Some types of jellyfish are dangerous and all should be treated with extreme caution whenever possible.

Jet lag

Children generally adjust much faster than adults, and jet lag is not usually a problem for them. A mild sedative such as Phenergan may help re-establish sleep patterns.

Malaria

Last year over 2250 people came back to Britain with malaria, the disease that kills more people worldwide than any other. Malaria occurs in over 100 countries and is spreading; so is resistance of the malarial parasite to conventional preventive drugs – chloroquine in particular. Increasing experience with alternative drugs revealed unacceptably high risks of toxic effects, so in countries with chloroquine-resistant malaria the most effective drugs now cannot be used.

All travellers to countries with malaria should take antimalarial tablets for the duration of their trip and for at least four weeks after leaving a malarial area; antimalarial drugs are not foolproof, and precautions against insect bites are also important to reduce the risk (see Insects). Malaria causes rapid deterioration and is potentially fatal; travellers to areas with resistant malaria should also now carry a treatment dose of antimalarial tablets with them at all times and take it if they develop a fever in circumstances where prompt medical care is not available. Advice on appropriate antimalarial drugs for the country you intend to visit is available from immunization centres; the Malaria Reference Laboratory (Tel: (071) 636 7921) can advise in case of particular difficulty.

Prickly heat

True prickly heat is a sweat rash occurring on the sweatier parts of the body. The rash consists of tiny blisters on sore, reddened, mildly inflamed skin; prevent with frequent showers and keeping the skin clean and dry. Treat with calamine lotion.

Rabies

Dog bites are common even in the UK; abroad, the risks are higher. Rabies occurs in most parts of the world and is uniformly fatal. Only the following areas are free of rabies at present: Britain and Ireland, Scandinavian countries (including Iceland and Finland but excluding Denmark), peninsular Malaysia, Taiwan, Japan, the Pacific Islands, Antarctica, Australia and New Zealand. It is spread by licks and scratches from infected animals, not just bites, so never handle a stray or strange dog abroad.

In the event of a bite, prompt and thorough cleansing of the wound is most important; if rabies is a risk, make sure that safe new Human Diploid Cell vaccine is given – it is expensive and not always widely available; some British and US embassies abroad hold stocks, or may help you obtain it. Otherwise, return home at once – most insurance companies accept this as a valid reason for curtailing a holiday. Anyone travelling to remote places in countries with rabies should consider prior vaccination, which is safe and effective.

Return home

A feature of malaria is a delay in onset of symptoms. In the UK many tragic deaths of returning travellers have occurred because the early symptoms of malaria – fever and headache – have been mistaken with those of flu; it is common for symptoms of malaria to be delayed perhaps for several weeks after return home. Symptoms of many other infectious and tropical diseases may also be much delayed, making diagnosis out of context much more difficult. If you or your children become unwell on return home, make sure your doctor knows that you have been abroad.

Stitches

Make sure that all needles to be used are sterile. Clean, gaping wounds can sometimes be held together with Steristrips, or similar adhesive tapes.

Sun

The British lobster-look abroad is a familiar sight, but those who are more at home with the sun treat it with greater respect. Acute sunburn not only adds to the long term risk of cancer and premature ageing, but is a miserable way to begin a holiday and results in a blotchy, uneven tan. Anyone not interested in a tan should cover up and use a high protection factor sunscreen; if you want a tan, achieve it very slowly. Sunburn is especially cruel for children, whose skin is easily damaged, and to whom a tan is anyway of little interest; waterproof sunscreens, now produced by almost all of the leading manufacturers, are more likely to provide effective protection.

Sunburn

Calamine lotion soothes affected areas and mild painkillers are often helpful. Stay out of the sun, or use a total block sunscreen, until the skin has healed.

Teeth

Make sure that all current dental problems have been attended to before departure with a dental check-up for all the family. Sugary drinks, sweets and ices become all the more tempting on holiday and in the heat. This temptation should be resisted – teeth are as vulnerable to decay abroad as they are at home. Make sure that water used for brushing teeth is safe – use bottled, boiled, or purified water if there is any doubt as to the safety of the local tap water.

Travel sickness

Many different brands of travel sickness remedy are available, including Kwells and Stugeron, the latter causing less drowsiness than most other brands. The important thing with any remedy is to take it *before* travel – once vomiting starts, it will be useless.

Vaccinations

These fall into two categories, those that are an essential requirement of entry – usually to protect the country you intend to visit – and those recommended for your own personal protection. The full DOH recommendations are given in DOH leaflet SA35 (available free from social security offices and some travel agents; GPs and immunization centres can also provide this information, as well as the vaccinations themselves, although GPs do not provide vaccination against yellow fever). A full course of vaccinations for travel outside Europe may take up to two months, so it is as well to plan ahead.

Most European countries have no formal vaccination requirements. Protection against typhoid, polio, and possibly hepatitis, is often advised for travellers to the Mediterranean. And all travellers should be immunized against tetanus.

Vaccinations provide valuable protection against important diseases, but are no substitute for other health precautions and are never enough on their own.

Vomiting

Specific treatment of vomiting in food-poisoning is not generally advised or considered necessary unless symptoms are so severe that skilled medical treatment is required.

Water

When there is the least doubt about hygiene, only drink water that you know is safe. Don't drink tap water or brush your teeth with it, stick to bottled or canned drinks – well-known brands are safe, but have bottled mineral waters opened in your presence and regard all ice as unsafe. Alcohol does not sterilize a drink!

Purify water by boiling, or with chlorine or iodine. For the latter, add four drops of 2 per cent tincture of iodine (a standard solution, available from any pharmacy) to each litre of water and allow to stand for 20 minutes before drinking.

Further reading

How to Stay Healthy Abroad, Dr Richard Darwood (Oxford University Press)
Traveller's Medical Resource, William W. Forgey, MD (ICS Books, Inc.)
Traveller's Self Care Manual, William W. Forgey, MD (ICS Books, Inc.)
Holidays and Travel Abroad – A Guide for Disabled People '90–'91 (RADAR)
The Traveller's Health Guide, Dr Anthony C. Turner (Roger Lascelles)
Nothing Ventured – Disabled People Travel the World (A Rough Guide special) (Harrap-Columbus)

 **Section 7**

Information and addresses

Useful addresses

AA: see Automobile Association

Aberdeen Airport
Dyce
Aberdeen AB2 0DU
Tel: (0224) 722331

ABTA: see Association of British Travel Agents

Ainsworth's Homoeopathic Pharmacy
(Mail-order service available)
38 New Cavendish Street
London W1M 7LH
Tel: (071) 935 5330

Air Travel Advisory Bureau
41–45 Goswell Road
London EC1V 7EH
Tel: (071) 636 5000
(See also Civil Aviation Authority)

Allergy Identification Tags: see Medic-Alert Foundation

Association of Breastfeeding Mothers
26 Homeshore Close
London SE26 4TH
Tel: (081) 778 4769
(See also La Leche League)

Association of British Travel Agents
(ABTA)
55–57 Newman Street
London W1P 4AH
Tel: (071) 637 2444
(See also Civil Aviation Authority)

Association of Pleasure Craft Operators
35a High Street
Newport
Shropshire TF10 8JW
Tel: (0952) 813572

Association Régionale de Tourisme Equestre (Riding in France)
Domaine du Volcelest
33830 Joué Belin–Beliet
Tel: 010 33 56 88 02 68

Automobile Association (AA)
Fanum House
Basingstoke
Hants RG21 2EA
Tel: (0256) 20123

Belfast Airport
Belfast
BT29 4AB
Tel: (08494) 22888

Birmingham International Airport
Birmingham
B26 3QJ
Tel: (021) 767 7145/6

Boats and Cruisers: see Association of Pleasure Craft Operators; British Waterways Board; Inland Waterways Association

Books, cassettes and videos: see CFL; Books for Children; The Red House; Stanfords; Tapeworm; Travel Bookshop Ltd

Books for Children
(Children's Bookshop)
97 Wandsworth Bridge Road
Fulham
London SW6 2TD
Tel: (071) 384 1821

Bournemouth Airport
Christchurch
Dorset BH23 6SE
Tel: (0202) 593939

Breastfeeding: see Association of Breastfeeding Mothers; and La Leche League

Information and addresses

Bristol Airport
Bristol
BS19 3DY
Tel: (0275) 474444

British Airways Immunisation Unit
156 Regent Street
London W1R 7HG
Tel: (071) 439 9584
(*See also* Thomas Cook Vaccination Centre)

British Consulates *see* CFL

British Rail (Motorail Reservations)
Tel: (0345) 090700
Fax: (031) 557 4795

British Waterways Board
Willow Grange
Church Road
Watford WD1 3QA
Tel: (0923) 226422
(*See also* Inland Waterways Association)

Bus and Coach Services (UK): *see* National Express

CAA: *see* Civil Aviation Authority

Cadw Welsh Historic Monuments
(For information on historic monuments of Wales)
Admissions Officer
Brunel House
2 Fitzalan Road
Cardiff CF2 1UY
Tel: (0222) 465511

Camping and Caravanning Club
Greenfields House
Westwood Way
Coventry CV4 8JH
Tel: (0203) 694 995
(*See also* National Caravan Council)

Cardiff–Wales Airport
Nr Cardiff
South Glamorgan
CF6 9BD
Tel: (0446) 711111

CFL Vision
PO Box 35
Wetherby
West Yorks LS23 7EX
Tel: (0937) 541010
('Get it right before you go . . . and while you're there'. Catalogue number UK 6178. Central Office of Information-produced VHS or Betamax video, available on free loan, giving information for travellers on consular protection overseas)

Civil Aviation Authority (CAA)
CAA House
45–59 Kingsway
London WC2B 6TE
Tel: (071) 379 7311

Comité National de Sentiers de Grande Randonée
(Walking in France)
92 rue Clignancourt
75883 Paris
Cedéx 15
France

Consumers' Association
2 Marylebone Road
London
NW1 4DX
Tel: (071) 486 5544

Cork Airport
Kinsale Road
Cork
Eire
Tel: 010 353 21313131

Corona Society: *see* Women's Corona Society

Countryside Commission
John Dower House
Crescent Place
Cheltenham
Glos GL50 3RA
Tel: (0242) 521381

Countryside Commission for Scotland: *see* Scottish Natural Heritage

Coventry Airport
Baginton
Coventry
Warwickshire
CV8 3AZ
Tel: (0203) 301717

Cyclists Touring Club
Cotteroll House
69 Meadrow
Godalming
Surrey GU7 5HS
Tel: (0483) 417714

DER Travel Service
(Motorail bookings for Germany, Belgium and Austria)
18 Conduit Street
London W1R 9TD
Tel: (071) 408 0111

Useful addresses

Dublin Airport
Co. Dublin
Eire
Tel: 010 353 1379900

East Midlands Airport
Castle Donnington
Derby DE7 2SA
Tel: (0332) 810621

Edinburgh Airport
Edinburgh
EH12 9DN
Tel: (031) 333 1000

English Heritage
(For information on historic monuments in England)
Marketing Department
Keysign House
429 Oxford Street
London W1R 2HD
Tel: (071) 973 3000

English Tourist Board
Thames Tower
Black's Road
Hammersmith
London W6 9EL
Tel: (081) 846 9000

Europ Assistance
(Insurance)
Europ Assistance
252 High Street
Croydon
Surrey CR0 1NF
Tel: (081) 680 1234

Exeter Airport
Exeter
Devon EX5 2BD
Tel: (0392) 67233

Gatwick Airport
West Sussex
RH6 0NP
Tel: (0293) 567675

Gingerbread
(Association for one-parent families; groups nationwide)
35 Wellington Street
London WC2E 7BN
Tel: (071) 240 0953

Gingerbread Services Ltd.
35 Wellington Street
London WC2E 7BN
(Send S.A.E. for details of two group holidays organized each year)
Tel: (071) 240 0953

Glasgow Airport
Paisley
PA3 2ST
Tel: (041) 887 1111/1807

Glasgow/Prestwick: see Prestwick

Guernsey Airport
La Villiaze
Forest
Guernsey CI
Tel: (0481) 37766

Heathrow Airport
Hounslow
Middx TW6 1JH
Tel: (081) 745 7702/4 (Terminal 1)
     (081) 745 7115/7 (Terminal 2)
     (081) 745 7412/4 (Terminal 3)
     (081) 745 4541 (Terminal 4)

Historic buildings, monuments and gardens: see Cadw Welsh Historic Monuments; English Heritage; Historic Scotland; National Trust; National Trust for Scotland

Historic Scotland
(Information on historic monuments in Scotland)
Marketing and Visitor Services
20 Brandon Street
Edinburgh EH3 5RA
Tel: (031) 244 3101

Holiday Care Service
(Holidays for single and lone parents)
2 Old Bank Chambers
Station Road
Horley
Surrey RH6 9HW
Tel: (0293) 774535

Homoeopathic Chemists with mail-order service; see Ainsworth's Homoeopathic Pharmacy

Horse Riding: see Association Régionale de Tourisme Equestre

Humberside Airport
Kirmington
South Humberside DN39 6YH
Tel: (0652) 688456

Immunization: see British Airways Immunization Unit; and Thomas Cook Vaccination Centre

Inland Waterways Association
114 Regent's Park Road
London NW1 8UQ
Tel: (071) 586 2510/2556
(*See also* Boats and Cruisers)

Insurance: *see* Europ Assistance

Intolerance to drugs: *see* Medic-Alert Foundation

Jersey Airport
St Peters
Jersey JE1 1BY
Tel: (0534) 46111

Kent International Airport
PO Box 500
Manston
Kent CT12 5BP
Tel: (0843) 823333

La Leche League
(Advice on breastfeeding)
BM 3424
London WC1N 3XX
Tel: (071) 242 1278
(*See also* Association of Breastfeeding Mothers)

Leeds-Bradford Airport
Yeadon
Leeds LS19 7TZ
Tel: (0532) 509696

Liverpool Airport
Liverpool
L24 1YD
Tel: (051) 486 8877

London: *see* Gatwick, Heathrow, London City Airport & Stansted

London City Airport
King George V Dock
Silvertown
London E16 2PX
Tel: (071) 474 5555

London-Luton Airport
Luton
Bedfordshire LU2 9LY
Tel: (0582) 405100

Medic-Alert Foundation
(Allergy identification bracelets and necklaces)
21 Bridge Wharf
156 Caledonian Road
London N1 9UU
Tel: (071) 833 3034

Manchester Airport
Wythenshawe
Manchester
M22 5PA
Tel: (061) 489 3000

Motorail: *see* British Rail (Motorail Reservations); DER Travel Service

National Caravan Council
Catherine House
Victoria Road
Aldershot
Hants GU11 1SS
Tel: (0252) 318251
(*See also* Camping and Caravanning Club)

National Express (Central)
Spencer House
Digbeth
Birmingham B5 6DQ
Tel: (021) 622 4373

National Express (North)
Coach Station
Chorlton Street
Manchester M1 3JR
Tel: (061) 228 3881

National Express (South)
23 Crawley Road
Luton LU1 1HX
Tel: (0582) 24407

National Express Scotland
(Caledonian Express)
Walnut Grove
Perth PH2 7LP
Tel: (0738) 31501

National Trust
(Details for properties in England, Wales & Ireland)
36 Queen Anne's Gate
London SW1H 9AS
Tel: (071) 222 9251

National Trust for Scotland
5 Charlotte Square
Edinburgh EH2 4DU
Tel: (031) 226 5922

Newcastle International Airport
Woolsington
Newcastle
NE13 8BZ
Tel: (091) 286 0966

Norwich Airport
Norwich
Norfolk NR6 6JA
Tel: (0603) 411923

Useful addresses

One-parent families: see SPLASH; Gingerbread; Gingerbread Services Ltd.; Holiday Care Service

Ordnance Survey
Romsey Road
Southampton SO9 4DH
Tel: (0703) 792765

Prestwick Airport
Prestwick
Ayrshire KA9 2PL
Tel: (0292) 79822

RAC: see Royal Automobile Club

Ramblers' Association
1–5 Wandsworth Road
London SW8 2XX
Tel: (071) 582 6878

The Red House
(Children's Bookclub)
Red House Books Ltd
Cotswold Business Park
Witney
Oxford OX8 5YF
Tel: (0993) 774171

RAC
RAC House
PO Box 100
7 Brighton Road
South Croydon
CR2 6XW
Tel: (081) 686 2525

Scottish Natural Heritage
Battleby
Redgorton
Perth PH1 3EW
Tel: (0738) 27921

Scottish Travel Centre
Scottish Tourist Board
23 Ravelston Terrace
Edinburgh EH4 3EU
Tel: (031) 332 2433

Scottish Youth Hostels Association
7 Glebe Crescent
Stirling FK8 2JA
Tel: (0786) 51181

Shannon Airport
Co. Clare
Eire
Tel: 010 353 6145556

Single Parents: see SPLASH; Gingerbread; Gingerbread Services Ltd.; Holiday Care Service

Ski Club of Great Britain
118 Eaton Square
London SW1 W9AF
Tel: (071) 245 1033

Southampton Airport
Southampton
SO9 1RH
Tel: (0703) 629600

Southend Airport
Southend
Essex SS2 6YF
Tel: (0702) 340201

SPLASH
(Holidays for lone parents)
19 North Street
Plymouth
Devon PL1 9AH
Tel: (0752) 674067

Stanfords
(Bookshop specializing in travel books & atlases)
12–14 Long Acre
London WC2E 9LP
Tel: (071) 836 1321

Stansted Airport
Enterprise House
Bassingbourn Road
Stansted
Essex CM24 1QW
Tel: (0279) 680500

Tapeworm
(Stories on cassette)
Aspoey House
Aspoey Road
New Morden,
Surrey KT3 3NJ
Tel: (081) 942 7788

Teesside Airport
Darlington
Co. Durham DL2 1LU
Tel: (0325) 332811

Thomas Cook Vaccination Centre
45 Berkeley Street
London W1A 1EB
Tel: (071) 408 4157
(See also British Airways Immunization Unit)

Travel Bookshop Ltd
(Specialist bookshop)
13 Blenheim Crescent
London W11 2EE
Tel: (071) 229 5260

Information and addresses

Vaccination: see British Airways Immunization Unit; and Thomas Cook Vaccination Centre

Wales Tourist Board
Brunel House
Cardiff CF2 1UY
Tel: (0222) 499909

Walking and rambling: see Comité National de Sentiers de Grande Randonée; Countryside Commission; Ordnance Survey; Ramblers' Association; Scottish Natural Heritage; Scottish Youth Hostels Association; Youth Hostels Association

Women's Corona Society
(Aid to families moving abroad or returning home)
Commonwealth House
18 Northumberland Avenue
London WC2N 5BJ
Tel: (071) 235 1230

Youth Hostels Association
(England and Wales)
Trevelyan House
8 St Stephen's Hill
St Albans
Herts AL1 2DY
Tel: (0727) 55215
(*See also* Scottish Youth Hostels Association)

Tourist Boards

Regional Tourist Boards

Cumbria Tourist Board
(Covering the county of Cumbria)
Ashleigh
Holly Road
Windermere
Cumbria LA23 2AQ
Tel: (05394) 4444
Fax: (05394) 4041

East Anglia Tourist Board
(Covering the counties of Cambridgeshire, Essex, Norfolk, and Suffolk)
Topplesfield Hall
Hadleigh
Suffolk IP7 5DN
Tel: (0473) 822922
Fax: (0473) 823063

East Midlands Tourist Board
(Covering the counties of Derbyshire, Leicestershire, Lincolnshire, Northamptonshire, and Nottinghamshire)
Exchequergate
Lincoln LN2 1PZ
Tel: (0522) 531521
Fax: (0522) 532501

Heart of England Tourist Board
(Covering the counties of Gloucestershire, Shropshire, Staffordshire, Warwickshire, West Midlands, and Worcestershire)
Woodside
Larkhill
Worcester WR5 2EF
Tel: (0905) 763436
Fax: (0905) 763450

London Tourist Board
(Covering the London Area)
26 Grosvenor Gardens
London SW1W 0DU
Tel: (071) 730 3450
Telex: 919041
Fax: (071) 730 9367

North West Tourist Board
(Covering the counties of Cheshire, Greater Manchester, Lancashire and Merseyside)
Swan House
Swan Meadow Road
Wigan Pier
Wigan WN3 5BB
Tel: (0942) 821222
Fax: (0942) 820002

Northumbria Tourist Board
(Covering the counties of Cleveland, Durham, Northumberland, and Tyne & Wear)
Aykley Heads
Durham DH1 5UX
Tel: (091) 384 6905
Fax: (091) 386 0899

South East England Tourist Board
(Covering the counties of East Sussex, Kent, Surrey, and West Sussex)
The Old Brew House
Warwick Park
Tunbridge Wells
Kent TN2 5TA
Tel: (0892) 540766
Telex: 95523
Fax: (0892) 511008

Southern Tourist Board
(Covering the counties of Hampshire, East Dorset, and the Isle of Wight)
40 Chamberlayne Road
Eastleigh
Hampshire SO5 5JH
Tel: (0703) 620006
Fax: (0703) 620010

West Country Tourist Board
(Covering the counties of Avon, Cornwall, Devon, Somerset, West Dorset, Wiltshire, and the Isles of Scilly)
60 St David's Hill
Exeter
EX4 4SY
Tel: (0392) 76351
Fax: (0392) 420891

282 Information and addresses

Yorkshire & Humberside Tourist Board
(Covering the counties of North Yorkshire, South Yorkshire, West Yorkshire, and Humberside)
312 Tadcaster Road
York YO2 2HF
Tel: (0904) 707961
Telex: 57715
Fax: (0904) 701414

National Tourist Boards

Northern Ireland Tourist Board
St Anne's Court
59 North Street
Belfast BT1 1NB
Tel: (0232) 231221
Telex: 748087
Fax: (0232) 240960

Scottish Tourist Board
23 Ravelston Terrace
Edinburgh EH4 3EU
Tel: (031) 332 2433
Telex: 72272
Fax: (031) 343 1513

Wales Tourist Board
Brunel House
2 Fitzalan Road
Cardiff CF2 1UY
Tel: (0222) 499909
Telex: 497269
Fax: (0222) 485031

States of Jersey Tourism Committee
Weighbridge
St Helier
Jersey
Channel Islands
Tel: (0534) 78000
Telex: 4192223
Fax: (0534) 35569

States of Guernsey Tourist Board
PO Box 23
White Rock
St Peter Port
Guernsey
Channel Islands
Tel: (04817) 726611
Fax: (04817) 721246

Isle of Man Dept of Tourism & Transport
Sea Terminal Building
Douglas
Isle of Man
Tel: (06246) 674323
Fax: (06246) 686800

The following Official Tourist Boards also have offices in London and deal with written, personal and telephone enquiries.

Scottish Tourist Board
19 Cockspur Street
SW1Y 5BL
Tel: (071) 930 8661

Northern Ireland Tourist Board
11 Berkeley Street
W1X 5AD
Tel: (071) 493 0601
Telex: 21839
Fax: (071) 499 3731
Free linkline 0800 282662

Jersey Tourist Information Office
35 Albemarle Street
W1X 3FB
Tel: (071) 493 5278
Fax: (071) 491 1565

British Tourist Authority
Thames Tower
Black's Road
Hammersmith
W6 1EL
(written enquiries only)

British Travel Centre
12 Regent Street
Piccadilly Circus
SW1Y 4PQ
(personal callers only)

Details of International Tourist Boards are given in Section 1.

Major mainline stations: British Rail

284 Information and addresses

British Rail station	Telephone enquiries	Car parking	Car hire available	Booking by telephone	Sleeper reservations
Aberdeen	0224 594222	●	●	0224 582005	0224 582005
Alfreton & Mansfield Parkway	0332 32051	●			
Banbury	0295 262256	●			
Basingstoke	0256 464966	●			0256 843483
Bath	0225 463075	●	●		
Bedford	0234 269686	●			
Berwick-upon-Tweed	0289 306771	●			
Birmingham New Street	021 643 2711	●	●	021 644 4285	021 644 4285
Birmingham International	021 643 2711	●	●	021 644 4475	
Blackpool	0772 59439	●	●	0253 20385	
Bournemouth	0202 292474	●			0202 554099
Bradford	0274 733994	●			
Bridgend	0222 228000	●			
Brighton	0273 206755	●	●		
Bristol Parkway	0272 294255	●	●		
Bristol Temple Meads	0272 294255	●	●	0272 294255	0272 294255
Bromley South	071 928 5100				
Burton-on-Trent	0332 32051	●			
Cambridge	0223 311999	●	●		
Canterbury East	0227 454411	●			
Cardiff	0222 228000	●			
Carlisle	0228 44711	●	●	0228 49433	0228 49433
Camarthen	0792 467777	●			
Chatham	0227 454411	●			
Chelmsford	0245 252111	●			
Cheltenham Spa	0452 529501	●	●		
Chester	0244 340170	●	●		
Chesterfield	0742 726411	●			
Chippenham	0793 536804	●			
Clapham Junction	071 928 5100				
Cleethorpes	0472 353556	●			
Colchester	0206 564777	●	●		
Coventry	0203 555211	●		0203 520012	
Crewe	0782 411411	●			0270 214343
Darlington	0325 355111	●	●		
Derby	0332 32051	●			
Didcot Parkway	0865 722333	●			
Doncaster	0302 340222	●	●		
Dover Priory	0227 454411	●	●		
Dumfries	0387 64105	●			
Dunbar	031 556 2451	●			
Dundee	0382 28046	●	●		0382 28776
Durham	091 232 6262	●			

Major mainline stations: British Rail 285

British Rail station	Telephone enquiries	Car parking	Car hire available	Booking by telephone	Sleeper reservations
East Croydon	071 928 5100				
Eastbourne	0273 206755	•			
Edinburgh	031 556 2451	•	•	031 556 5633	031 556 5633
Exeter St Davids	0392 433551	•	•	0392 411154	0392 56908
Folkestone Central	0227 454411	•			
Fort William	041 204 2844	•			0397 703791
Gatwick Airport	0403 62218		•		
Glasgow Central	041 204 2844		•	041 221 2305	041 221 2305
Glasgow Queen Street	041 204 2844	•	•		
Gloucester	0452 529501	•			
Grantham	0476 64135	•			
Great Yarmouth	0603 632055	•			
Grimsby	0472 353556	•			
Guildford	0483 755905	•	•		
Harrogate	0532 448133	•	•		
Hartford	051 709 9696	•			
Harwich Parkeston Quay	0206 564777	•			
Haverfordwest	0792 467777				
Haywards Heath	0273 206755	•			
Hereford	0452 29501	•			
High Wycombe	0494 441561	•			
Holyhead	0407 769222	•			
Hull	0482 26033	•	•		
Inverness	0463 238924	•	•	0463 242124	0463 242124
Ipswich	0473 693396	•	•		
Kettering	0533 629811	•			
Lancaster	0524 32333	•	•		
Leamington Spa	0203 555211				
Leeds	0532 448133	•	•		
Leicester	0533 629811	•			
Lincoln	0522 539502	•			
Liverpool Lime Street	051 709 9696		•	051 709 2894	
LONDON					
Euston	071 387 7070	•	•	071 387 8541	071 388 6061
Kensington Olympia	071 387 7070	•	•		
Kings Cross	071 278 2477	•	•	071 278 9431	
Liverpool Street	071 928 5100				
Paddington	071 262 6767	•	•	071 922 4372	071 922 4372
St Pancras	071 387 7070	•		071 837 5483	
Victoria	071 928 5100		•		
Waterloo	071 928 5100	•	•		
Loughborough	0533 629811	•			
Luton	0582 27612	•			

286 Information and addresses

British Rail station	Telephone enquiries	Car parking	Car hire available	Booking by telephone	Sleeper reservations
Macclesfield	061 832 8353	•			
Manchester Picadilly	061 832 8353	•	•	061 236 6091	
Market Harborough	0533 629811	•			
Middlesborough	0642 225535	•	•		
Milton Keynes	0908 370883	•	•		
Motherwell	041 204 2844	•			
Neath	0792 467777	•			
Newark North Gate	0636 704491	•			
Newcastle-upon-Tyne	091 232 6262	•	•	091 232 6262	
Newport (Gwent)	0633 842222	•	•	0633 257271	
Newton Abbot	0392 433551	•	•		
Northallerton	0325 355111	•			
Northampton	0788 560116	•			
Norwich	0603 632055	•	•		
Nottingham	0332 32051	•	•		
Nuneaton	0788 560116	•			
Oxenholme	0539 720397	•			
Oxford	0865 722333	•	•		0865 722333
Paignton	0392 433551	•			
Penrith	0768 62466	•			
Penzance	0872 76244	•			0872 76244
Perth	0738 37117	•	•		0738 37228
Peterborough	0733 68181	•	•		
Plymouth	0752 221300	•	•	0752 225141	0752 225141
Poole	0202 292474	•			0202 554099
Portsmouth & Southsea	0705 825771	•	•		
Portsmouth Harbour	0705 825771				
Port Talbot Parkway	0222 228000	•			
Preston	0772 59439	•	•	0772 823335	0772 823335
Reading	0734 595911	•	•	0734 585579	0734 587751
Retford	0302 340222	•			
Rugby	0788 560116	•			
Runcorn	051 709 9696	•	•		
St Austell	0872 76244	•			0872 76244
Salisbury	0722 27591	•	•		
Sandwell & Dudley	021 643 2711	•			
Scarborough	0723 373486	•			
Sheffield	0742 726411	•	•	0742 721780	
Shipley	0274 733994	•			
Shrewsbury	0743 364041	•			
Slough	0753 538621	•	•		
Southampton	0703 229393	•	•		0202 554099
Southend Victoria	0702 611811				

Major mainline stations: British Rail

British Rail station	Telephone enquiries	Car parking	Car hire available	Booking by telephone	Sleeper reservations
Stafford	0782 411411	•			
Stevenage	071 278 2477	•	•		
Stirling	0786 64754	•			0786 73812
Stockport	061 832 8353	•		061 480 4482	
Stoke-on-Trent	0782 411411	•			
Sunderland	091 232 6262				
Swansea	0792 467777	•	•	0792 467777	
Swindon	0793 536804	•	•		
Tamworth	021 643 2711	•			
Taunton	0823 283444	•	•	0823 256537	0823 256537
Telford	0743 64041	•			
Tiverton Parkway	0823 283444	•			
Tonbridge	0732 770111	•			
Torquay	0392 433551	•	•		
Totnes	0392 433551	•			
Truro	0872 76244	•	•	0872 76244	0872 76244
Tunbridge Wells	0732 770111	•			
Wakefield Westgate	0532 448133	•	•		
Warrington	061 832 8353	•			
Watford Junction	0923 245001	•	•		
Wellingborough	0234 60230	•			
Westbury	0793 536804	•			
Weston-super-Mare	0272 294255	•			
Weymouth	0305 785501	•			
Wigan	0772 59439	•			
Wilmslow	061 832 8353	•			
Winchester	0703 229393	•			0202 554099
Wolverhampton	021 643 2711	•		0902 456311	
Worcester Shrub Hill	0452 529501	•			
Worcester Foregate Street	0452 529501				
York	0904 642155	•	•		

Traffic and weather news

Accurate traffic and weather information is often difficult to find. The information meted out by the national radio stations is all too often restricted to generalities, tending to ignore anywhere north of Watford or west of Windsor. Information is *far* better than instinct – the listings that follow will help you to be prepared for anything the weather or motorway maintenance may throw at you, wherever you may be.

AA Roadwatch

The AA operate a number of telephone information lines, offering up-to-date traffic reports.

National

West Country	0836 401 481
Wales	0836 401 482
Midlands	0836 401 483
East Anglia	0836 401 484
North-West England	0836 401 485
North-East England	0836 401 486
Scotland	0836 401 487
National Motorway Network	0836 401 488

London and the South East

Central London (inside North/South Circulars)   0836 401 779

Motorways/roads between
M4 & M1   0836 401 780
Motorways/roads between
M1 & Dartford Tunnel   0836 401 781
Motorways/roads between
Dartford Tunnel and M23   0836 401 782
Motorways/roads between
M23 and M4   0836 401 783
M25 London Orbital only   0836 401 784

N.B. These phone lines are charged at a higher rate than the standard charge.

AA Weatherwatch

The AA operate a number of telephone information lines, offering an up-to-date weather report and a two-day forecast.

South-East England	0836 401 785
West Country	0836 401 786
Wales	0836 401 787
Midlands	0836 401 788
East Anglia	0836 401 789
North-West England	0836 401 790
North-East England	0836 401 791
Scotland	0836 401 792

N.B. These phone lines are charged at a higher rate than the standard charge.

Local radio

One of the easiest ways of obtaining an up-to-date forecast of the local travel or weather conditions is by tuning in to the local radio network. Armed with an accurate description of what lies ahead of you, routes can quickly be replanned and journey breaks can be scheduled.

The established network of local radio stations is split into two sectors – those companies that are run by the BBC and those which operate as independent stations.

We have listed below the main local stations and the radio frequencies on which they can be found. Please note that this type of information is often subject to change and we apologize in advance for any inaccuracies you may encounter.

BBC local radio

	MW service	VHF service
Radio Bedfordshire	1161 kHz 630 kHz	95.5 MHz 103.8 MHz
Radio Bristol	1548 kHz 1323 kHz	94.9 MHz 95.5 MHz 104.6 MHz
Radio Cambridgeshire	1026 kHz 1449 kHz	96.0 MHz 95.7 MHz
Radio Cleveland	1548 kHz	95.0 MHz 95.8 MHz
Radio Cornwall	657 kHz 630 kHz	95.2 MHz 96.0 MHz 103.9 MHz
Radio Cumbria	756 kHz 1458 kHz	95.6 MHz
Radio Derby	1116 kHz	94.2 MHz 104.5 MHz 95.3 MHz
Radio Devon	801 kHz 990 kHz 855 kHz 1458 kHz	94.8 MHz 103.4 MHz 96.0 MHz 95.8 MHz
BBC Essex	1530 kHz 765 kHz 729 kHz	103.5 MHz 95.3 MHz
Radio Furness	837 kHz	96.1 MHz 95.2 MHz 104.2 MHz
Radio Guernsey	1116 kHz	93.2 MHz
Radio Humberside	1485 kHz	95.9 MHz
Radio Jersey	1026 kHz	88.8 MHz
Radio Kent	1035 kHz 774 kHz 1602 kHz	104.2 MHz 96.7 MHz
Radio Lancashire	1557 kHz 855 kHz	95.5 MHz 104.5 MHz 103.9 MHz
Radio Leeds	774 kHz	92.4 MHz 95.3 MHz
Radio Leicester	837 kHz	95.1 MHz
Radio Lincolnshire	1368 kHz	94.9 MHz
Radio London	1458 kHz	94.9 MHz
Radio Manchester	1458 kHz	95.1 MHz
Radio Merseyside	1485 kHz	95.8 MHz
Radio Newcastle	1458 kHz	96.0 MHz 95.4 MHz 104.4 MHz
Radio Norfolk	855 kHz 873 kHz	104.4 MHz 95.1 MHz

290 Information and addresses

Radio Northampton	1107 kHz	103.6 MHz
		104.2 MHz
Radio Nottingham	1584 kHz	103.8 MHz
	1521 kHz	95.5 MHz
Radio Oxford	1485 kHz	95.2 MHz
Radio Sheffield	1035 kHz	104.1 MHz
		88.6 MHz
Radio Shropshire	756 kHz	95.0 MHz
	1584 kHz	96.0 MHz
Radio Solent	1359 kHz	96.1 MHz
	999 kHz	
Radio Stoke-on-Trent	1503 kHz	94.6 MHz
		104.6 MHz
Radio Sussex	1161 kHz	95.3 MHz
	1485 kHz	104.5 MHz
	1368 kHz	104.0 MHz
		95.0 MHz
Radio WM	1458 kHz	95.6 MHz
	828 kHz	
Radio York	666 kHz	103.7 MHz
	1260 kHz	95.5 MHz
		104.3 MHz

Independent local radio

	MW service	VHF service
Scotland & Northern Ireland		
Moray Firth Radio (Inverness)	1107 kHz	97.4 MHz
Northsound Radio (Aberdeen)	1035 kHz	96.9 MHz
Radio Tay (Dundee)	1161 kHz	102.8 MHz
	1584 kHz	96.4 MHz
Radio Forth (Edinburgh)	1548 kHz	97.3 MHz
Radio Clyde	1548 kHz	102.5 MHz
West Sound (Ayr)	1035 kHz	96.7 MHz
		97.5 MHz
Downtown Radio (Newtownards)	1026 kHz	96.4 MHz
		96.6 MHz
		97.4 MHz
		102.4 MHz

North East		
Metro Radio (Newcastle)	1152 kHz	97.1 MHz
Tim Radio	1170 kHz	96.6 MHz
Yorkshire & North West		
Red Rose Radio (Preston)	999 kHz	97.4 MHz
Viking Radio (Hull)	1161 kHz	96.9 MHz
Radio Aire	828 kHz	96.3 MHz
Pennine Radio (Bradford)	1278 kHz	97.5 MHz
	1530 kHz	102.5 MHz
Piccadilly Radio	1152 kHz	103.0 MHz
Radio Hallam (Sheffield)	1548 kHz	97.4 MHz
	990 kHz	96.1 MHz
	1305 kHz	103.4 MHz
		102.9 MHz
Radio City (Liverpool)	1548 kHz	96.7 MHz
Marcher Sound (Wrexham)	1260 kHz	103.4 MHz
Midlands		
Radio Trent (Nottingham)	999 kHz	96.2 MHz
	945 kHz	102.8 MHz
Signal Radio (Stoke-on-Trent)	1170 kHz	102.6 MHz
Beacon Radio (Wolverhampton)	990 kHz	97.2 MHz
		103.1 MHz
Leicester Sound	1260 kHz	103.2 MHz
BRMB Radio	1152 kHz	96.4 MHz
Mercia Sound (Coventry)	1359 kHz	97.0 MHz
Radio Wyvern (Worcester)	954 kHz	97.6 MHz
	1530 kHz	102.8 MHz
East Anglia		
Hereward Radio (Peterborough)	1332 kHz	102.7 MHz
Radio Broadland (Norwich)	1152 kHz	102.4 MHz
Saxon Radio (Ipswich)	1251 kHz	96.4 MHz
Chiltern Radio Network (Dunstable)	828 kHz	97.6 MHz
	792 kHz	96.9 MHz
	1557 kHz	96.6 MHz

Radio Orwell (Ipswich)	1170 kHz 97.1 MHz	Radio 210 (Reading)	1431 kHz 97.0 MHz 102.9 MHz
Essex Radio	1431 kHz 96.3 MHz 1359 kHz 102.6 MHz	Invicta Radio (Canterbury)	1242 kHz 103.1 MHz 603 kHz 102.8 MHz 95.9 MHz 97.0 MHz 96.1 MHz
Wales & West			
Severn Sound (Gloucester)	774 kHz 102.4 MHz		
GWR Radio (Bristol)	1260 kHz 96.3 MHz 1161 kHz 97.2 MHz 936 kHz 102.6 MHz 103.0 MHz	County Sound Radio (Guildford)	1476 kHz 96.4 MHz
		Radio Mercury (Crawley)	1521 kHz 102.7 MHz 97.5 MHz
Swansea Sound	1170 kHz 96.4 MHz		
Red Dragon Radio (Cardiff)	1359 kHz 103.2 MHz 1305 kHz 97.4 MHz	Ocean Sound (Fareham)	1557 kHz 102.2 MHz 1170 kHz 97.5 MHz 96.7 MHz
DevonAir Radio (Exeter)	666 kHz 97.0 MHz 954 kHz 96.4 MHz		
Plymouth Sound	1152 kHz 97.0 MHz 96.6 MHz	Southern Sound (Brighton)	1323 kHz 103.5 MHz
London & South East		Two Counties Radio (Bournemouth)	828 kHz 97.2 MHz
Capital Radio	1548 kHz 95.8 MHz		
LBC News Radio	1152 kHz 97.3 MHz		

Information and addresses

Diseases and the precautions you should take

Disease	Risk areas	How caught	Vaccination	Vaccination certificate needed	Other precautions
AIDS	Worldwide.	From having sex with an infected person; from injections with infected blood or needles.	None available.	No, but some countries have introduced HIV antibody testing for some visitors (or require an HIV antibody test certificate for some visitors. See note below).	Using a condom (rubber or sheath) during sex gives some protection. Take a travel kit for use in medical emergencies.
Cholera	Africa, Asia, Middle East, especially in conditions of poor hygiene and sanitation.	From contaminated food or water.	Usually 2 injections by your doctor.	Some countries may require evidence of vaccination. Certificate valid for 6 months.	Vaccination gives modest protection only, so take scrupulous care over food and drink.
Viral hepatitis A	Most parts of the world but especially in conditions of poor hygiene and sanitation.	From contaminated food or water.	Immunoglobulin if not already immune.	No.	Take scrupulous care over food and drink.
Viral hepatitis B	Worldwide.	By intimate contact with an infected person; from injections with infected blood or needles (as AIDS).	Your doctor will advise on the need for vaccination.	No.	Avoid casual sexual or other intimate contact (as for AIDS).
Malaria	Africa, Asia, Central and South America.	Bite from infected mosquito.	None, but anti-malarial tablets are available.	No.	Avoid mosquito bites.
Poliomyelitis	Everywhere except Australia, New Zealand, Europe and North America.	Direct contact with an infected person; rarely by contaminated water or food.	Drops by mouth in 3 doses (spacing depends upon age). Reinforcing dose advised after 10 years.	No.	Take scrupulous care over food and drink.
Rabies	Many parts of the world.	Bite or scratch from an infected animal.	Vaccination may be advised after a bite. Get advice from a doctor immediately.	No.	
Tetanus	Worldwide but particularly dangerous in places where medical facilities not readily available.	Any skin-penetrating wound, especially if soiled.	Vaccination is safe, effective and gives long-lasting protection.	No.	Wash the wound thoroughly and consult a doctor without delay.
Tuberculosis	Asia, Africa, Central and South America.	Airborne from infectious person.	Skin test and injection at least 2 months before travel.	No.	Seek medical advice for chest pain, persistent cough or sputum, especially if bloodstained.
Typhoid	Everywhere except Australia, New Zealand, Europe, North America, in conditions of poor hygiene and sanitation.	Contaminated food, water or milk.	2 injections from your doctor, 4–6 weeks apart. Revaccination by 1 injection usually after 3 years.	No.	Take scrupulous care over food and drink.
Yellow fever	Africa and South America.	Bite from infected mosquito.	1 injection at a yellow fever vaccination centre at least 10 days before you go abroad.	Yes. Certificate valid for 10 years.	Avoid mosquito bites, as for malaria.

Diseases and precautions

Smallpox has been eradicated worldwide and there is NO requirement for the vaccination of travellers.

Note: If in doubt about HIV antibody test certificate requirements, check the current position with the Embassy or High Commission in London of the country concerned.

Before you travel abroad, remember that you are at risk from a number of diseases. You should always take necessary precautions.

Remember too, if you are unsure about any health requirements, always consult your Doctor. You should make arrangements for any vaccinations at least two months before departure.

What to do in a medical emergency

- Make sure you have all your documents with you – your passport, proof of UK residence (e.g. driver's licence, NHS card) and vaccination certificates – before seeking treatment. Where you are eligible for free or reduced-cost treatment, these documents will help prove your entitlement.
- Read your insurance document before treatment so that you know what will be paid for by the insurance company.
- Make sure you are not eligible for free or reduced-cost treatment. Leaflet T3 (available free from the Health Literature Line on 0800 555777) will give you all relevant information.
- Contact the local representative of your travel company.
- Tell the doctor if you are taking any medicine or tablets and, if you know it, give him the name of the drug (as on the prescription label), not just the brand name.
- Also tell the doctor if you have been to another country before becoming ill.
- Keep all receipts, special proofs of purchase, price tags and labels for all payments made for treatment or drugs, if you intend to claim a refund or claim on your insurance.
- The British Consular officials will be able to contact friends or relatives in the UK to ask for help so make sure their names and addresses are in your passport.
- If you need to return to the UK quickly, the British Consular officials may be able to arrange it if no one else can help you, but you will have to pay for this.

What to do when you return home

- If you were given any tablets or medicine when abroad it may not be legal to bring these back into the UK. If in doubt declare the drugs at Customs when you return.
- If you have been taking malaria tablets while abroad, continue taking them for a month after you return.
- If you become ill, tell your doctor which countries you have stayed in or travelled through.
- Even if you have received treatment and feel well, always consult your UK doctor if you have:
    - been bitten by an animal or
    - risked catching a sexually transmitted disease.
- When going to donate blood, tell the blood transfusion staff if you have been abroad outside Europe or had medical treatment abroad.
- If you had medical treatment abroad claim on your insurance as soon as you return.

Diseases and precautions 295

Vaccination recommendations
As issued by the Department of Health

Country	Malaria	Cholera	Typhoid	Polio	Yellow fever
Afghanistan	●	●	●	●	★
Albania					★
Algeria	●		●	●	★
Angola	●	●	●	●	★●
Antigua/Barbuda					★
Argentina	●		●	●	
Australia					★
Austria					
Azores					★
Bahamas			●	●	★
Bahrain		●	●	●	★
Bali	●	●	●	●	★
Bangladesh	●	●	●	●	★
Barbados			●	●	★
Belize	●		●	●	★
Benin	●	●	●	●	■
Bermuda					
Bhutan	●	●	●	●	★
Bolivia	●		●	●	★●
Botswana	●	●	●	●	
Brazil	●		●	●	★●
Brunei		●	●	●	★
Bulgaria					
Burkina Faso	●	●	●	●	■
Burundi	●	●	●	●	★
Cameroon	●	●	●	●	■
Canada					
Canary Islands					
Cape Verde	●		●	●	★
Cayman Islands				●	
Central African Republic	●	●	●	●	■
Chad	●	●	●	●	■
Chile			●	●	

296 Information and addresses

Country	Malaria	Cholera	Typhoid	Polio	Yellow fever
China	●		●	●	★
CIS					
Colombia	●		●	●	●·
Comoros	●		●	●	
Congo	●	●	●	●	■
Cook Islands			●	●	
Corsica					
Costa Rica	●		●	●	
Crete					★
Cuba			●	●	
Cyprus					
Czechoslovakia					
Djibouti	●	●	●	●	★
Dominica			●	●	★
Dominican Republic	●		●	●	
Ecuador	●	●	●	●	★
Egypt	●	●	●	●	★
El Salvador	●		●	●	★
Equatorial Guinea	●	●	●	●	★●
Ethiopia	●	●	●	●	★●
Falkland Islands					●
Fiji			●	●	★
Finland					
Gabon	●	●	●	●	■
The Gambia	●	●	●	●	■
Ghana	●	●	●	●	■
Greenland					
Grenada			●	●	★
Guam			●	●	★
Guatemala	●		●	●	★
Guiana, French	●		●	●	■
Guinea	●	●	●	●	★●
Guinea-Bissau	●	●	●	●	★●

Diseases and precautions 297

Country	Malaria	Cholera	Typhoid	Polio	Yellow fever
Guyana	●		●	●	★●
Haiti	●		●	●	★
Honduras	●		●	●	★
Hong Kong			●	●	
Hungary					
Iceland					
India	●	●	●	●	★
Indonesia	●	●	●	●	★
Iran	●		●	●	★
Iraq	●	●	●	●	★
Israel			●	●	
Ivory Coast	●	●	●	●	■
Jamaica			●	●	★
Japan			●	●	
Jordan		●	●	●	
Kampuchea	●	●	●	●	★
Kenya	●	●	●	●	★●
Kiribati			●	●	★
Korea (North & South)		●	●	●	
Kuwait		●	●	●	
Laos	●	●	●	●	★
Lebanon		●	●	●	★
Lesotho		★●	●	●	★
Liberia	●	●	●	●	■
Libya	●	●	●	●	★
Madagascar	●	●	●	●	★
Madeira					★
Malawi	●		●	●	★
Malaysia	●	●	●	●	★
Maldives	●	●	●	●	★
Mali	●	●	●	●	■
Malta					●
Mauritania	●	●	●	●	▲●

Country	Malaria	Cholera	Typhoid	Polio	Yellow fever
Mauritius	●		●		★
Mexico	●		●	●	★
Monaco					
Mongolia			●	●	
Montserrat			●	●	★
Morocco	●	●	●	●	★
Mozambique	●	●	●	●	★
Myanmar	●	●	●	●	★
Namibia	●	●	●	●	★●
Nauru			●	●	★
Nepal	●	●	●	●	★
Netherlands, Antilles			●	●	★
New Caledonia			●	●	★
New Zealand					
Nicaragua	●		●	●	★
Niger	●	●	●	●	■
Nigeria	●	●	●	●	★●
Niue			●	●	★
Norway					
Oman	●	●	●	●	★
Pakistan	●	★●	●	●	★
Panama	●		●	●	■
Papua New Guinea	●	●	●	●	★
Paraguay	●		●	●	★
Peru	●	●	●	●	★●
Philippines	●	●	●	●	★
Pitcairn Islands		★	●	●	★
Poland					
Polynesia, French (Tahiti)			●	●	★
Puerto Rico			●	●	
Qatar		●	●	●	★
Réunion			●		★
Romania					

Diseases and precautions

Country	Malaria	Cholera	Typhoid	Polio	Yellow fever
Rwanda	●	●	●	●	■
St Helena			●	●	
St Kitts and Nevis			●	●	★
St Lucia			●	●	★
St Vincent and Grenadines			●	●	★
Samoa			●	●	★
Sao Tome and Principe	●	●	●	●	▲
Saudi Arabia	●	●	●	●	★
Senegal	●	●	●	●	■
Seychelles			●	●	
Sierra Leone	●	●	●	●	★●
Singapore		●	●	●	★
Solomon Islands	●		●	●	★
Somalia	●	★●	●	●	★●
South Africa	●	●	●	●	★
Sri Lanka	●	●	●	●	★
Sudan	●	★●	●	●	★●
Surinam	●		●	●	★
Swaziland	●	●	●	●	★
Sweden					
Switzerland					
Syria	●	●	●	●	★
Tahiti			●	●	★
Taiwan		●	●	●	★
Tanzania	●	◆	●	●	◆
Thailand	●	●	●	●	★
Togo	●	●	●	●	■
Tonga			●	●	★
Trinidad and Tobago			●	●	★
Tunisia		●	●	●	★
Turkey	●	●	●	●	
Tuvalu			●	●	★
Uganda	●	●	●	●	★●

Information and addresses

Country	Malaria	Cholera	Typhoid	Polio	Yellow fever
United Arab Emirates	●	●	●	●	★
USA					
Uruguay			●	●	
Vanuatu	●		●	●	
Venezuela	●		●	●	
Vietnam	●	●	●	●	★
Virgin Islands			●	●	
West Indies Associated States			●	●	★
French West Indies			●	●	★
Yemen Arab Rep. (North) Yemen Dem. Rep. (South)	●	●	●	●	★
Yugoslavia					
Zaire	●	●	●	●	★●
Zambia	●	●	●	●	★●
Zimbabwe	●	●	●	●	★

This list is correct at the time and date of printing. Changes are notified on PRESTEL page 50063.

This alphabetical list shows for each country:
- ● = Vaccinations or tablets recommended for protection against disease.
- ◆ = Vaccinations which are an essential requirement for entry to the country concerned and for which you will require a certificate.

For Cholera only, the following symbols apply:
- ● = Cholera vaccination recommended to give improved protection against the disease especially for those who may not be able to avoid the risk of contaminated food or water.
- ★ = These countries consider vaccination essential for those who have visited or passed through any zone where cholera is present. A vaccination certificate (valid for six months) should be held. In the presence of a cholera epidemic, more countries may insist on a certificate.

For Yellow Fever only, the following symbols apply:
- ● = Vaccination recommended for protection against the disease, but note that pregnant women and infants under nine months should not normally be vaccinated and therefore should not be exposed to the disease.
- ■ = Vaccination essential except for infants under one year (but note the advice above).
- ▲ = Vaccination essential except for infants under one year and except for travellers arriving from non-infected areas and staying for less than two weeks. The UK is a non-infected area, but if travelling via equatorial Africa or South America, seek medical advice.
- ★ = Vaccination essential if the traveller arrives from an infected counry – i.e. where yellow fever is present. This will not apply if your journey is direct from the UK.

International SOS

We assume that you will be taking a standard phrase book with you and will therefore have the basics of the language to hand (guide to grammar, pronunciation etc.). But there are times when you will probably need a more specifically child-orientated vocabulary and would not expect to find the Spanish for teat, French for dummy or Italian for bib in your average Berlitz!

We have chosen the words in this section seeing two main uses: one so you can go to the chemist for minor ailments, such as antiseptic for insect bites; the other helps you ask the local people for children's needs, such as a baby-sitter or a cot. Major medical terms have been omitted on the assumption that if your child is very ill you will go straight to the experts.

French, German, Italian, Portuguese and Spanish are the languages given here as they are the ones generally in use at the popular holiday destinations.

International food vocabulary

Ordering food for children can be difficult enough at the best of times, but when you are relying on inspired guesswork to interpret a menu you frankly don't understand it's easy to make costly and frustrating mistakes.

This International Food Vocabulary will help you to find your way around a menu in French, German, Italian or Spanish. And when the kids demand sausage, egg and chips . . . at least you will know how to order them.

	French	**German**
accident	accident (*m*)	Unfall (*m*)
allergy	allergie (*f*)	Allergie (*f*)
antibiotic	antibiotique (*m*)	Antibiotikum (*n*)
appetite	appétit (*m*)	Appetit (*m*)
arm	bras (*m*)	Arm (*m*)
asthma	asthme (m)	Asthma (*n*)
baby	bébé (*m/f*)	baby (*n*)
baby-sitter	garde d'enfant (*f*)	Babysitter (*m*)
bib	bavette (*f*)	Latz (*m*)
bites	piqûres (*f*)	Stiche (*m*)
bleed	saigner (*v*)	Blutung (*f*)
blister	ampoule (*f*)	Blasen (*f*)
boil	clou (*m*) abcès (*m*)	Furunkel (*m*)
bone	os (*m*)	Knochen (*m*)
breathe	respirer (*v*)	atmen (*v*)
bronchitis	bronchite (*f*)	Bronchitis (*f*)
bruise	bleu (*m*) meurtrissure (*f*)	Bluterguss (*m*)
burn	brûlure (*f*)	Verbrennung (*f*)
child	enfant (*m/f*)	Kind (*n*)
choke	étouffer (*v*)	erstickung (*v*)
cold	rhume (*f*)	Erkältung (*f*)
colic	colique (*f*)	Kolik (*f*)
constipation	constipation (*f*)	Vestopfung (*f*)
cot	lit d'enfant (*m*)	Kinderbett (*n*)

Italian	Portuguese	Spanish
incidente (*m*)	acident (*m*)	accidente (m)
allergia (*f*)	alergia (*f*)	alergia (*f*)
antibiotico (*m*)	antibiótico (*m*)	antibiotico (*m*)
appetito (*m*)	apetite (*m*)	apetito (*m*)
braccio (*m*)	braço (*m*)	brazo (*m*)
asma (*f*)	asma (*f*)	asma (*f*)
bambino (*m*)	bebê (*m*)	bebé (*m*)
babysitter (*m/f*)	babá (*f*)	canguro (*m*) cuidar para los niños (*m*)
bavaglino (*m*)	babador (*m*)	babero (*m*)
puntura (*f*)	mordida/picada (*f*)	mordiscos (*m*)
sanguinare (*v*)	sangrar (*v*)	sangrar (*v*)
pustola (*f*)	bolha (*f*)	ampolla (*f*)
foruncolo (*m*)	furunculo (*m*)	tumor (*m*)
osso (*m*)	osso (*f*)	hueso (*m*)
respirare (*v*)	respirar (*v*)	respirar (*v*)
bronchite (*f*)	bronquite (*f*)	bronquitis (*f*)
contusione (*f*)	machucads (*m*)	cardenal (*m*)
scottare (*v*)	queimadura (*f*)	quemadura (*f*)
ragazzo (*m*)	crianca (*f*)	niño (*m*)
strozzare (*v*)	sufocada (*v*)	atragantarse (*v*)
raffreddore (*m*)	frio (*f*)	resfriado (*m*)
colica (*f*)	cólica (*f*)	colico (*m*)
constipazione intestinale (*f*)	prisão de ventre (*f*)	estreñimiento
lettino (*m*)	berco (*m*)	cuna (*f*)

	French	**German**
dehydration	déshydration (f)	Dehydierung (f)
dentist	dentiste (m/f)	Zahnarzt (m)
diarrhoea	diarrhée (f)	Durchfall (m)
doctor	médecin (m)	Arzt (m)
drug	médicament (m)	Medizin (f)
dummy	sucette (f)	Schnuller (m)
ear	oreille (f)	Ohr (n)
earache	mal à l'oreille (m)	Ohrenschmerz (m)
electric shock	secousse électrique (f)	Elektroschock (m)
emergency	urgence (f)	Not (f)
eye	oeil (m) yeux (pl)	Auge (n)
faeces	selles (f) fèces (f)	Stuhl (m)
fainting	s'évanouir (v) evanouissement (m)	ohnmächtig werden (v)
fever	fièvre (f)	Fieber (n)
finger	doigt (m)	Finger (m)
food poisoning	intoxication alimentaire (f)	Vergiftung (f)
foot	pied (m)	Fuss (m)
frost-bite	gelure (f)	Erfrierung (f)
genitals	organs sexuels (m)	Genitalien (pl)
graze	eraflure (f)	Kratzung (m)
hand	main (f)	Hand (f)
head	tête (f)	Kopf (m)

Italian	Portuguese	Spanish
disidratazione (f)	desidratação (f)	deshidratación
dentista (m)	dentista (m/f)	dentista (m)
diarrea (f)	diarréia (f)	diarrea (f)
dottore (m)	médico (m/f)	médico (m)
medicina (f)	remédio (m)	droga (f)
ciucciotto (m)	chupeta (f)	chupete (m)
orecchio (m)	ouvido (m)	oido (m)
mal d'orrechio (m)	dor de ouvido (f)	dolor de oido (m)
scossa elettrica (f)	choque eléctrico (m)	calambre (m)
emergenza (f)	emergência (f)	emergencia (f)
occhio (m)	olho (m)	ojo (m)
feci (f)	fezes (f)	excrementos (m)
svenire (v)	desmaiar (v)	desmayar (v) desmayo (m)
febbre (f)	febre (f)	fiebre (f)
dita (f)	dedo (m)	dedo (m)
intossicazione alimentari (f)	comida estragada (f)	envenenamiento (m)
piede (m)	pé (m)	pie (m)
congelamento (m)	ulceração produzida pelo frio (f)	congelación (f)
genitali (m)	genitais (m)	genitales (m)
scalfire (v)	arranhar esfolar (v)	rozar (v) rozadura (f)
mano (f)	haõ (f)	mano (m)
testa (f)	cabeça (f)	cabeza (f)

	French	**German**
heat stroke	coup de chaleur (*m*)	Hitzschlag (*m*)
high chair	chaise d'enfant (*f*)	Stuhl für Kind (*m*)
hospital	hôpital (*m*)	Krankenanstalt (*f*)
ill	malade (*adj*)	krank (*adj*)
infection	infection (*f*)	Ansteckung (*f*)
injection	piqûre (*f*)	Einspritzung (*f*)
knee	genou (*m*)	Knie (*n*)
leg	jambe (*f*)	Bein (*n*)
medicine	médicament (*m*)	Arzneimittel (*n*)
mouth	bouche (*f*)	Mund (*m*)
nap	somme (*m*)	Schlaf (*m*)
nappy	couche (à jeter) (*f*)	Windel (*f*)
neck	cou (*m*)	Hals (*m*)
nightmare	cauchemar (*m*)	Alptraum (*m*)
nose	nez (*m*)	Nase (*f*)
nose-bleed	saignement de nez (*m*)	Nasenblutung (*f*)
pain	douleur (*f*)	Schmerz (*m*)
play	jouer (*v*)	spiel (*v*)
potty	pot de chambre (*m*)	Topf (*m*)
pram	landau (*m*)	Kinderwagen (*m*)
pregnant	enceinte (*adj*)	schwanger (*adj*)
prescription	ordonnance (*f*)	Verschreibung (*f*)

Italian	Portuguese	Spanish
colpo di sole (m)	insolação (f)	insolación (f)
sedia da bambino (f)	cadeirá alta (f)	sillita de niño (f)
ospedale (m)	hospital (m)	hospital (m)
ammalato (m)	doente (m/f)	enfermo (m)
infezione (f)	infecção (f)	infección (f)
iniezione (f)	injeção (f)	inyección (f)
ginocchio (m)	soelho (f)	rodilla (f)
gamba (f)	perna (f)	pierna (f)
medicina (f)	remédio (m)	medicina (f)
bocca (f)	boca (f)	boca (f)
sonnino (m)	cochilo (m) dormida (f)	siesta (f)
pannolino (m)	fralda (f)	pañal (m)
collo (m)	pescoço (m)	cuello (m)
incubo (m)	pesadelo (m)	pesadilla (f)
naso (m)	nariz (m)	nariz (m)
sanguinare dal naso (v)	nariz sangranda (m)	hemorragia nasal (f)
dolore (m)	dor (f)	dolor (m)
giocare (v)	brincar (v)	jugar (v)
orinale (m)	pinico (m)	orinal (m)
incinta (adj)	grávida (adj)	embarazada (adj)
ricetta (f)	receita médica (f)	receta (f)
carrozzina (per bambini) (f)	carrinho de bebé (m)	cochecito de niño (m)

	French	**German**
shiver	frissonner (v)	zittern (v)
shock	choque (m)	Schock (m)
skin	peau (f)	Haut (f)
sore	avoir mal (v)	Wunde (f)
sterilize	stériliser (v)	sterilisieren (v)
sting	piqûre (f)	Stich (m)
stitches	agrafes (f)	Stich (m)
stomach	estomac (m)	Magen (m)
sunburn	coup de soleil (m)	Sonnenbrand (m)
sweat	suer (v)	schwitzen (v)
swelling	enflure (f)	Schwellung (m)
teat	tétine (f)	Schnuller (m)
teeth	dents (f)	Zähne (f)
temperature	température (f)	Temperatur (f)
throat	gorge (f)	Hals (m)
tired	fatigué (adj)	müde (adj)
toe	orteil (m)	Zeh (m)
toothache	mal de dent (m)	Zahnschmerz (m)
toys	jouets (m)	Spielzeug (n)
urine	urine (f)	Urin (m)
vaccination	vaccination (f)	Impfung (f)
vomit	vomir (v)	brechen (v)

(m) = masculine (f) = feminine (v) = verb (adj) = adjective

	indefinite articles:	indefinite articles:
	(m) un	(m) ein
	(f) une	(f) eine (n) ein

Italian	Portuguese	Spanish
pelle (f)	pele (f)	piel (m)
tremare (v)	calafrio/tremedeira (v)	temblar (v)
schock (m)	choque (m)	susto (m)
piaga (f)	inflamação (f)	llaga (f)
sterilizare (v)	esterelizar (v)	esterilizar (v)
puntura (f)	picada (f)	picadura (f)
stomaco (m)	estômago (m)	estómago (m)
punti (m)	pontos (m)	puntos (m)
scottatura (f)	queimadura de sol (f)	quemadura de sol (f)
sudare (v)	suor (v)	sudar (v)
rigonfiamento (m)	inchação (f)	hinchazón (f)
ciuccio (m)	bico de seio (m)	tetina (f)
temperatura (f)	êle tem febre	temperatura (f)
denti (m)	dentes (m)	dientes (m)
stanco (adj)	cansado (adj)	cansado (adj)
gola (f)	garganta (f)	garganta (f)
dito (m)	dedo do pé (m)	dedo del pie (m)
mal di dente (m)	dor de dente (m)	dolor de muelas (m)
gioccattoli (m)	brinquedos (m)	juguetes (m)
orina (f)	urina (f)	orina (f)
vaccinazione (f)	vacina (f)	vacuna (f)
vomitaire (v)	vomitar (v)	vomitar (v)

(*m*) = masculine   (*f*) = feminine   (*v*) = verb   (*adj*) = adjective

indefinite articles:	indefinite articles:	indefinite articles:
(*m*) un, uno	(*m*) um, umos	(*m*) un, unos
(*f*) una, un	(*f*) uma, umas	(*f*) una, unas

Meats

	French	**Italian**
Bacon	lard (*m*)	pancetta
Bacon & eggs	œufs (*mpl*) au lard	nova (*fpl*)/al prosciutto con la pancetta
Beef	bœuf (*m*)	manzo (*m*)
Beef sausage	saucisse (*f*) de bœuf/ Strasbourg	salsiccia di manzo
Chicken	poulet (*m*)	pollo (*m*)
Ham	jambon (*m*)	prosciutto (*m*)
Hamburger	hamburger (*m*)	hamburger (*m*)
Lamb	agneau (*m*)	agnello (*m*)
Lamb chop	côtelette (*f*) d'agneau	cotoletta (*f*) d'agnello
Pork	porc (*m*)	(carne (*f*) di) maiale (*m*)
Pork/mutton chop	côtelette de porc/ mouton	braciola (or) costoletta di maiale
Pork pie	pâté (*m*) en croûte	pasticcio di maiale in crosta
Pork sausage	saucisse (*f*) de porc	salsiccia di maiale
Roast beef	rôti (*m*) de bœuf rosbif (*m*)	arrosto di manzo
Sausage	saucisse (*f*) [uncooked]	salsiccia (to be cooked)
Sausage (pre-cooked)	saucisson (*m*)	salame (*m*)
Steak	bifteck (*m*) steak (*m*)	carne (*f*) di manzo bistecca
Veal	veau (*m*)	vitello (*m*)
Veal cutlet	escalope (*f*) de veau	cotoletta di vitello

Meats

Spanish	German
tocino (*m*)	Schinkenspeck (*m*)
tocino (*m*) con huevos	Eier mit Speck
carne (*f*) de vaca [or] de res	Rindfleisch (*nt*)
salchicha (*f*) de carne de vacuno	Rindswürstchen (*nt*)
pollo (*m*)	Huhn (*nt*)
jamón (*m*) pernil (*m*)	Schinken (*m*)
hamburguesa (*f*)	Hamburger (*m*)
(carne (*f*) de) cordero (*m*)	Lamm(fleisch) (*nt*)
chuleta (*f*) de cordero	Lammkotelett (*nt*)
(carne (*f*) de) cerdo/chancho (*m*)	Schweinefleisch (*nt*)
chuleta (*f*) de cerdo	Schweinskotelett (*nt*)
pastel (*m*) de carne de cerdo	Schweinepastete (*f*)
salchicha (*f*) de cerdo	Schweinewurst (*f*)
carne (*f*) de vaca asado	Roastbeef (*nt*)
embutido (*m*)	Wurst (*f*)
embutido (*m*)	Wurst (*f*)
biftec (*m*) bistec (*m*) bife (*m*)	Beefsteak (*nt*)
ternera (*f*)	Kalb(fleisch) (*nt*)
chuleta (*f*) de ternera (*f*)	Kalbsschnitzel (*nt*)

Vegetables

	French	**Italian**
Artichoke	Artichaut (*m*)	Carciofi (*m*)
Asparagus	Asperges (*fpl*)	Asparago (*m*)
Bean, French or Runner	Haricots verts (*mpl*)	Fagiolini (*mpl*)
Beetroot	Betterave (*f*)	Barbabictoia (*f*)
Broccoli	Brocoli (*m*)	Broccoli (*mpl*)
Brussels sprout	Chou de Bruxelles (*m*)	Cavoli di Bruxelles (*mpl*)
Cabbage	Chou (*m*)	Cavolo (*m*)
Carrot	Carrote (*f*)	Carota (*f*)
Cauliflower	Chou-fleur (*m*)	Cavolfiore (*m*)
Celeriac	Céleri-rave (*m*)	Sedano capa (*f*)
Courgette/Marrow	Courgette (*f*)	Zuchini (*mpl*)
Cucumber	Concombre (*m*)	Cetriolo (*m*)
Endive	Chicorée (*f*)	Cicoria (*f*)
Fennel	Fenouil (*m*)	Finocchio
Leek	Poireau (*m*)	Porro (*m*)
Lentil	Lentille (*f*)	Lente (*f*)
Lettuce	Laitue (*f*)	Lattuga (*f*)
Mushroom	Champignon (*m*)	Pilz-fungo (*m*)
Onion	Oignon (*m*)	Cipolla (*f*)
Parsnip	Panais (*m*)	Pastinaca (*f*)
Pea	Petit pois (*m*)	Piselli freschi (*fpl*)
Pepper, green	Poivron vert (*m*)	Peperone (*m*)
Potato	Pomme de terre (*f*)	Patata (*f*)
Sweetcorn	Mais (*m*)	Granturco (*m*)
Tomato	Tomate (*f*)	Pomodoro (*m*)
Turnip	Navet (*m*)	Rapa (*f*)

International food vocabulary 315

Vegetables

Spanish	German
Alcachofa (f)	Artischocke (m)
Espárragos (mpl)	Spargel (m)
Judias (fpl)	Zwergbohne (f)
Remolacha (f)	Crote Rube (f)
Brecol (m)	Spargelkohl (f)
Col (f) de Bruselas	Rosenkohl (m)
Col (f)	Kohl (m)
Zanahoria (f)	Karotte (m)
Coliflor (f)	Blumenkohl (m)
Apio-nabo (m)	Knollensellerie (m)
Calabacin (m)	Eierkurbis (m)
Pepino (m)	Gurke (f)
Escarola (f)	Endive (f)
Hinojo	Fenchel (m)
Puerro (m)	Lauch (m)
Lenteja (f)	Linse (f)
Lechuga (f)	Lattich (m)
Champiñon (m)	Champignon (m)
Cebolla (f)	Zwiebel (f)
Chirivia (f)	Pastinake (f)
Guisante (m)	Erbse (f)
Pimiento verde (m)	Spanischer Pfeffer (m)
Patata (f)	Kartoffel (f)
Maiz (m)	Mais (m)
Tomate (m)	Tomate (f)
Nabo (m)	Weiße Rübe (f)

Fish

	French	**Italian**
Anchovy	Anchois (*m*)	Accuiga (*f*)
Bass	Bar (*m*)/loup de mer (*m*)	Spigola (*f*)
Brill	Barbue (*f*)	
Caviar	Caviare (*m*)	Caviale (*m*) Not common
Cod, fresh	Cabillaud (*m*)	Merluzzo (*m*)
Cod, salted	Morue (*f*)	Baccala (*f*)
Crab	Crabe (*m*)	Granchio (*m*)
Crayfish:		
Freshwater	Ecrevisse (*f*)	Astaco (*m*), gamberi di fiume (*mpl*)
Saltwater	Langouste (*f*)	Scampo (i)
Haddock	Merlan (*m*)	Merluzzo (*m*)
Hake	Colin (*m*)	Merluzzo/Nasello (*m*)
Halibut	Flétan (*m*)	Sogliola Atlantica (*f*)
Herring	Hareng (*m*)	Aringe (*f*)
John Dory	St-Pierre (*m*)	Piesce San Pietro (*m*)
Lobster	Homard (*m*)	Aragosta (*f*)
Mackerel	Maquereau (*m*)	Sgombro (*m*)
Monkfish (Angler fish)	Lotte (*f*)	Coda di rospo (*f*)
Mussel	Moule (*f*)	Cozze (fpl)
Oyster	Huitre (*f*)	Ostrica (*f*)
Pike	Brochet (*m*)	Luccio (*m*)
Plaice	Flet (*m*)	Passerino (*m*)
Prawn		
Small	Ecrivette (*f*)	Gamberetti (*mpl*)
Larger	Langoustine (*f*)	Gamberetti/Scampi (*mpl*)
Ray/Skate	Raie (*f*)	Razza (*f*)
Red mullet	Rouget (*m*)	Triglia (*f*)

International food vocabulary

Fish

Spanish	German
Anchoa (f)	Sardelle (f)
Lubina (f)	Seebarsch (m)
	Seebutt (m)
Caviar (m)	Kaviar (m)
Merluza (f)	Dorsch (m)
Bacalao (m)	Stockfisch (m)
Buey cangrejo (m)	Krabbe (f)
Cigala (f)	Flußbrebr (m)
Cigala (f)/Langostino (m)	Languste (f)
Merlango (m)	Schellfisch (m)
Merluza (f)	Hechtdorsch (m)
Halibut (m)	Heilbutt (m)
Arengue (m)	Hering (m)
Pez de San Pedro	Not common
Langosta (f) (large) bogavante (m)	Hummer (m)
Caballa (f)	Makrele (f)
Pejesapo (m)	Meerteufel (m)
Mejillón (m)	Muschel (f)
Ostra (f)	Austern (f)
Lucio (m)	Hecht (m)
Platija (f)	Scholle (f)
Camaron (m) Gamba (f)	Krabbe (f) Hummerkrabe (f)
Raya (f)	Glattroche (m)
Salmonette de fango (m)	Rote Meerbarbe (f)

Fish cont.

	French	**Italian**
Salmon	Saumon (f)	Salmone (m)
Salmon, smoked	Saumon fumé (m)	Salmone affumicato (m)
Sardine	Sardine (f)	Sardina (f)
Scallop	Coquille-Saint-Jacques	Ventaglio (m)
Shad	Alose (f)	Alosa (f)
Sole, Dover	Sole (f)	Sogliola (f)
Trout	Truite (f)	Trota (f)
Tuna	Thon (m)	Tonno (m)
Turbot	Turbot (m)	Rombo (m)

Fruit

	French	**Italian**
Apple	Pomme (f)	Mela (f)
Apricot	Abricot (m)	Albicocca (f)
Avocado	Poire d'avocat (f)	Avocado (m)
Banana	Banane (f)	Banana (f)
Blackberry	Mûre (m)	Moro (m)
Blackcurrant	Cassis (m)	Ribes nero (m)
Blueberry/Bilberry	Myrtille (m)	Mirtillo (m)
Cherry: Red	Cerise (f)	Ciliegia (f)
Black	Cerise noire (f)	Ciliegia nera (f)
Date	Datte (f)	Dattero (m)
Fig	Figrie (f)	Fico (m)
Gooseberry	Groseille à maquereau (f)	Uvaspina (f)
Grape	Raisin (m)	Uva (fpl)
Grapefruit	Pamplemousse (m)	Pompelmo (m)
Lemon	Citron (m)	Limone (m)
Lime	Limon	Lima (f)

International food vocabulary 319

Fish cont.

Spanish	German
Salmon (m)	Lachs (m)
Salmon fumaro (m)	Raucherlachs (m)
Sardina (f)	Sardine (f)
Vierra concha de peregrino (f)	Jackobsmuschel (f)
Alosa (f)	Alse (f)
Lenguado (m)	Seezunge (f)
Trucha (f)	Forelle (f)
Atún (m)	Thunfisch (m)
Rodaballo (m)	Steinbutt (m)

Fruit

Spanish	German
Manzana (f)	Apfel (m)
Albaricoque (m)	Aprikose (f)
Aguacate (m)	Avacatobirne (f)
Plántano (m)	Banane (f)
Zarzamola (f)	Brombeere (f)
Grosella negra (f)	Korinth (f)
Azuramore (f)	Blaubeere (f)
Cereza (f)	Kirsche (f)
Cereza negra (f)	Schwarze Kirsche (f)
Datil (m)	Dattel (f)
Higo (m)	Feige (f)
Grosella espinoza (f)	Stachelbeere (f)
Uva (f)	Weintrabe (f)
Pomelo (m)	Pampelmuse (f)
Limon (m)	Zitrone (f)
Lima (f)	Limone (f)

Fruit cont.

	French	Italian
Loganberry	Ronce-framboise (f)	Frutta che somiglia alquanto al lampone (f)
Mango	Mangue (f)	Mango (m)
Melon	Melon (m)	Melone (m)
Nectarine	Brugnon (m)	Pesco-nettarovie (f)
Orange	Orange (f)	Arancia (f)
Papaya	Papaye (f)	Papaia (f)
Peach	Pêche (f)	Pesca (f)
Passion fruit	Fruit de la passiflore (m)	Granadilla (f)
Pear	Poire (f)	Pera (f)
Pineapple	Ananas (m)	Ananas (m)
Plum	Prune (f)	Susina (f)
Raspberry	Framboise (f)	Lampona (m)
Redcurrant	Groseille (f)	Ribes rosso (m)
Strawberry:		
Ordinary	Fraise (f)	Fragola (f)
Wild	Fraise de bois (f)	Fragola di campo (f)
Tangerine	Manderine (f)	Mandariono (m)

Herbs and spices

	French	Italian
Allspice	Toute-épice (f) Piment Jamaïque	Pimento (m)
Aniseed	Anis (m)	Anice (m)
Basil	Basilic (m)	Basilico (m)
Bay Leaf	Laurier (m)	Lauro (m)
Caraway	Carvi (m)	Comino (m)
Cardamon	Cardamome (m)	Cardamomo (m)
Cayenne	Cayenne (f)	Pepe di Caienna (m)
Chervil	Cefeuil (m)	Cerfoglio (m)
Chili	Piment fort (m)	Diavoletto (m)

Fruit cont.

Spanish	German
Fruta obtenido del cruce de la zarzamora y la frambuesa (f)	Kreuzung von Himbeere und Brombeere (f)
Mango (m)	Mangopflanze (f)
Melón (m)	Melone (f)
Nectarina (f)	Nektarine (f)
Naranja (f)	Orange (f)
Papaya (f)	Melononbaum (m)
Melocoton (m)	Pfersich (m)
Granadilla (f)	Eßbare Passionsblume (f)
Pera (f)	Birne (f)
Ananas (m)	Ananas (f)
Ciruela (f)	Pflaume (f)
Frambuesa (f)	Himbeere (f)
Grosella (roja) (f)	Johannisbeere (f)
Fresa (f)	Erdbeere (f)
Fresa silvestra (f)	Wilde Erdbeere (f)
Mandarina (f)	Mandarine (f)

Herbs and spices

Spanish	German
Fruita del pimiento de jamaica (f)	Jamaikapfeffer (m)
Anis (m)	Anis (m)
Alabega (f)	Basilienkraut (m)
Laurel (m)	Lorbeer (m)
Alcarvea (f)	Kümmel (m)
Cardamomo (m)	Kardamome (f)
Cayena inglesa (f)	Cayennepfeffer (m)
Perifollo (m)	Kerbel (m)
Chile (m)	Roter Pfeffer (m)

Herbs and spices cont.

	French	**Italian**
Cinnamon	Cannelle (f)	Cannella (f)
Clove	Clou de girofle (m)	Chiodo de garofano (m)
Coriander	Coriandre (f)	Coriandolo (m)
Cumin	Cumin (m)	Comino (m)
Fennel	Fenouil (m)	Finocchio (m)
Garlic	Ail (m)	Aglio (m)
Ginger	Gingembre (m)	Zenzero (m)
Mace	Fleur de Muscade (f)	Macis (m)
Marjoram	Marjolaine (f)	Maggiorana (f)
Mint	Menthe (f)	Menta (f)
Mustard	Moutarde (f)	Senape (f)
Nutmeg	Muscade (f)	Noce moscata (f)
Oregano	Oregan (m)	Origano (m)
Paprika	Paprika de Hongrie (f)	Paprika (f)
Parsley	Persil (m)	Prezzemolo (m)
Pepper	Poivre (m)	Pepe nero (m)
Rosemary	Romarin (m)	Rosmarino (m)
Saffron	Safran (m)	Zafferano (m)
Sage	Sauge (f)	Salvia (f)
Tarragon	Estragon (m)	Dragoncello (m)
Thyme	Thym (m)	Timo (m)
Turmeric	Curcuma (m)	Curcuma (f)

Herbs and spices cont.

Spanish	German
Canela (f)	Kaneel (m)
Clavo (m)	Givürznelke (f)
Coriandro/Cilantro (m)	Koriander (m)
Comino (m)	Kreuzkümmel (m)
Hinojo (m)	Fenchel (m)
Ajo (m)	Knoblauch (m)
Jengibre (m)	Ingwer (m)
Macia (f)	Muskatblüte (f)
Mejorana (f)	Majoran (m)
Hierbabuera (f)	Minze (f)
Mostaza (f)	Senf (m)
Moscada (f)	Muskat (m)
Oregano (m)	Oregano (m)
Paprika (f)	Paprika (f)
Perejil (m)	Petersilie (f)
Pimienta negra (f)	Pfeffer (m)
Romero (m)	Rosmarein (m)
Azafrán (m)	Safran (m)
Salvia (f)	Salbei (m or f)
Estragán (m)	Estragon (m)
Tomillo (m)	Romischer Quendel (m)
Curcuma (f)	Gelbwürz (f)

Contributors

In addition to the many contributors to *Section 1*, the following have written for *How to Have Stress-Free Family Holidays*.

Wayne Jackman has been a children's TV presenter for ten years, has written many children's books and has three children himself.

Ernest R. Jones is the chairman and chief executive of Mercury Insurance Services Ltd., a specialist emergency assistance company.

Kathy Rooney lives in London and is the mother of two long-suffering children.

Sheila Sang is Consumer Editor for *Essentials* magazine, has worked for the Consumers Association, and also freelances as a researcher and writer.

Index

Abercrombie & Kent Travel 159, 168
Accidents 263
Acorn Activities 172
Addresses 275–82
Adventure Cruisers 99
Adventure International 172
Aer Lingus Holidays 135
Africa Exclusive 159
Air France Holidays 135
Airlines 198–200
Airlink Holidays 135
Airport facilities for families (Britain) 189–97
Airtours PLC 135
Air travel 189, 198, 224, 247, 249, 263
Albany Tours 135
Allergy 264
Allez France 108
Anglo Dutch Sports 172
Animal bites 264
Antibiotics 264
Antigua 74
Antihistamines 264
Aquasun Holidays 136
Arctic Experience Ltd 136
Aultbea Highland Lodges 108
Australia 4
Austria 5
Auto Plan Holidays Ltd 176
Avis 102

Baby-sitting 249
B+I Line UK Ltd 108
Bales Tours Ltd 159
Balkan Holidays 136, 168
Barbados 74
Bargain breaks and short-stay holidays 80–90
Barrowfield Hotel 128
Bath Holiday Homes 108
Beach Villas 177
Becks Holidays 92
Bee stings 264
Belgium 6
BelleAir Holidays 136
Bell-Ingram Self-Catering Holidays 108
Best Travel Ltd 137
Best Western Hotels 81
Beverley Park Holidays 129
Blakes Country Cottages 109
Blakes Holidays 99
Blakes Villas 177

Blisters 264
Blisworth Tunnel Boats Ltd 99
Blood transfusion 265
Books for children 227–32
Boredom ix, 221, 227, 250
Bosham Sailing 163
Bowhill's 109
Brazil 8
Breastfeeding 251, 265
Bridgewater Boats 99
Britain 9
British Airways Fly/Drive 125
British Rail mainline stations 284–7
Brittany Caravan Hire 92
Brittany Direct Holidays 109
Brittany Ferries Gîtes Holidays 109
Bulgaria 11
Butlins Holiday Worlds and Hotels 129, 236–9
Butterfield's Indian Railway Tours 159

Cabervans 92
Calotels 81
Camping, caravan sites and mobile homes 91–7
Canada 12
Canal holidays 98–101
Canaries 60
Cant Farm 109
Canvas Holidays Ltd 93
Caprice Holidays Ltd 137
Car rental 102–3
Caribbean Connection 137
Caribbean Villas and Hideaway Hotels 137, 177
Carisma Holidays Ltd 93
Casas Cantabricas 110
Castaways 138
Celebrity Holidays & Travel 138
Celtik Holidays 138
Center Parcs 129
Cerbid's Quality Cottages 110
Chandris Ltd 122
Character Cottages 110
Chateau Welcome 177
China 14
China Travel Service (UK) Ltd 160
C.I.E. Tours International 138
CIS 55
Citalia 138
Clansman Monarch Holidays 104

326 Index

Claymoore Navigation Ltd 100
Clothing 251
Club Cantabrica 139
Club Méditerranée 139
Coach travel 104–6, 210, 225, 252
Coast and Country Holidays 110
Coastal Cottages of Pembrokeshire 110
Colds 265
Color Line Holidays 139
Consort Hotels 81
Constipation 266
Contraception 265
Thomas Cook Faraway Holidays 160
Coping in cold weather 251, 252, 265
Coping in hot weather 251, 252, 268
Corfu a la Carte 178
Cornish Traditional Cottages Ltd 111
Corona Holidays 178
Corsair Cruisers Ltd 100
Corton Beach Holiday Village 130
Cosmosair PLC 105
Cottage in the Country, A 111
Cottages, gîtes and farmhouses 108–21
Countryside Cottages 111
Countrywide Holidays 172
Courtlands Centre 173
Creative Tours Ltd 160
Creeping eruption 266
Cresta Holidays 139
Cruises 122–4
CTC Lines 123
Cunard 123
CV Travel 178
Cycling for Softies 173
Cyprair Holidays 140
Cyprus 16

Dales Holiday Cottages 111
Davies and Newman Travel Ltd 140
Dehydration 266
Denmark 17
Diarrhoea 266
Discover Britain Holidays 111
Disease precautions 292–300
Dogs 266
Dominique's Villas 178
Drugs and medicines abroad 267
Dunaird Cabins 111
Dysentery 267

Earache 267
Eating on the move 215–7
Egypt 19
Eire 20
Elmsworth Sailing School 164
Embassy Leisure Breaks 140
England 9
English Country Cottages Ltd 112
Enterprise Summersun 141
Equity Cruises 123
Eurocamp Travel Ltd 93
Eurodollar Rent a Car Ltd 103
Evan Evans Tours 105

Falcon Family Holidays 141

Farm and Cottage Holidays 112
Felindre 112
Fermanagh Lakeland 112
Ferries 201–9, 225, 253
Fever 267
Finlandia Travel Agency Ltd 141
Finnchalet Holidays Ltd 113
First-aid kits 267
Flamingo Land Holiday Village 130
Florida Home Owners' Association 179
Fly/drive 125–7
Food safety 268
Forestdale Hotels 82
Forest Holidays 119
Forte Hotels 83
Forte Travelodge 85
Four Pillars Group 85
Frames Rickards 105
France 21
Freedom Holidays Ltd 113
French Affair 113
French Country Cottages 113
French Country Cruises 100
French Villa Centre 179
Friendly Hotels 85

Germany 23
Getaway America 126
Getting lost 253
Golden Gateways 142
Gordon Holiday Cottages 113
Granada Hotels and Lodges 86
Great British Holidays 142
Greece 24
Greek Islands Club 179
Greek Islands Sailing Club 164
Greyhound World Travel 105

Haven 94
Haven France and Spain 94
Haywood-Amaro Holidays 114
Heatherwood Park 114
Heat stroke 268
Hepatitis 268
Hertz (UK) Ltd 103
HF Holidays 184
Highlife London and UK Breaks 142
Holiday Club Pontins 130
Holiday Cottages 114
Holiday Houses Dumfries and Galloway 114
Holiday reading for kids 227
Holiday Scandinavia Ltd 114
Holidays in Lakeland 114
Holimarine 94
Holland 48
Home From Home 115
Home swapping 133
Homoeopathic medicine 253
Hong Kong 26
Horizon Holidays 142
Horning Pleasurecraft Ltd 100
Hoseasons Holidays Ltd 100, 115
Hotel holidays 134–51

Index

Hotels of the Cinque Ports 86
Hungary 27

Ice 268
Ilios Island Holidays 179
India 28
Inghams Travel 168
Inghams Travel (Canary Travel) 143
Inghams Travel Eurobreak 87
Inghams Travel (Lakes and Mountains) 143
Injections 269
Inntravel 143
Insects 269
Insurance 240–3, 269
Interhome 168, 180
International Chapters 177
Intervac International House Exchange Service 133
Intourist Travel Ltd 143
Island Holidays 144
Israel 30
Italy 31

Jamaica 34
Japan 35
Jasmin Tours Ltd 160
Jellyfish 269
Jet lag 270
Jetsave 126, 144
Jordan 37
Jubilee Sailing Trust 164
Just France 115, 144
Just Pedalling 173

Kenya 38

Laskarina Holidays 180
Local radio 289–91
Long-haul travel 254
Lorne Leader 164
Lotus Supertravel 168

Mackay's Agency 115
Made to Measure Holidays 165, 169
Magic of Italy 126, 180
Majorca 61
Malaria 270
Malaysia 40
Malta 41
Mann's Holidays 116
Manos Holidays 144
Matthews Holidays 94
Mauritius 42
Medical emergencies 240, 255, 263, 303–11
Meon Villa Holidays 180
Meridian Holidays 145
Metak Holidays 145
Mexico 44
Milkbere Holidays 116
Millfield Village of Education 174
Minorca Sailing Holidays 165
Minotels 87
Morocco 46
Motorail 255

Motoring 102, 125, 211, 223, 255
Mount Holiday Park, The 95
Multitours 145
Mundi Color Holidays 126

Nappies 256
National Express Ltd 105
National Trust for Scotland 116
Naturist Holidays 174
Nepal 47
Netherlands, The 48
Nevis 74
New Century Holidays 180
Newman's (David) European Collection 116, 145
New Zealand 50
Northern Ireland 20
North Norfolk Holiday Homes 116
North Sea Ferries Holidays 145
Northumbria Horse Holidays 174
North Wales Holiday Cottages and Farmhouses 117
Northwest Fly/Drive USA 127
Norway 51

Olympic Holidays 146
Open spaces 257
Osborne (Vikki) 146
Owners Holiday Letting Consortium 117

PAB Travel 146
Page and Moy Ltd 146
Palmer & Parker 181
P & O Cruises 124
Parks 257
Passage to South America 160
Passports 258
Peak and Moorland Farm Holidays 117
Peregor Travel 127
PGL Family Adventure 174
Philippines, The 52
Pontins 130
Portland Holidays 147
Portugal 53
Portuguese Property Bureau, The 181
Powell's Cottage Holidays 117
Precautions against diseases 292–300
Preparation 258
Preston Holiday 147
Prickly heat 270
Proprietaires de l'Ouest, Les 117
Pullman Holidays UK Ltd 147

Queens Moat Houses 87

Rabies 270
Radfords Country Hotel 131
Radio (local) 289–91
Rail travel 152–7, 224, 258, 284
Ramblers Holidays 185
Rank Hotels 88
Recommended Cottage Holidays 118
Regent Holidays (UK) Ltd 161
Rendezvous France 118

328 Index

Renting somewhere to stay 108, 176, 233
Resort Hotels 89
Return home 270
Rockley Point Sailing School 165
Romany Caravan Holidays 95
Russia 55

Safaris, Treks and exotic tours 158–62
Sailing 163–6
Sandpiper Camping Holidays 95
Saudi Arabia 56
Saunton Sands Hotel, The 131
Scandinavian Seaways 147
Scotland 9
Seasun Holidays Ltd 95
Service stations 211–4
Seymour Hotels and Holidays 148
S.F.V. Holidays 118
Shamrock Cottages 118
Shaw's Holidays 118
Shopping 257
Sicily 33
Sightseeing 259
Silk Cut Travel Ltd 161
Simolda Ltd 101
Simply Simon Holidays 148
Simply Turkey 181
Singapore 40
Skiathos Travel 181
Ski Chamois 169
Ski Esprit 148, 169
Skiing 167, 259
Ski Thomson 169
Ski-Val 170
Skytours Holidays 148
Solaire International Holidays 96
Sonata Travel 106
South Africa 57
Southern Voyages 119
Sovereign Italia 182
Sovereign Just Turkey 182
Sovereign Sailing 165
Sovereign Scanscape Holidays 148
Spain 59
Steepwest Holidays Ltd 149
Stitches 271
Stratton Creber 119
Summer Cottages Ltd 119
Sun Blessed Holidays 149
Sunburn 271
Sun Esprit 119, 149, 182
Sunsail/Sunsail Clubs 166
Sunselect Villas 182
Sunspot Tours Ltd 149
Sunvil Travel 150
Sunvista Holidays Ltd 119
Surfrider Activity Holidays 175
Sweden 62
Swiss Ski 170
Switzerland 63

Taber Holidays 150
Teeth 271
Thailand 65
Thomson Holidays 150
Thomson Tour Operations 183
Timescape Holidays Ltd 106
Toad Hall Cottages 119
Tourist boards 281–2
Tracks Africa/Europe Ltd 161
Traffic news 288–91
Travelbag PLC 161
Travel-cots 260
Travelodge *see* Forte Travelodge
Travelscene Ltd 150
Travel sickness 223, 271
Treble B Holiday Centre 96
Tunisia 66
Turkey 67
Twickers World of the Red Sea 175

UK Express 151
Uley Carriage Hire 96
Unijet 127
United Kingdom 9
United States 70

Vacances en Campagne/Vacances in Italia 120
Vaccinations 271, 292–300
Valley USA 127
Vauxhall Holiday Park 132
VFB Holidays Ltd 120, 151
Viking Afloat 101
Villas and apartments 176–83
Virgin Holidays Ltd 151
Vomiting 271
Voyages Jules Verne 162

Wales 9
Walking holidays 184–5, 260
Wallace Arnold 106
Water 272
Weather news 288–91
Welcome Caravan Company 96
Welcome Lodge 90
Welsh Holidays 120
Welsh Wayfaring Holidays 185
Westents Ltd 97
West Indies 74
Wildblood (Patricia) 183
Windermere Lake District Holidays 121
Worldwide Journeys and Expeditions 162

YMCA National Centre 175

Zimbabwe 75